P9-CIU-889

EARTH AND OTHER PLANETS

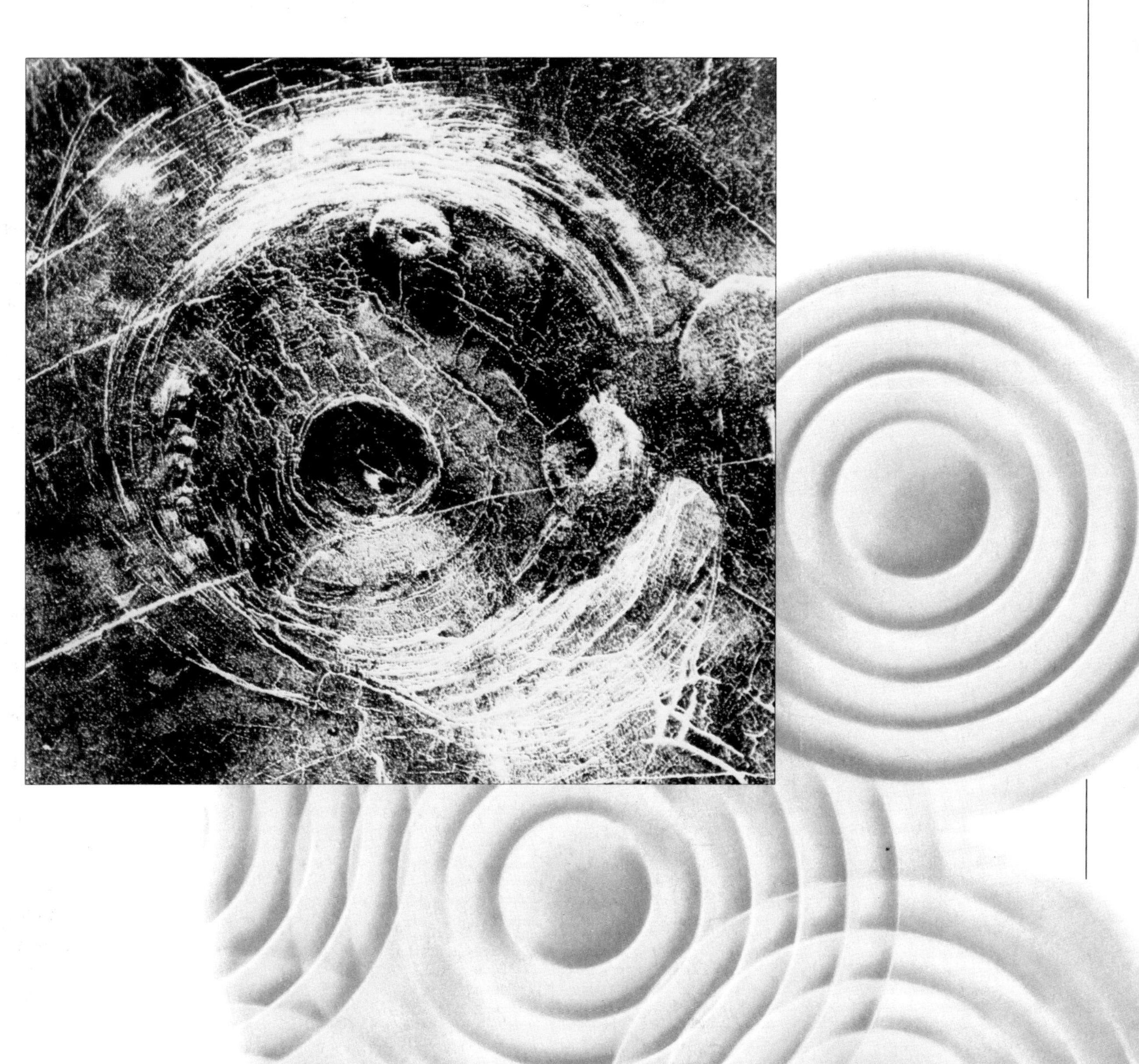

EARTH AND OTHER PLANETS

Geology and Space Research

PETER CATTERMOLE

OXFORD UNIVERSITY PRESS

New York 1995

CONTENTS

Introduction 6

Knowledge Map 8

Timechart 12

Geology Keywords 16

1 *Out of the Cosmos* 48

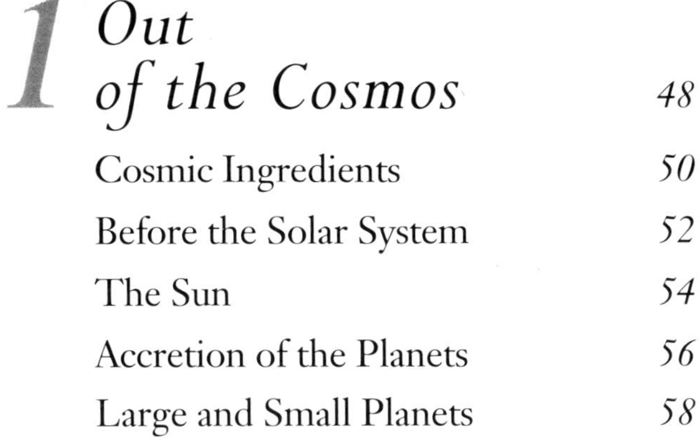

Cosmic Ingredients 50

Before the Solar System 52

The Sun 54

Accretion of the Planets 56

Large and Small Planets 58

2 *The Sun's Family* 60

Planets and Their Orbits 62

Earth and Moon 64

The Inner Planets 66

Distant Companions 68

Bits and Pieces 70

Beyond the Fringe 72

3 *Heat Engines* 74

The Planets Heat Up 76

Forming Cores 78

Segregating the Elements 80

Magnetic Fields 82

How Atmospheres Evolved 84

Internal Clocks 86

4 *Dynamic Planets* 88

The First Crusts 90

The Rise of Magmas 92

Volcanoes 94

Seismic Waves 96

Present-day Atmospheres 98

Earth's Oceans 100

Early Continents 102

The Ice Ages 104

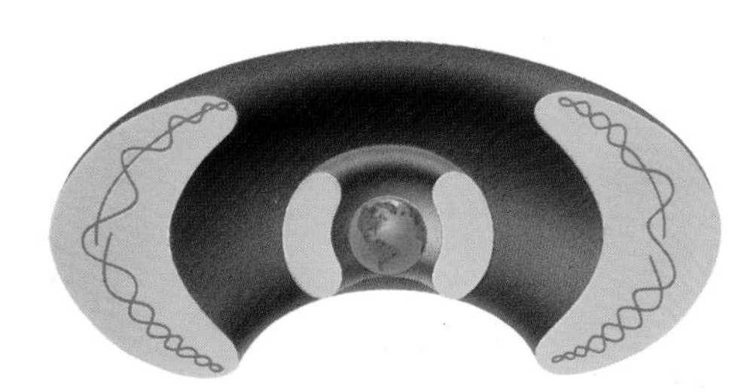

Project editor	Peter Furtado
Editors	Lauren Bourque, John Clark
Editorial assistant	Marian Dreier
Art editor	Ayala Kingsley
Visualization and artwork	Ted McCausland/ Siena Artworks
Senior designer	Martin Anderson
Designer	Roger Hutchins
Picture manager	Jo Rapley
Picture research	David Pratt
Production	Clive Sparling

Planned and produced by
Andromeda Oxford Ltd
9-15 The Vineyard
Abingdon
Oxfordshire OX14 3PX

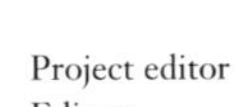

Published in the United States of America by
Oxford University Press, Inc.,
198 Madison Avenue
New York, NY 10016

Oxford is a resitered trademark of Oxford University Press

5 *Geological Jigsaws* 106

Mobile and Stable Zones 108
Wandering Continents 110
Plates and Plumes 112
Beneath the Ocean Floor 114
Island Arcs 116
Mountains from the Sea 118
Rift Valleys 120

6 *Changing Worlds* 122

Nothing is Forever 124
The Work of Rivers 126
Coasts and Oceans 128
Deserts and Winds 130
Glaciers and Ice 132

7 *Beginnings and Endings* 134

Life on Earth 136
Natural Catastrophes 138
Man Leaves the Earth 140
Geological Stories 142

Factfile 144

Metric Prefixes 144
Conversion Factors 144
SI Units 145
The Solar System 146
The Major Moons of the Planets 147
Earth Data 148
Rocks and Minerals 150
Metallic Minerals 152
Rock Dating 152
Geological Eras 153
Earthquake Scales 153
Space Probes 154

Further Reading 154
Index 156
Acknowledgments 160

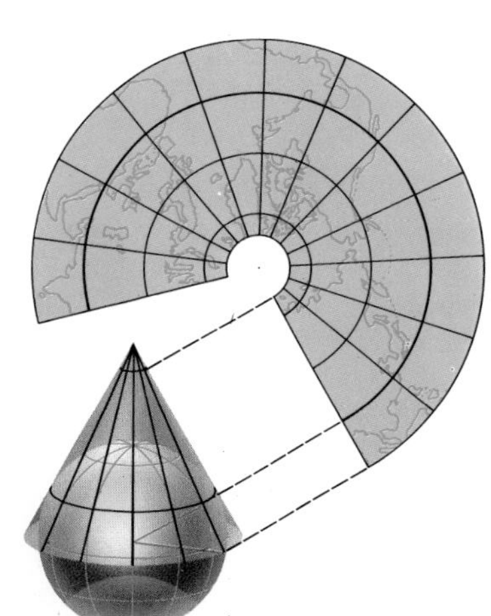

Library of Congress Cataloging-in-Publication Data

Cattermole, Peter John
Earth and other planets : geology and space research / by Peter Cattermole.
160pp. 29 x 23cm - (New encyclopedia of science)
Includes bibliographical references and index
ISBN 0-19-521138-3 (alk. paper)
1. Earth 2. Planets – Geology I. Title II. Series
Q8631.C37 1995
550 – dc20 95-17138

Printed in Spain by Graficromo SA, Córdoba

INTRODUCTION

GEOLOGY is the science of the Earth. Our planet, however, is but part of a system of worlds that comprise the Sun's family, including planets, comets, meteoroids and asteroids, which were all created, together with the Sun itself, about 4.6 billion years ago out of a cloud of gases. Scientific study of the Earth began in Europe in the 17th and 18th centuries, and since about the same dates scientists have also been able to study the Moon and planets through telescopes. Understanding of the structure and composition of the planets, and of the Earth itself, have both dergone vast changes since the beginning of the 1960s.

With the start of the space age, it became possible to explore some of these distant worlds by sending spacecraft to photograph them as they flew close by or even to land on their surfaces. The first journey by humans to another world was to the Moon, in July 1969. The Moon remains the only other body in the Solar System to have been explored by astronauts, who collected samples to take back to Earth for chemical analysis and for dating. Because of the distances involved, such data cannot be obtained for any other world, although meteorites – stray cosmic particles that hit the Earth – provide free samples of parts of planets and other ancient objects disrupted long ago. They therefore offer unique insights into the primitive material from which the Earth was formed.

Even the distant regions of the Solar System have been explored, by means of long-distance space probes. The Magellan radar-mapping mission to Venus, which was completed in 1994, was one of the most successful of all time and penetrated the planet's thick cloud cover to provide geologists with an abundance of fascinating images, many of them featuring landforms not found on Earth. In the 1980s and 1990s there has also been a dramatic increase in the study of space objects by Earth-based and orbiting telescopes, including the Hubble Space Telescope. As astronomers have learned to make use of a wide range of wavelengths, the wealth of data obtained has also grown.

At the same time as this revolution in our knowledge of the planets, understanding of how the Earth itself is constructed has improved immeasurably since the 1960s, when the concept of plate tectonics was first proposed to explain continental drift, thed distribution of regions vulnerable to earthquakes and volcanoes, and other phenomena. This theory has given geologists new insights into how the rigid outer surfaces of some planets adjust to the internal movements of their mantles and cores. It has become appreciated that the Earth's solid mantle is able to convect, inducing movements in the brittle layers above, and that these movements account for many of the largest features of the planet on which we live, its oceans and continents, mountain ranges and zones vulnerable to earthquakes and volcanoes. Scientists have learned how the Earth's crust is continually being generated and destroyed at certain sites, where huge slabs of the lithosphere jostle against one another.

By understanding better the forces at work within the Earth, geologists have been able to make more meaningful comparisons with the other planets. It seems that no other planet has such a dynamic structure as the Earth's, but several exhibit features such as volcanoes, rift valleys and ice-caps, about which terrrestrial knowledge provides important insights. Conversely, some baffling characteristics of Earth may be explained in terms of those known to exist on other planets. This two-way study is the basis of the science of comparative planetology. This book seeks excitingly to demonstrate the many ways in which the study of our own planet, and that of the rest of the Solar System, have become intertwined.

THIS BOOK aims to make all this information available to the whole family, from students studying for exams and projects to adults wanting to bring their scientific knowledge up to date. To achieve this, the book is organized in such a way as to provide readers with a quick answer to a specific query, or allow them to follow a more detailed account of a particular topic.

At the heart of the book is a 96-page thematic section, made up of 48 major narrative topics, each one richly illustrated to tell the story of a central theme of the book. The strong graphic presentation and the style of writing are designed to make this section the ideal point of departure for the less well-informed reader. Sets of Keywords highlighted on each topic spread point the reader to the second major section, a 32-page alphabetic mini-encyclopedia of the subject, which contains some 400 entries. This section, too, leads the reader back to the thematic topics.

No part of modern science can be separated cleanly from other fields. Earth Science merges into astronomy on the one hand, chemistry on the other and environmental studies in other respects. It has also been affected by biology and physics. The Knowledge Map, immediately following this Introduction, maps out the entire field of modern science, shows how each area of science interacts with another, and defines the major fields. This is followed by a Timechart, which traces the development of the subject through great discoveries.

Finally, to ensure that the volume is of value for reference as well as for browsing, the Factfile provides hard data, tables and statistics, covering aspects of the Earth, the planets, geology and space research.

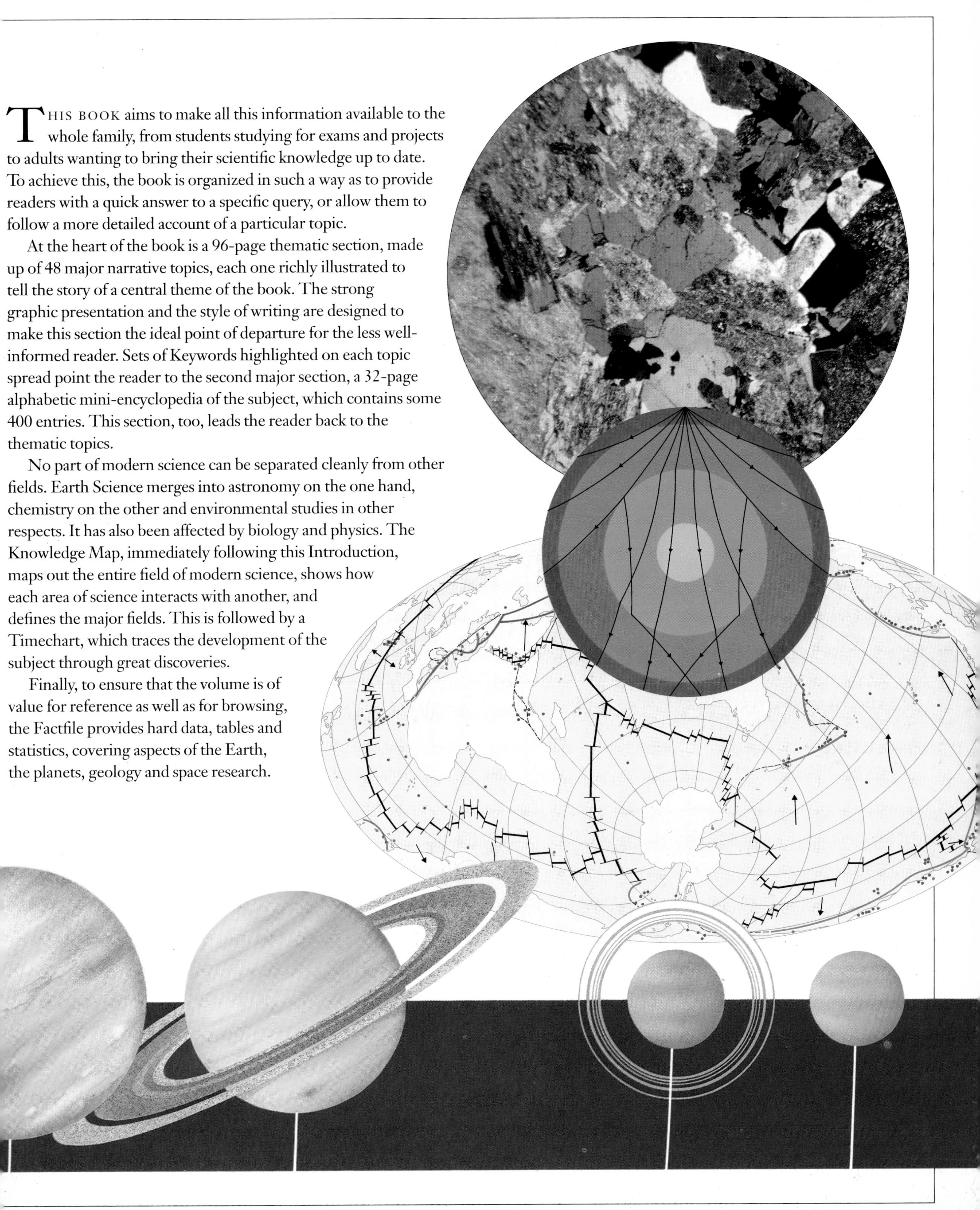

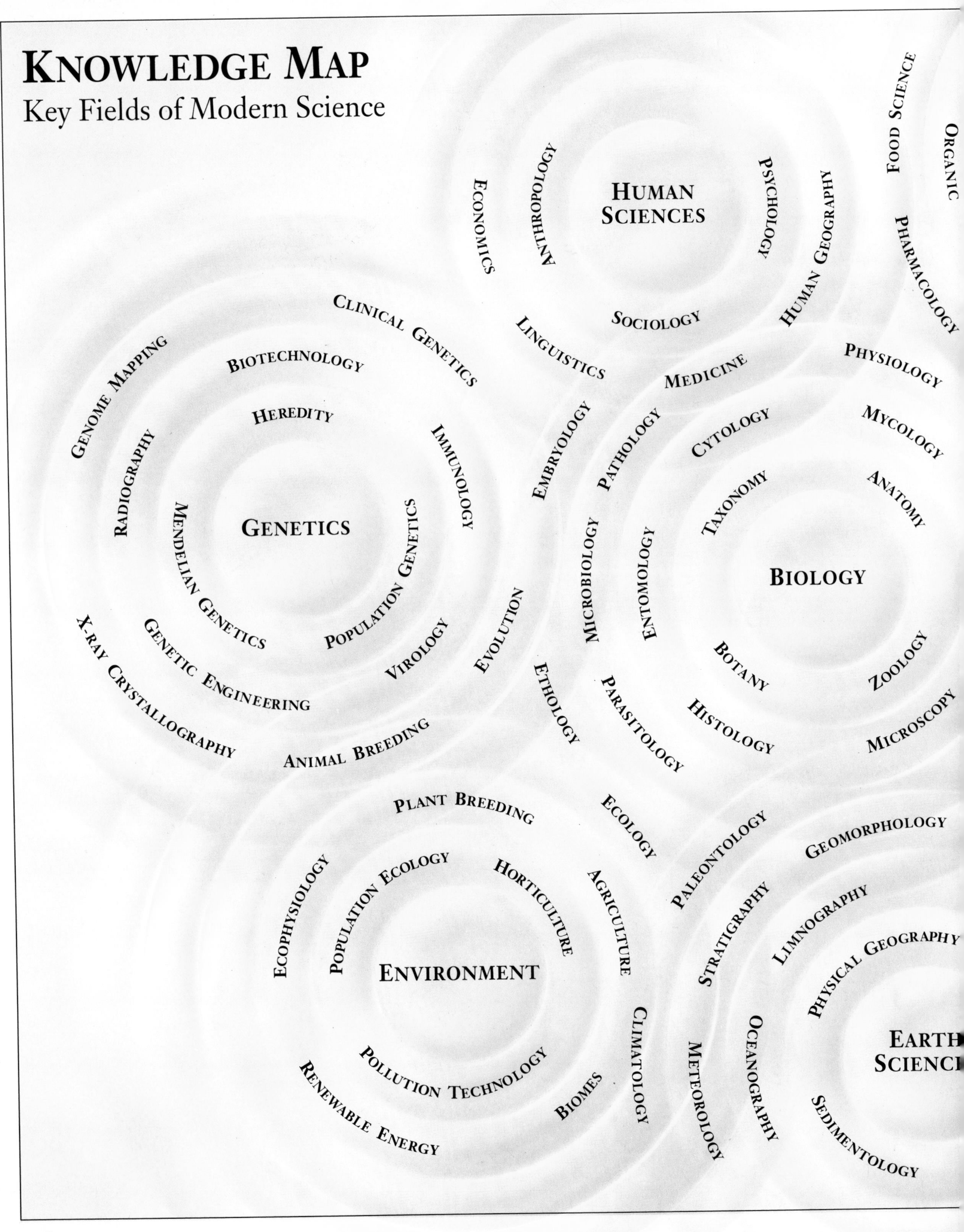

Knowledge Map
Key Fields of Modern Science
Human Sciences
Economics
Anthropology
Psychology
Human Geography
Sociology
Linguistics
Food Science
Organic
Pharmacology
Physiology
Genetics
Genome Mapping
Biotechnology
Clinical Genetics
Heredity
Radiography
Immunology
Mendelian Genetics
Population Genetics
X-ray Crystallography
Genetic Engineering
Virology
Evolution
Animal Breeding
Biology
Medicine
Embryology
Pathology
Cytology
Mycology
Taxonomy
Anatomy
Microbiology
Entomology
Botany
Zoology
Ethology
Parasitology
Histology
Microscopy
Environment
Plant Breeding
Ecology
Ecophysiology
Population Ecology
Horticulture
Agriculture
Paleontology
Geomorphology
Stratigraphy
Limnography
Physical Geography
Pollution Technology
Renewable Energy
Biomes
Climatology
Meteorology
Oceanography
Sedimentology
Earth

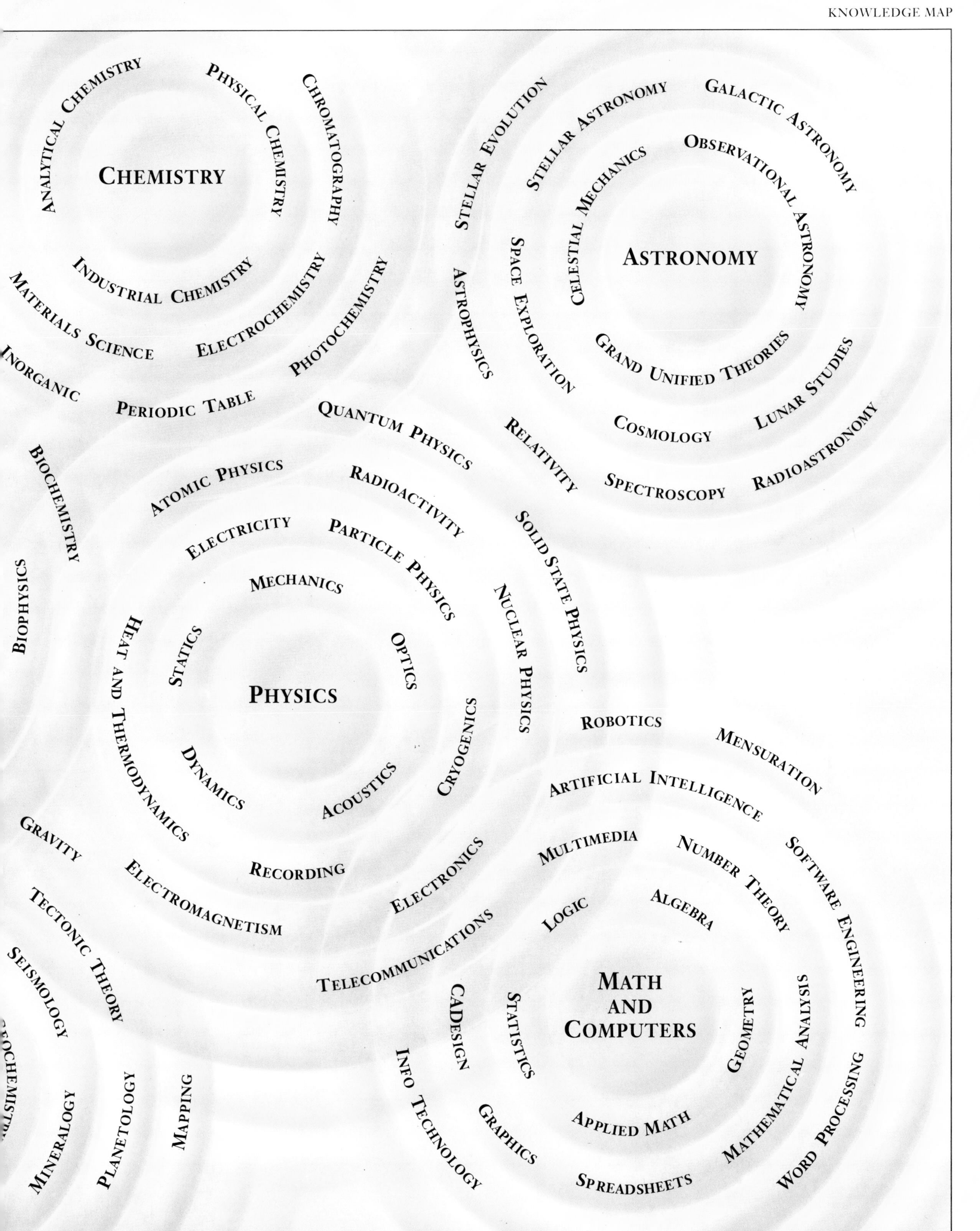

Chemistry
Analytical Chemistry
Physical Chemistry
Chromatography
Industrial Chemistry
Electrochemistry
Photochemistry
Materials Science
Inorganic
Periodic Table
Biochemistry
Biophysics
Astronomy
Stellar Evolution
Stellar Astronomy
Galactic Astronomy
Celestial Mechanics
Observational Astronomy
Space Exploration
Astrophysics
Grand Unified Theories
Lunar Studies
Cosmology
Relativity
Spectroscopy
Radioastronomy
Physics
Quantum Physics
Atomic Physics
Radioactivity
Electricity
Particle Physics
Mechanics
Solid State Physics
Nuclear Physics
Heat and Thermodynamics
Statics
Optics
Cryogenics
Dynamics
Acoustics
Recording
Gravity
Electromagnetism
Electronics
Telecommunications
Tectonic Theory
Seismology
Mineralogy
Planetology
Mapping
Math and Computers
Robotics
Mensuration
Artificial Intelligence
Multimedia
Number Theory
Software Engineering
Logic
Algebra
CADesign
Statistics
Geometry
Mathematical Analysis
Info Technology
Graphics
Applied Math
Spreadsheets
Word Processing

KNOWLEDGE MAP

Modern Geology

GEOMORPHOLOGY

The study of the evolution of landscape features, and of the fundamental physical processes that generate the variety of landforms that are found in nature. While most of these landforms are caused by erosion, others are constructional – for example, spits, bars and moraines.

OCEANOGRAPHY

The study of the Earth's oceans, including the gross structure of the ocean basins; ocean currents; marine life forms the morphology and distribution of submarine landforms; and the changing forms of these as plate tectonic processes modify them.

PHYSICAL GEOGRAPHY

The branch of geography which deal s with the development and distribution of natural features found at the Earth's surface. A practical subject, it includes the study of relief, weather and simple geology, and also some aspects of meteorology and climatology. Mapwork and statistical studies are an integral part of it.

PLANETOLOGY

This relatively recent branch of geology has developed with the exploration of space. Essentially, planetology is the study of the surfaces, atmospheres and interiors of planets and their satellites. As such, it has become a comparative discipline, in which features recognized on other planets are compared with each other and with Earth's.

GEOCHEMISTRY

The science concerned with the chemistry of the Earth as a whole, and of its constituent parts. It deals with the distribution and movement of chemical elements inside the Earth in both time and space. While the solid crust is the principal focus, interactions between it, the hydrosphere and atmosphere are a vital part of it.

SEDIMENTOLOGY

The study of the formation, movement and deposition of sedimentary materials, together with diagenetic changes – physical and chemical changes which occur after sediment is buried. This necessarily involves an understanding of mechanical and chemical activity at the Earth's surface, along with the analysis of different sedimentary environments and of clastic materials, carbonates and evaporites.

MINERALOGY

The study of minerals – the naturally-occurring crystalline substances, both silicate and non-silicate, that make up the rocks of the Earth and the other solid planets. Mineralogy seeks to understand the structure, physical and chemical behavior that distinguish the 3000 different known minerals, and the ways in which they react together in natural rock systems. It encompasses several aspects of geology, chemistry and physics.

SEISMOLOGY

The branch of geophysics that deals with earthquakes – that is, with seismic activity. Recording of seismic waves by instruments called seismographs allows seismologists not only to monitor geological activity along faults and at sites of volcanicity, but also to map out the Earth's internal structure.

PETROLOGY

The study of different kinds of rocks – their origins, interactions and distribution. It may involve analysis of geochemical data, thin sections of rock samples, laboratory experimental data and field mapping. It is potentially applicable to many geological problems.

GEOMAGNETICS

The study of remnant magnetization in rocks. Magnetic minerals fossilize polarity and the direction of the magnetic pole when they crystallize, which allows scientists to map out polarity reversals.

STRATIGRAPHY

A branch of geology whose name is derived from *stratum*, a bed or layer, which refers to the basic form of a sedimentary rock unit. Stratigraphy studies the succession of such layers and interprets them as historical records of the processes which have operated over billions of years at the surface of the planets.

CLIMATOLOGY

The analysis of average weather conditions in different regions of the Earth, over long periods of time. This discipline has importance to ethnographers, agriculturalists, meteorologists and to issues of trade and health.

LIMNOGRAPHY

The study of the physical phenomena associated with lakes and other bodies of fresh water. Limnographic data can be useful in dating recent geological events, in understanding climatic changes and in paleontology.

PALEONTOLOGY

The study of the plant and animal e which has left its records as fossils ne stratigraphic record – the layers of and sediment laid down over geolog- time. It includes analyses of plant and mal form, growth and development; f evolutionary changes; and of the relationship of all these factors to environmental variations with time.

TECTONICS

The study of geological structures: the features produced by the deformation of a planet's crust, either by internal movements or by impacts. In recent years a unifying model of the deformation of the Earth's crust has been developed; this is known as plate tectonics. Tectonic theory invokes movements of the mantle to explain the largescale physical features, such as mountain ranges, that are observed at the planet's surface.

MAPPING

The study of geological structures and of the relationships between different rock types and ages. This enables geologists to construct models of three-dimensional sections of the Earth, either from the ground or using remote sensing techniques.

REMOTE SENSING

The use of orbiting spacecraft (in addition to traditional aircraft or balloons) for the purpose of data collection. A wide variety of different kinds of mapping can be accomplished by such means. Use of long wavelengths allows a surface (such as that of a planet) to be inspected even through layers of clouds.

GRAVITY MAPPING

The measurement of the distribution of mass within a planet by spacecraft or by the effects that the planet has on natural satellites. The presence of dense material near to the surface will speed up an orbiting object, and vice-versa. Gravity mapping enables geophysicists to understand planetary interiors.

METEOROLOGY

The study of weather. It involves the recording of changes in temperature and pressure of air masses on a day-to-day basis with a view to the accurate prediction of expected weather patterns over different regions of the Earth's surface with time. In recent years, the science of meteorology has been extended to the atmospheres of other planets, such as Mars, in an attempt to understand these planets better.

TIMECHART

QUESTIONS about the age and size of the Earth, and its relation to the Sun, Moon, planets and stars, have challenged the human mind since time immemorial. Space exploration in the late 20th century enabled people, for the first time, to see their planet in its true perspective.

The ancient Greeks made profound contributions to Earth science. Even in the 6th century BC, Pythagoras recognized that the Earth is a sphere and that the tilt of its axis causes the seasons. This realization was based, in part, on observations by sailors who noted that as they traveled to the north or south, the elevations of stars changed. It was correctly interpreted that sailors were in fact traveling along a curved surface. By 200 BC Eratosthenes estimated the circumference of the Earth remarkably accurately.

The Greeks imagined the planets, Sun and Moon as separate spheres moving in orbits around the Earth. Aristarchus of Samos had argued that the Sun is at the center of the Universe, but most people continued to believe in an Earth-centered system, until the 16th-century Polish astronomer Nicolas Copernicus changed

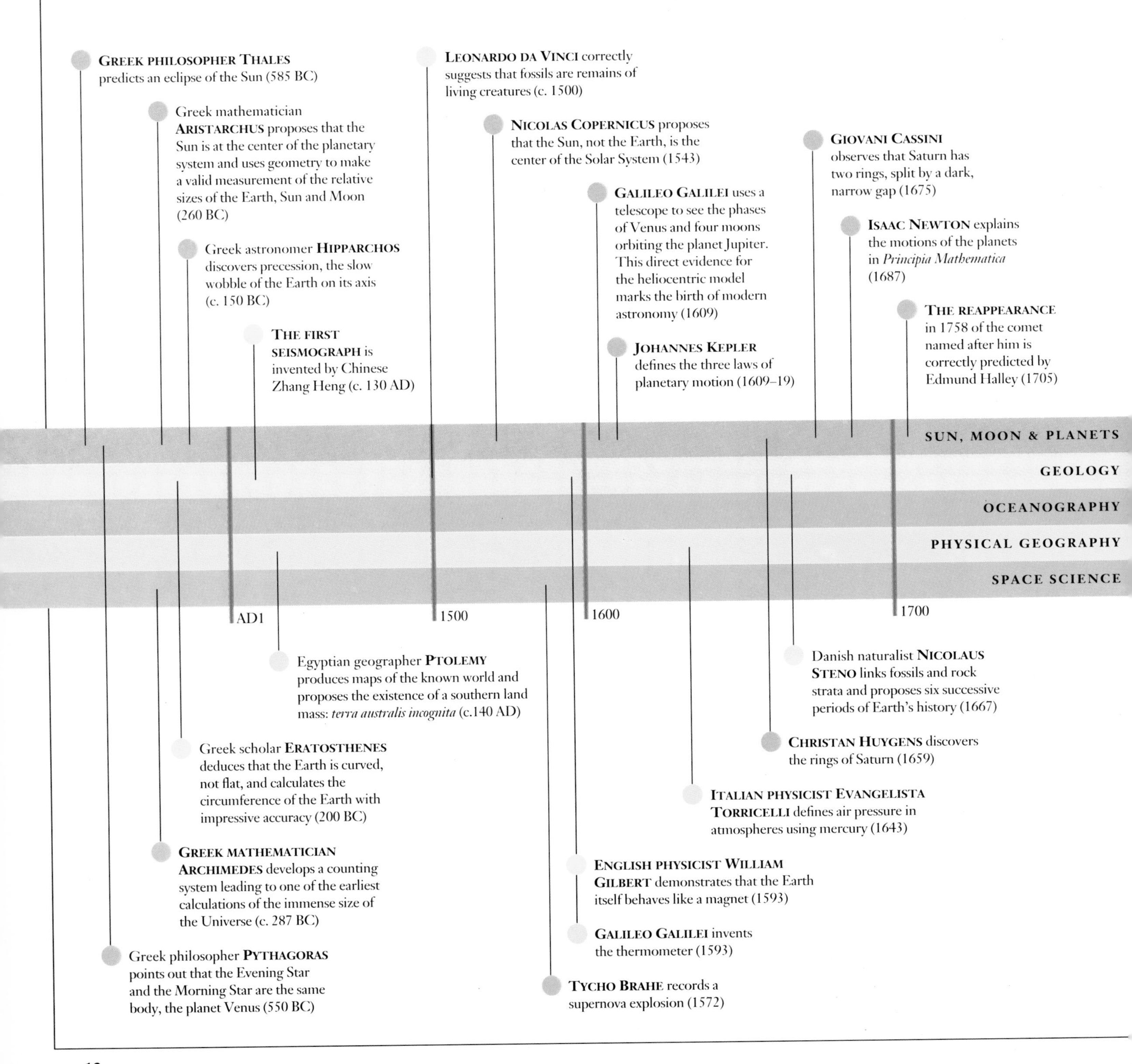

astronomical thinking. His model of the Universe, with the Sun at the center of the Solar System and the Earth and planets revolving around it, was widely regarded as heretical at the time, but led to the image of the Solar System we know today.

During subsequent years, Tycho Brahe and Johannes Kepler refined the Copernican mode, and Kepler formulated the laws of planetary motion. Galileo Galilei confirmed the Copernican concept with the help of a new invention, the telescope, and Isaac Newton gave a theoretical explanation of the physical character of the Solar System. The known planets – Mercury, Venus, Mars, Jupiter and Saturn – were, by then, established telescopically as more or less spherical bodies.

Distinct investigation of the Earth itself – geology – dates back a mere 200 years. Scientific thinking about the Earth grew out of other traditions, such as philosophy, geography and mineralogy. Aristotle had considered that the world was eternal, and had drawn attention to natural processes that changed its surface features. Knowledge on minerals, gems, fossils, metals, crystals

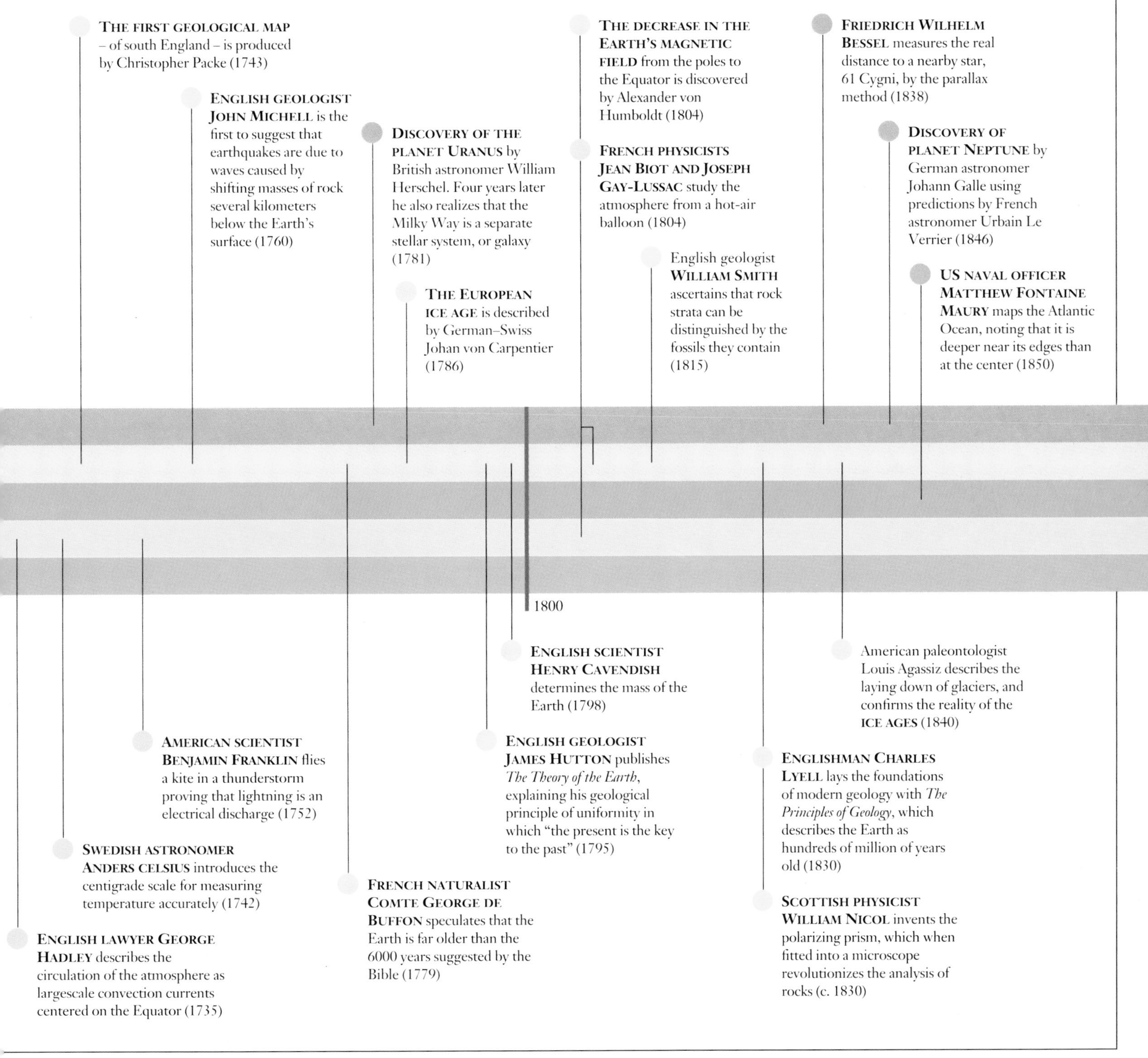

and useful chemicals was collected in encyclopedic works by natural historians such as Pliny (AD 23–79), Isidore of Seville (560–636) and Conrad Gesner (1516–1565).

Biblical accounts of Creation shaped beliefs about the Earth's age, which Bishop Ussher estimated in 1660 to be less than 6,000 years. It was not until the late 18th and 19th centuries that geologists, biologists and physicists grappled with the question of the Earth's age. James Hutton (1726–97), the "father of geology", used stratigraphical evidence to argue that the Earth was far older. In the early 20th century the discovery of radioactive isotopes allowed the measurement of the age of rocks. Radioactive dating of the oldest rocks on Earth, meteorites and lunar rocks, led to the conclusion that the age of the Earth is 4.6 billion years.

The concept of the continents adrift on the surface of the Earth was first put forward by Alfred Wegener in 1912. Few scientists supported him. But during the 1940s Victor Vacquier's discovery that the ocean floor is imprinted with a regular pattern of remanent magnetism indicated that the continents do indeed

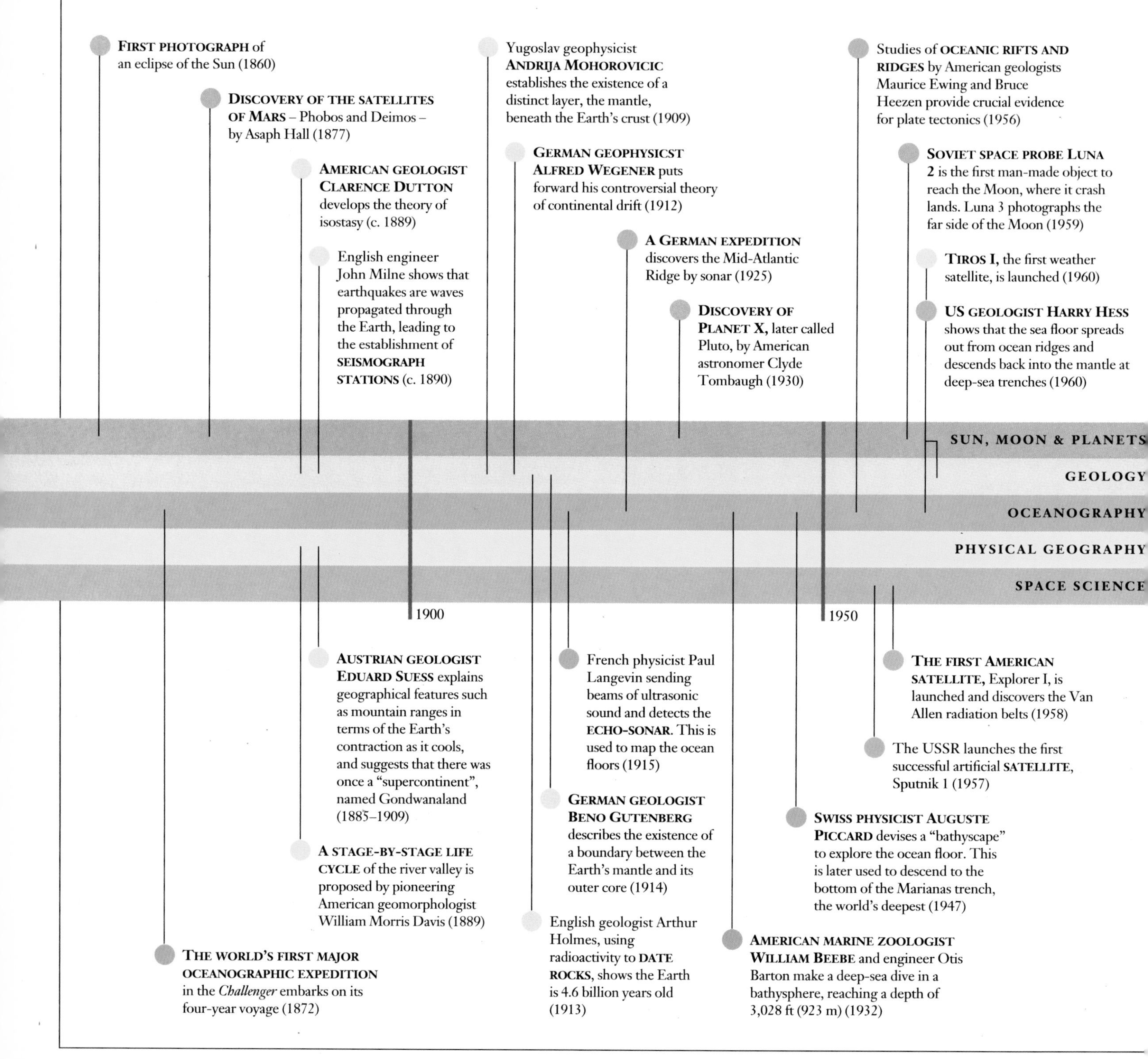

move. During the 1960s and 1970s, a wealth of evidence emerged to confirm the theory now called plate tectonics. The theory as a whole explains the occurrence of many planetary features such as volcanoes, earthquakes and the configuration of the oceans and continents, and led to a new view of the evolution of the Earth.

Information from unmanned spacecraft has been hugely revealing. Robotic spacecraft – the Mariner, Venera, Viking, Voyager and Giotto missions – have introduced us to the features of eight planets, more than 60 moons and planetary rings and comets, and have made discoveries vital to the understanding of our own planet. We now know, for example, that oceans may have existed on Venus until a carbon dioxide buildup created a massive "greenhouse effect" and very high temperatures. Such knowledge may be useful in understanding the greenhouse effect on Earth.

Planetary exploration has confirmed many ideas about our own planet. The Earth has much in common with other planets – volcanoes, hurricanes and a moon – yet retains unique features, such as drifting continents, liquid water and living beings.

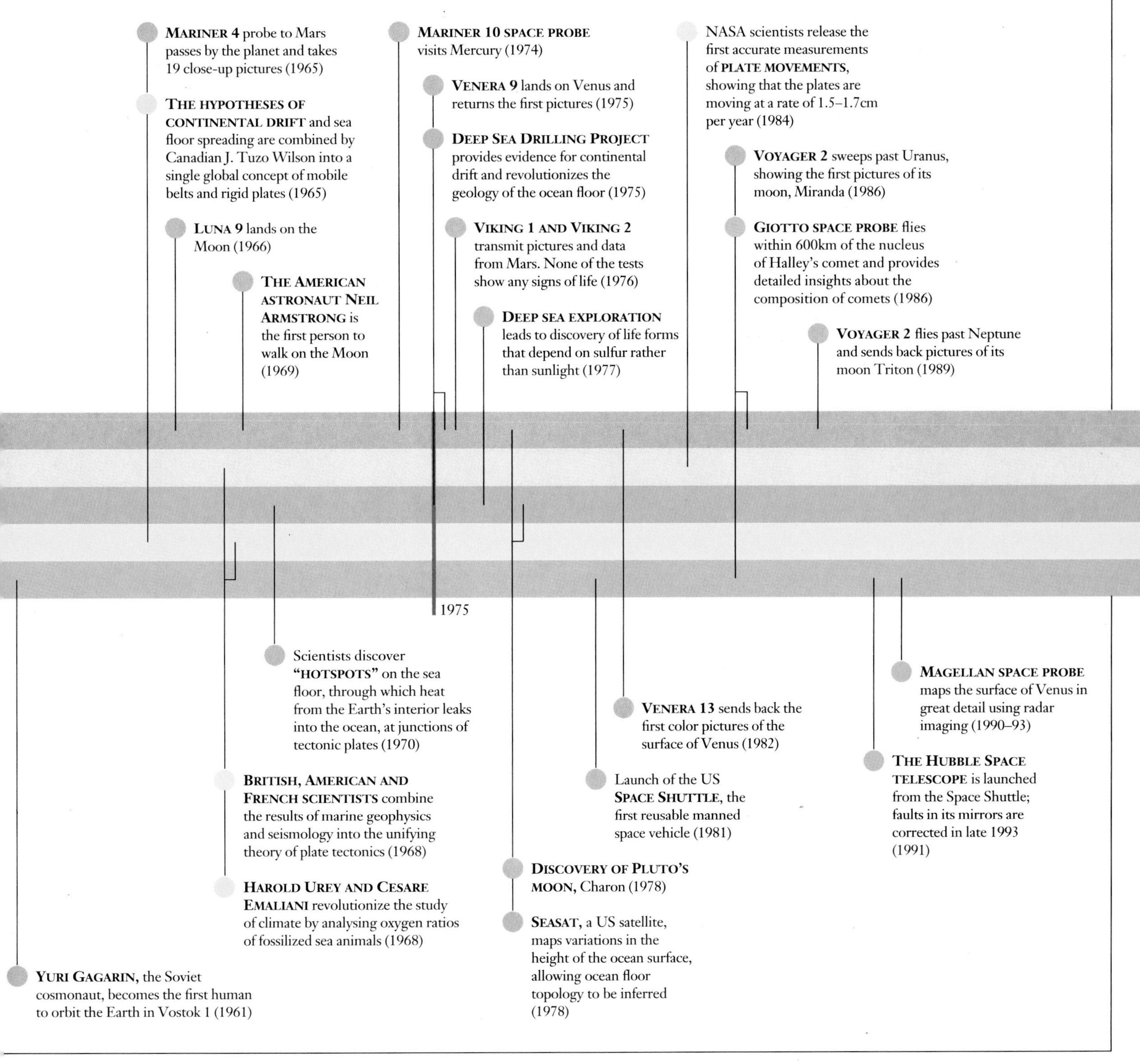

Geology KEYWORDS

aa
A type of volcanic lava, typically found in Hawaii, that solidifies and takes the form of rough blocks. *See also* **pahoehoe.**

ablation
The loss of ice from the surface of a glacier as a result of melting or sublimation (in which ice turns directly into water vapor without first melting).

absolute age
The age of a geological event as derived from **radiometric dating**. The technique depends on the fact that **isotopes** of certain elements are radioactive and decay to a daughter isotope in a known time. When a magma (molten rock) consolidates and the isotopes are fixed into it, decay begins. Calculation of the proportion of the isotope that has decayed within the sample allows the date of consolidation to be estimated.

abyssal plain
The extensive flat region of the deep ocean floor that extends oceanward from the base of the **continental slope**. It is covered mainly by **pelagic** sediments.

abyssal sediment
Sediment on the ocean floor, consisting of very fine-grained material, including settled airborne dust, calcite shell material (calcareous **ooze**) and sediment derived from diatom skeletons (siliceous ooze). The most widespread type is chocolate brown clay, which covers nearly 70 million square kilometers of the Pacific floor. In fact, abyssal sediment is generally abundant wherever the water depth exceeds 4.5 kilometers. It is derived from fine-grained volcanic material, much of it of submarine origin. Compared with shallow water sediment, abyssal sediment accumulates very slowly.

accretion
The enlargement of a continent by the tectonic addition of other crustal fragments. The term also describes the process by which **planetesimals** grow into planets by collisions with other objects, both larger and smaller than themselves.

CONNECTIONS

THE FORMATION OF THE SOLAR SYSTEM **48**
THE PLANETS FORM **56**
THE INNER PLANETS **66**
THE OUTER PLANETS **68**
THE PLANETS HEAT UP **76**
ISLAND ARCS **116**

accretional energy
The energy transferred to an object by collisions with others. The energy of **meteorite** or **asteroid** impact is converted to heat, some of which is stored inside the object.

adiabatic heating
The natural increase in temperature in rocks produced by growing **hydrostatic pressure** with increasing depth. On Earth, before core formation, temperatures of around 2500°C would have been encountered at depths of about 1000 kilometers.

aerolite
Another name for a stony **meteorite**.

air-fall deposit
Any sediment produced by fallout from a volcanic eruption cloud, as opposed to the action of a lava flow.

albedo
The proportion of the incident light that is reflected from the surface of an object, expressed as a percentage. The average albedo of the Moon, although it appears silvery bright to observers on Earth, is a mere 7 percent.

alluvium
Unconsolidated gravels, sands and clays deposited by rivers on their flood plains or in deltas and estuaries.

amino acid
Any of a group of organic acids derived from ammonia which are the basic building blocks of proteins. The development of amino acids might have allowed the start of life on Earth.

ammonia
A pungent colorless gas with the formula NH_3. It is one of several gases found in abundance in the outer Solar System.

AMMONITE

ammonite
A type of extinct cephalopod mollusk that thrived in the Mesozoic era (from 248 million to 65 million years ago). Ammonites lived in shallow seas in many-chambered coiled shells that are abundant as fossils, particularly from the **Jurassic** period (213–144 million years ago).

amphibian
An animal with the ability to live both on land and in water. Modern representatives are frogs, toads and salamanders. Life first moved onto the land, from the water, about 395 million years ago. Amphibians were the first four-legged animals to develop. Reptiles evolved from amphibians.

amphibole
A dark rock-forming silicate mineral, such as hornblende.

andesite
A fine-grained volcanic rock that is typically generated at **subduction zones**. With a silica content of between 55 and 60 percent, its essential constituents are plagioclase **feldspar** and **amphibole**.

angular momentum
A property of a rotating object that depends on its velocity and the distribution of its mass around its axis of rotation. For any system, the total angular momentum is constant. For example, if a slowly rotating **nebula** contracts, its rotational velocity has to increase to conserve angular momentum.

anion
An ion with a negative electric charge.

anorthosite
An igneous rock composed largely of anorthite, a calcium-rich plagioclase **feldspar**. It is rich in calcium and aluminum, and is the predominant constituent of the ancient highland crust of the Moon.

anticline
An upfold in deformed rock strata, with the oldest beds at the center of the structure. Anticlines are usually separated from each other by downfolds or **synclines**.

aphelion
The point in the orbit of a planet that is farthest from the Sun.

Apollo missions
A series of United States space missions to land astronauts on the surface of the Moon and to return rock samples to Earth. The missions were initiated in the early 1960s and culminated in the landing of Apollo 11 on the surface of the Moon (on Mare Tranquillitatis) on 16 July 1969. Each Apollo spacecraft could carry three astronauts, and comprised a command module (designed to remain in lunar orbit), and a lunar module, which could land on the lunar surface, carrying two astronauts. Of the 17 Apollo missions, six were unmanned; of the 11 manned missions, two were Earth-orbital, two lunar orbital, one aborted by in-flight accident and six made lunar landings. The final Apollo (17) lifted off from the Taurus-Littrow Valley of the Moon in December 1972. By the time it returned to Earth, 380.8 kilograms of lunar rock samples had been collected.

CONNECTIONS

THE EARTH AND THE MOON **64**
MAN LEAVES THE EARTH **140**

aquifer
A layer of permeable rock in which groundwater can be stored and through which it can pass in sufficient volumes to feed wells.

Archean
The earliest part of **Precambrian** time, extending from the birth of the Earth as a planet to 2500 million years ago. Rocks of this age do not generally contain fossil remains. During this lengthy period there was a gradual thickening of the Earth's crust and growth of continental **cratons**.

argon
An inert gas present in small quantities in the atmospheres of the terrestrial planets. Certain isotopes of argon (Ar-40) are used in **radiometric dating**.

Ariel
The fourth largest moon of **Uranus**, measuring 1160 kilometers across and orbiting at a mean distance of 130,000 kilometers from its parent planet. Ariel is also the name of a series of British scientific/astronomical satellites launched by NASA in the 1970s as part of a joint British–United States program.

arkose
A **feldspar**-bearing sandstone that is usually derived from the relatively rapid weathering of granitic rocks.

ash flow deposit
A volcanic deposit derived from a *nuée ardente* or "glowing avalanche". Ash flow sediments may be blocky in part, but may also include tuffs that show varying degrees of welding together of the glassy lava shards which the flow carried along.

asteroid
Also known as a "minor planet", any of the rocky bodies that orbit the Sun in paths that mostly lie between those of Mars and Jupiter. This region is called the asteroid belt. Apollo asteroids, however, have Earth-crossing orbits, whereas Amor asteroids move in orbits reaching between Mars and the Earth. Aten objects have orbits that lie largely inside that of the Earth. The largest

ASTEROID

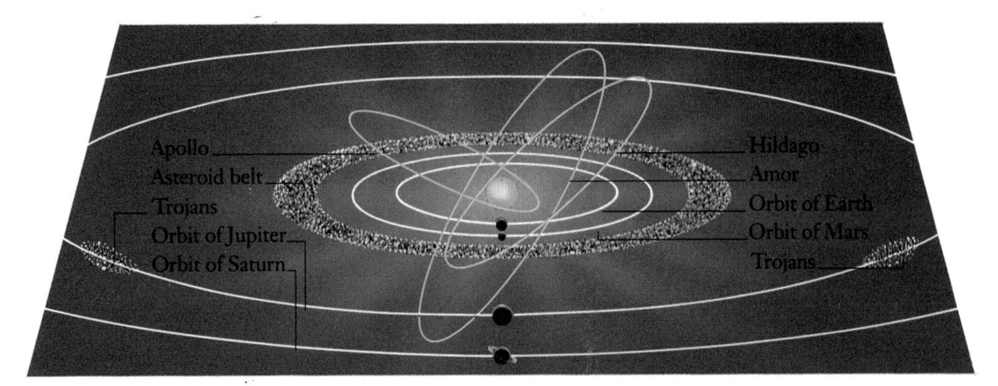

ATMOSPHERE

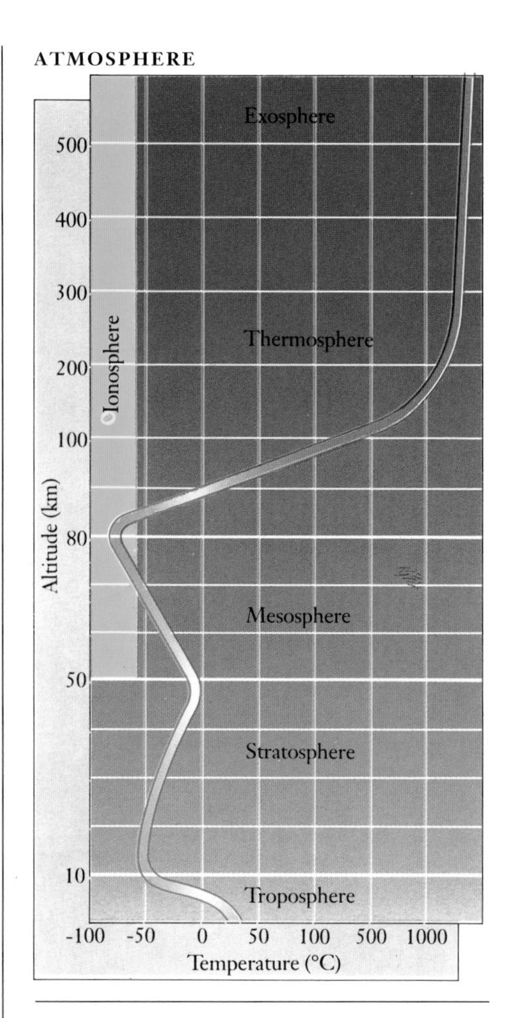

asteroid, Ceres, has a diameter of 1000 kilometers. Only one (Vesta) is visible to the naked eye. Asteroids are of various chemical types and are thought to be the parent bodies of **meteorites**. Although once believed to be the remains of a fragmented planet or planets, most asteroids are now believed to have been **planetesimals** that were prevented from accreting into larger objects by the strong gravitational effect of Jupiter.

CONNECTIONS

LARGE AND SMALL PLANETS **58**
THE SUN'S FAMILY **60**
ASTEROIDS AND METEORITES **70**

asteroid belt
See **asteroid**

asthenosphere
The layer inside the Earth that lies immediately below the **lithosphere**. The asthenosphere is characterized by low seismic wave velocities (that is, a high degree of seismic attenuation) and low rigidity. It is thought to be a zone in which **partial melting** of rock in the Earth's **mantle** occurs.

astronomical unit (AU)
A unit of length commonly used by astronomers to measure distances within the Solar System. One astronomical unit is equal to the mean distance between the Earth and the Sun (149,599,000 kilometers, about 93 million miles). For example, the solar nebula – the mass of rotating gas and dust from which the Sun and planets formed – is estimated to have been about 20 astronomical units across.

atmophile elements
Gaseous elements that tend to be concentrated in **atmospheres** in the uncombined state; for example, argon and nitrogen.

atmosphere
The layer of gases that surround a planetary object. Earth's atmosphere is divisible into a number of layers with different temperature and pressures. The weather is confined to the lowest layer, the **troposphere**. The term is also used as a unit of pressure (101,325 Newtons per square meter).

atoll
A continuous or discontinuous reef of coral encircling a central lagoon.

atom
The smallest particle of a chemical element that can take part in chemical reactions yet still keep its identity. Each atom consists of a central nucleus (comprising protons and neutrons) surrounded by one or more **electrons**. With an increasing number of particles in the nucleus, the relative atomic mass (atomic weight) increases.

aurora
A diffuse but striking glow with streamers that is sometimes observed in the sky at high latitudes near both north and south poles. Aurorae are believed to be caused by charged particles that enter the upper atmosphere along magnetic lines of force, interact with it and fluoresce. Particularly intense aurorae are associated with high solar activity.

barchan
A crescent-shaped sand dune in which the convex face lies upwind and the concave face downwind.

basalt
A dark gray, fine-grained volcanic rock with a silica content of between 44 and 50 percent, composed primarily of plagioclase **feldspar** and a **pyroxene**, usually augite. Basalt is the most widespread volcanic rock found on the surfaces of the terrestrial planets. On the Earth basalt is formed by the **partial melting** of peridotitic upper mantle materials, and it is the dominant constituent of the **oceanic crust**.

CONNECTIONS

THE EARTH AND THE MOON **64**
VOLCANOES **94**
BENEATH THE OCEAN FLOOR **114**

batholith
A large, often irregularly shaped body of intrusive igneous rocks, usually granitic or syenitic in composition, with an exposed surface area greater than 100 square kilometers. Batholiths descend to great depths and are typical of **orogenic belts**, in which they occur in association with metamorphic and other highly deformed rocks.

bedding plane
The surface that separates one sedimentary stratum from another.

Benioff zone
The steeply inclined zone of seismic activity that extends downward from an oceanic trench toward the **asthenosphere**. Named for H. Benioff, a designer of musical instruments and seismographs, such zones mark the path of a tectonic plate being subducted at a destructive plate margin (*see* **subduction**). The foci of earthquakes become deeper toward the non-subducting plate, reaching more than 600 kilometers deep.

benthic
A term describing organisms that live on the floor of a sea or lake. *See also* **pelagic**.

Beta Regio
A major volcanic rise in the western hemisphere of the planet Venus. Beta Regio is a region of crustal rifting and volcanism comparable with the East African Rift on Earth.

Big Bang
The cosmological theory which proposes that the Universe (including space itself, time, matter and energy, as well as the laws of physics) began about 15 billion years ago, in an extremely hot, dense, body which exploded, sending material radially outward; this material cooled and diluted to form galaxies and stars. The expansion of the Universe, as observed by scientists, is a direct result of the initial explosion.

binary star
A pair of stars that move around their shared center of gravity. Binary stars are extremely common.

black smoker
A jet of very hot mineralized water rising from the ocean floor. Seven black smokers were discovered by the submersible *Alvin* along the crest of the East Pacific Rise in 1979. Their black color is caused by dissolved sulfides of iron, zinc, manganese and copper. Smokers emerge where oceanic rifts are particularly active, and may be redeposited as black, sulfide-coated chimneys surrounding the active vents.

blue-green alga
A primitive cellular organism which occurs either singly or in groups. Most of these algae use sunlight to produce oxygen, and are sometimes called cyanobacteria. Their remains range from **Archean** times to the present day.

Bode's law
See **Titius–Bode rule**

boulder clay
A mixture of rock and finely pulverized rock flour that formed when it was dragged along at the base of a moving glacier and then left behind after the ice had melted. Boulder clay is also known as till.

bow shock
The interface at which high-energy particles (**plasma**) streaming outward from the Sun interact with a planet's **magnetic field**. The turbulent region immediately inside is called the **magnetosheath**.

brachiopod
A type of marine shellfish that first evolved in the early **Cambrian** period (about 550 million years ago) before the arrival of true mollusks.

caldera
A large depression, usually located at the summit of a shield volcano, that results from the withdrawal of magma from beneath the structure. Some calderas are formed partly by explosive activity and partly by subsidence. Most Hawaiian calderas measure just a few kilometers across and are characterized by lava lake activity during eruptive periods. Calderas also occur on both Mars and Venus, where they reach diameters of the order of 100 kilometers or more.

Callisto
The largest moon of **Jupiter**. Callisto was discovered in 1610 by the Italian Galileo Galilei (using his first primitive astronomical telescope). It is 4820 kilometers across and orbits Jupiter at an average distance of 1,800,000 kilometers.

CELESTIAL SPHERE

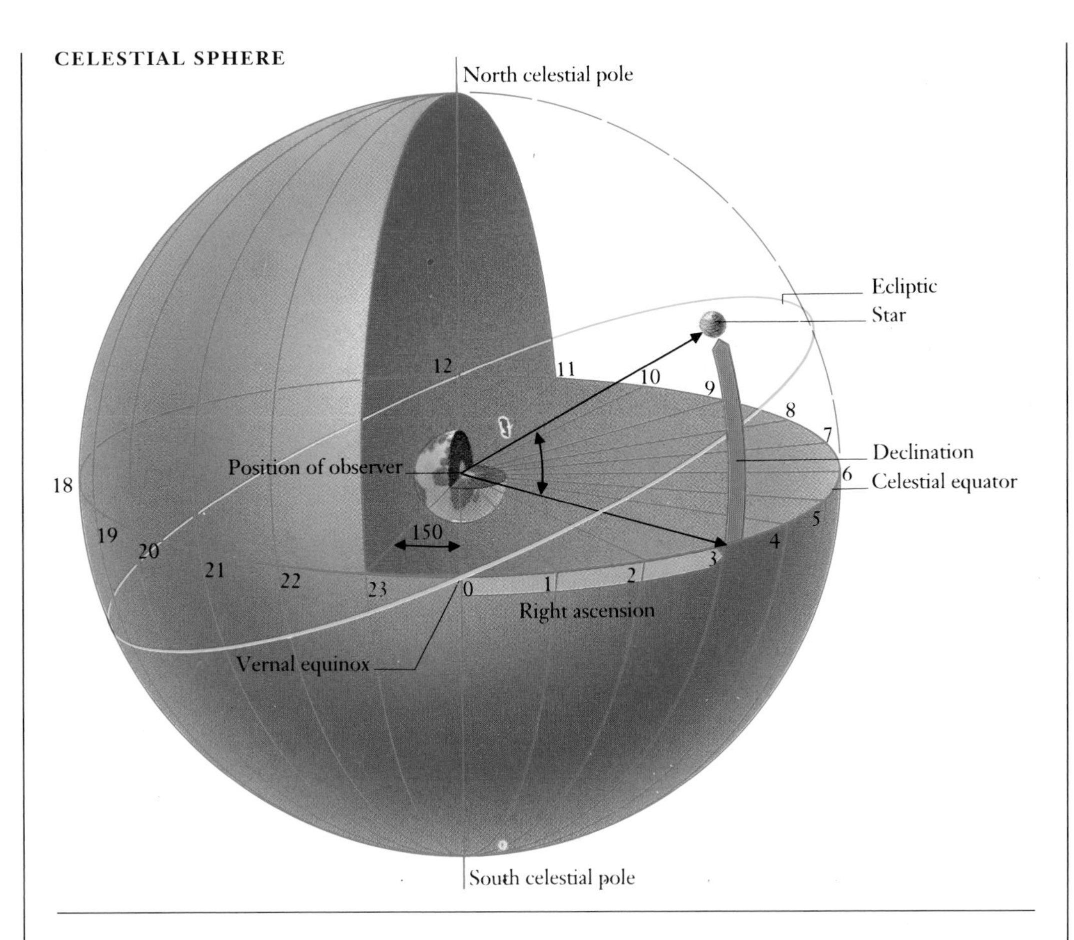

Cambrian
In the history of the Earth, the Cambrian period is the earliest part of the Paleozoic era, and is the period between about 590 million and 505 million years ago. It was at this time that shelled marine creatures first made their appearance. They included brachiopods, echinoderms, graptolites, mollusks and the dominant trilobites.

captured rotation
When the axial rotational period of an object is the same as its orbital period, it is said to have captured rotation. This is true of the Moon, which as a result always keeps the same face toward the Earth. It is also known as synchronous rotation.

carbonaceous chondrite
A type of primitive stony **meteorite** containing carbonaceous compounds and hydrated silicates. Hydrated silicates indicate that some of the original mineral grains have altered. Such meteorites are thought to be primordial Solar System debris

carbonatite
A rare **igneous rock** of extreme composition, consisting chiefly of carbonate minerals. Carbonatites are found in the East African Rift and are associated with explosively produced alkaline igneous rocks.

Carboniferous
The period in the history of the Earth from about 360 million to 286 million years ago, toward the end of the Paleozoic era. In North America the Carboniferous period is divided into the Mississippian and Pennsylvanian, which in Europe are known respectively as the Lower and Upper Carboniferous. At the beginning of the period, brachiopods and corals flourished in the shallow seas. Their fossil remains are common in limestones dating from that time. Later in the Carboniferous period, sandstone and shale were deposited in river mouths, on top of which beds of coal (from submerged swamp forests of tree ferns), sandstone, shale and clay formed the coal measures. *See also* **brachiopod**, **fossil** and **Paleozoic**.

celestial sphere
The imaginary sphere, centered on the Earth, upon which it is supposed that the stars and other celestial objects are fixed. The great circle on the sphere where it is cut by an extension of the plane of the Earth's equator is known as the celestial equator, and celestial north and south poles are similarly extensions of the terrestrial poles. The ecliptic is the projection of the Sun's apparent path onto the celestial sphere. *See also* **latitude** and **longitude.**

cement
Precipitated mineral matter between the grains of **clastic rocks** which may hold the rocks together. Cement production is a product of **diagenesis**. Silica and carbonate cements are the most common forms.

Cenozoic
The most recent geological era in the history of the Earth, following the Mesozoic and extending from about 65 million years ago to the present, and divided into the Tertiary and Quaternary periods. It is the time during which major mountains such as the Alps and Himalayas were formed, and mammals – along with birds – became the dominant animals on the planet.

Ceres
The largest **asteroid** and the first to be discovered (in 1801). Measuring 1000 kilometers across, it orbits – between Mars and Jupiter – at an average distance of about 414 million kilometers from the Sun, taking 4.6 years to complete one orbit.

chalcophile element
A chemical element that has an affinity for sulfur. Examples include copper (Cu), silver (Ag), zinc (Zn) and lead (Pb).

chalk
A white, fine-grained sedimentary rock (a type of **limestone**), composed of calcium carbonate ($CaCO_3$) and made up mainly of the skeletons of millions of microscopic sea creatures. Chalk is usually soft and easily eroded by weathering or by the sea.

Charon
The only moon of the planet **Pluto**, not discovered until 1978. It is 1200 kilometers across (nearly half as big as Pluto itself) and orbits at a mean distance of about 20,000 kilometers from its parent planet; its orbital period of 6.4 days is the same as the rotation period of Pluto.

CONNECTIONS
THE SUN'S FAMILY 60
THE OUTER PLANETS 68

chert
A sedimentary precipitate made up of very finely crystallized hydrous silica. It is formed on the sea floor and can be a replacement of limestone. Flint is one of its forms.

chondrite
A type of stony **meteorite** containing spherical inclusions known as chondrules. The latter are composed largely of the silicate minerals **olivine** and **pyroxene**, often with some glass, and show the effects of rapid cooling from a melt. The average composition of a chondrite is 12 percent nickel-iron, 46 percent olivine, 25 percent pyroxene and 11 percent plagioclase **feldspar**. Chondrites make up more than 90 percent of stony meteorites (aerolites). They are thought to be primordial Solar System debris.

CONNECTIONS
LARGE AND SMALL PLANETS 58
ASTEROIDS AND METEORITES 70

chondrule
See **chondrite**.

cirque
Also known (in Britain) as a corrie or a cwm, a cirque is a steep hollow carved out of the ground by the erosive action of a glacier. Many cirques are filled with water and create a common form of lake found in northern latitudes.

clastic rock
A rock made up of fragments of previously existing rocks, or their minerals (such as quartz).

clay
A fine-grained sediment composed predominantly of clay minerals – that is, hydrated alumino-silicates. The grain size of a typical clay is less than 1/256 millimeters. After burial, clay converts to shale or mudrock, principally by the squeezing out of the pore water and compaction of the tiny grains.

cleavage plane
A planar weakness within any mineral crystal, located along layers of weakest atomic bonding. The same term is also used to describe the tectonically-formed planes within slate and similar **metamorphic rocks**.

cloud belt
A persistent band of cloud in a planet's atmosphere, associated with the planets Jupiter, Saturn, Uranus and Neptune. Like Earth, the weather of these planets is dominated by cyclonic activity, but the rapid rotation of these non-solid worlds dominates atmospheric circulation so that strong parallel banding of the clouds can be seen (because the atmosphere rotates at different speeds at different latitudes). Unlike the Earth, where there is a distinct temperature gradient between the poles and the equator, there is little difference on the outer gaseous planets.

COMET

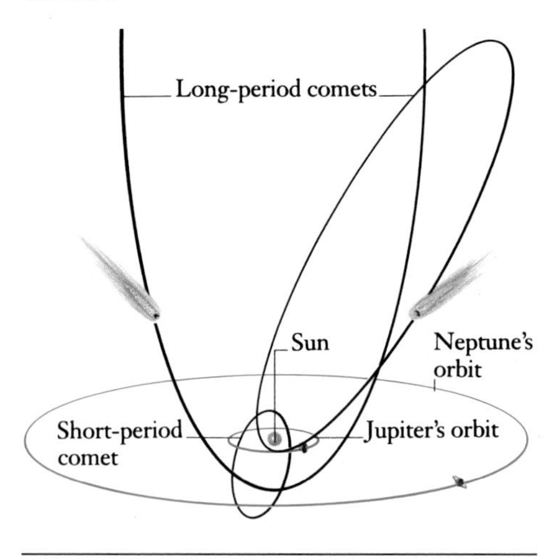

coastline of emergence
A shoreline that shows landforms such as raised beach terraces, fossil wave-cut platforms and clifflines, which indicate that there has been a relative rise in the level of the land with respect to the sea. A typical example is the northwestern seaboard of Scotland.

coastline of submergence
A shoreline that shows features such as a coast with many bays and active marine erosion concentrated at headlands. This type of coastline indicates a relative fall of the land surface with respect to sea level.

coesite
A high-pressure form of quartz (SiO_2) produced at pressures above 38,000 atmospheres. It is found in rocks impacted by large meteorites.

collision zone
A mobile zone of Earth's surface in which two or more **lithospheric plates** have come into contact. Also known as a convergence zone, it may involve oceanic-continental convergence, oceanic-oceanic convergence (eventually forming an **island arc**), or continent-continent convergence. In the last case, because of the buoyant nature of both plates, subduction does not occur; rather the rocks at the preceeding edges of the plates are buckled and thrust together as tectonic slices forming mountain ranges. *See also* **convergent margin** and **subduction**.

coma
The cloud of dust and gas that surrounds the nucleus of a **comet**. It usually forms only when a comet passes inside the orbit of Mars, when the Sun provides sufficient energy for ices within the nucleus to sublime (pass directly from a solid to a vapor).

comet
A primitive icy body of low mass that circles the Sun in a highly elliptical orbit. When a comet approaches the Sun, it may form a bright tail as ices in the nucleus vaporize or sublime, being carried away by the energetic solar wind. The nucleus – usually less than 20 kilometers across – takes the form of a "dirty snowball". The nucleus of Halley's comet, as revealed in 1986, consists of a cratered irregular mass 16 kilometers long and 8 kilometers wide, loosely held together and coated by dark carbonaceous matter. Volatile elements recorded on Halley's comet included hydrogen, nitrogen, carbon and sodium. A large number of small particles similar in type to carbonaceous chondrites were also discovered.

CONNECTIONS
THE SUN'S FAMILY **60**
ASTEROIDS AND METEORITES **70**
THE EVOLUTION OF ATMOSPHERES **84**

condensation
A physical change in which a solid crystallizes from a liquid or a liquid precipitates from a gas. The condensation of elements from the planetary nebula was the first stage in the formation of the Solar System.

conduction
A method of energy transfer in which heat is transferred through solids by molecular impact. *See also* **convection**.

continent
A body of relatively buoyant terrestrial crust composed of less-dense granitic rocks (density roughly 2,700 kilograms per cubic meter). The Earth's continents lie on average 4.6 kilometers above the ocean floor and range in thickness between 20 and 60 kilometers. The oldest continental-type rocks found so far have an age of 4 billion years. The cores of ancient continents are termed **shields**. The growth of the Earth's continents appears to have taken place largely by the accretion of successive island arcs during plate convergence.

CONNECTIONS
MOBILE AND STABLE ZONES **108**
WANDERING CONTINENTS **110**
RIFT VALLEYS **120**

continental drift
A theory, generally attributed to the German meteorologist Alfred Wegener, postulating the early existence of a single ancient supercontinent which eventually broke up, beginning to drift apart about 200 million years ago. Until his book was translated into English, in 1924, Wegener's ideas received little publicity. After that time, however, much hostility arose. Modern research has vindicated the idea and it has been established that drift is the result of **sea-floor spreading**, probably driven by movement within the Earth's mantle. *See also* **continent, Gondwanaland** and **Pangea**.

continental shelf
A gently shelving submerged part of a continent's margin that extends from the coastline to the top of the **continental slope**. Most sedimentation occurs on this part of the ocean floor.

continental slope
The relatively steep region beyond a continent's **continental shelf** which descends toward the **abyssal plain**. Along tectonically unstable coastlines, sediment may be transferred from the shelf to the abyssal plain by **turbidity currents**, which help to cut deep submarine canyons along which sediment flows. The shelf-derived sediment then builds out great fans at the foot of the continental slope.

convection
The mechanism of heat transfer within a flowing material, in which hot material from lower levels rises because it is less dense. The movement is complemented by the sinking of cooler material from near the surface. The overall motion thus generated comprises what is a convection cell.

convection cell
See **convection**.

convergence zone
See **convergent margin** and **collision zone**.

convergent margin
A region of the **lithosphere** in which lithospheric plates are driven together and crustal surface area is lost. It may be caused either by subduction, in which lithosphere is consumed into the mantle, or by crustal shortening or thickening, when slices of lithosphere are stacked upon each other as thrust slices. *See also* **subduction zone** and **collision zone**.

core
The central region of a planet in which density and temperature are highest and where the chemical elements of which it is composed tend to exist in the metallic state. The Earth's core lies below a depth of 2900 kilometers and the outer part is molten; the inner part is made up largely of iron-nickel alloy. In contrast, Jupiter's core is believed to be composed of metallic hydrogen.

CONNECTIONS
THE EARTH AND THE MOON **64**
THE INNER PLANETS **66**
THE PLANETS HEAT UP **76**
CORE, MANTLE AND CRUST **80**
PLANETARY MAGNETIC FIELDS **82**
DYNAMIC PLANETS **88**

corona
1 A circular volcano-tectonic structure on Venus, believed to have formed above a mantle **plume**. Such structures have an annulus of fractures surrounding a structural moat, radial faults and extensive volcanic flows associated with domes and shields. **2** The outer atmosphere of the Sun or other star. The term is also used for the halo seen

CONTINENTAL SHELF

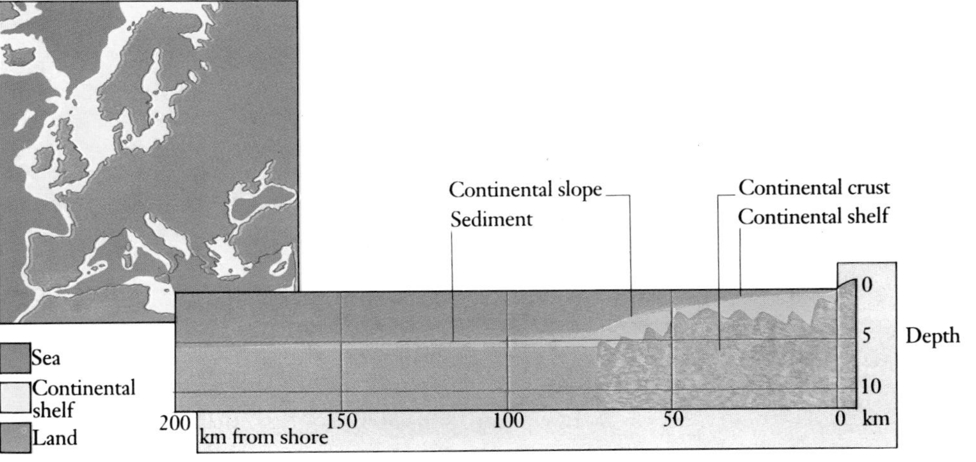

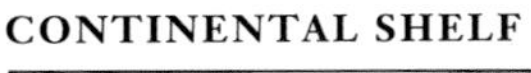

around a celestial body, due to the diffraction of its light by water droplets in thin clouds of the Earth's atmosphere.

cosmic rays
Very high-speed particles, mainly protons and atomic nuclei, which cannot penetrate the Earth's atmosphere but which interact with particles in the upper atmosphere, forming secondary particles (such as neutrons, muons, etc.).

crater
A generally circular depression formed during meteorite impact or by volcanic explosion. In the former type, the depression has a floor depressed below the exterior surface, whereas in the latter the crater floor is usually near or at the summit of a topographic ridge or volcanic cone. Impact craters are surrounded by an **ejecta blanket**.

CONNECTIONS
THE EARTH AND THE MOON **64**
THE INNER PLANETS **66**

craton
A region of ancient crust at the heart of a continent, which has evaded tectonic deformation for a protracted period.

CONNECTIONS
MOBILE AND STABLE ZONES **108**
WANDERING CONTINENTS **110**

Cretaceous
In the history of the Earth, the period between about 144 and 65 million years ago which formed the end of the Mesozoic era. During the Cretaceous, large deposits of chalk were laid down in western Europe at a time of extensive flooding. On dry land, flowering plants (angiosperms) began to appear, but by the end of the period the early dominance of dinosaur-type reptiles came to a sudden end.

critical point
The unique temperature and pressure at which a substance can co-exist in two phases simultaneously (solid, liquid or gas).

cross-bedding
Inclined planes in sedimentary rocks caused by strong currents of water or wind during deposition. For example, in a typical delta, where a flowing river drops its sediment load on reaching the deeper water of the sea, there are more or less horizontal or very gently shelving topset beds; inclined foreset beds (the delta front); and gently sloping bottomset beds which meet the flat-lying sea floor in front of the delta. Current flows in the direction of the downward-sloping strata. A similar pattern develops where dunes are formed by the wind in a desert environment. The phenomenon is also known as current bedding.

crust
The outermost layer of a planet's **lithosphere**. Compared with the mantle materials below, its rocks are relatively brittle and of lower density.

CONNECTIONS
THE INNER PLANETS **66**
CORE, MANTLE AND CRUST **80**
THE FIRST CRUSTS **90**
EARTHQUAKES AND SEISMIC WAVES **96**
EARTH'S FIRST OCEANS **100**
EARTH'S FIRST CONTINENTS **102**

cryovolcanism
A type of volcanic outbreak produced by vaporization of liquid nitrogen or methane at very low temperatures. It is the type of volcanism developed on outer planet moons, such as Neptune's moon Triton, on which geyserlike activity appears to have occurred.

crystallization
In rocks, the process whereby crystals with ordered atomic arrangements developed from atomically disordered molten magma during cooling, or as a result of **metamorphism**. Crystallization occurs in order of the decreasing melting temperatures of the solid phases (minerals) in the system.

Curie point
The critical temperature above which a ferromagnetic substance loses its permanent magnetization.

daughter isotope
An isotope of an element produced by the radioactive decay of another isotope.

decay sequence
The sequence of transformations by which a parent isotope is converted to other elements by spontaneous radioactive decay. For instance, the isotope U-238 decays to Pb-206, the process having a **half-life** of 4.5 billion years.

deep-focus earthquake
An earthquake with a focus at a depth greater than 300 kilometers. Most earthquakes of this type are found along the deeper parts of **Benioff zones**.

Deimos
The smaller, outer moon of Mars, measuring 8 kilometers across, which orbits Mars every 30.3 hours at an average distance of 23,500 kilometers.

delta
The wedge-shaped body of sediment deposited by a river at its point of entry into the sea or a lake.

desert
A dry barren region in which evaporation exceeds rainfall and there is little or no vegetation. Desert environments are dominated by aeolian processes, that is, characterized by dune development, mainly mechanical rock breakdown and development of ephemeral watercourses and lakes.

CONNECTIONS
THE WORK OF RIVERS **126**
DESERTS AND WINDS **130**

deuterium
A relatively rare natural isotope of hydrogen (abbreviated to D) that has twice the mass of the more common isotope.

Devonian
A period in the geological history of the Earth between about 408 million and 360 million years ago, in the middle of the Paleozoic era. Marine sedimentary rocks of the period contain fossils of invertebrates such as ammonites, brachiopods and corals, while the sandstones now characteristic of continental rocks contain fossils of fish and the early land plants.

diagenesis
The physical and chemical processes that affect a sediment after it becomes buried. It does not, however, include **metamorphism**.. Diagenetic products are the crystalline cements that weld many sedimentary rocks together, post-depositional concretions, and the conversion of some clays to others more stable at elevated temperatures and pressures.

diapir
A bulbous vertical body of **igneous rock** which rises into the Earth's **crust** because of its lower density compared with the surrounding rocks. Salt domes are also diapiric in nature – that is, they rise because they are lighter than surrounding rock.

differentiation
An important process by which a variety of rock types are produced by **partial melting** or **fractional crystallization** of a parent magma. In this way the various chemical components of a magma may be partitioned between different rocks. The primary differentiation of the Earth (and other planets) was achieved in a similar way.

Dione
At 1120 kilometers across, the third-largest moon of **Saturn**. Dione orbits its parent planet once every 2.7 days at an average distance of 377,000 kilometers. It is denser than Saturn's other inner moons and probably contains up to 40 percent of rocky material (the remainder being water ice).

dipole
An object that has opposite electric or magnetic poles at its ends.

divergent plate margin
A lithospheric plate boundary at which two plates are moving apart and along which there is upwelling of mantle-derived material to create new oceanic crust. It is also known as a divergence zone and is associated with such margins as mid-oceanic ridges, axial rifts and active submarine volcanism.

CONNECTIONS
EARTH'S FIRST OCEANS **100**
MOBILE AND STABLE ZONES **108**
GEOLOGICAL PLATE MOVEMENTS **112**

dreikanter
A faceted pebble sculpted by the activity of wind in a desert environment. Such pebbles are generated because they are too heavy to be lifted by the wind and are merely moved back and forth along the desert floor.

drumlin
An elongated hillock composed of sediment deposited by a glacier, with its longer axis parallel to the direction of ice movement.

ductility
The property of a solid that allows it to be drawn out, like a metal that can be drawn into wire.

dune
An elongated mound of sand constructed by wind. Dune formation generally involves the development of cross-bedded strata. Crescent dunes are known as barchans, and linear forms are seif dunes. In regions of variable wind, star-shaped dunes may occur.

dune-bedding
The **cross-bedding** resulting from the depositional action of wind during dune construction in desert areas.

dust (extraterrestrial)
Very fine-grained material that reaches the Earth's surface from outer space. Its existence was first reported by the Swedish scientist A.E. Nordenskiöld in 1874. Most is believed to result from the ablation of meteorites as they pass through the Earth's atmosphere; dust is also known to be widespread in interplanetary space.

dust tail
The part of a comet's tail that consists of solids derived from the nucleus.

dyke
A small sheetlike intrusion of volcanic or high-level plutonic rock which cuts across pre-existing strata.

Earth
The largest of the inner terrestrial planets of the Solar System. It is thought to be the only planet that supports life, and is probably the only planet with multiple tectonic plates in its crust. The Earth is a slightly flattened sphere with an equatorial diameter of 12,756.4 kilometers. It rotates on its axis, completing one rotation every 23 hours 56 minutes and 4.1 seconds (the sidereal day), and completes one orbit around the Sun every 365 days 6 hours 9 minutes and 9.5 seconds (the sidereal year). The Earth's axis is tilted at an angle of 23.5 degrees, resulting in changing day length and seasonal changes during the course of a year in higher latitudes. The planet has three zones: the **atmosphere**, the hydrosphere (the Earth's waters), and the **lithosphere** (the mobile layer above the atmosphere).

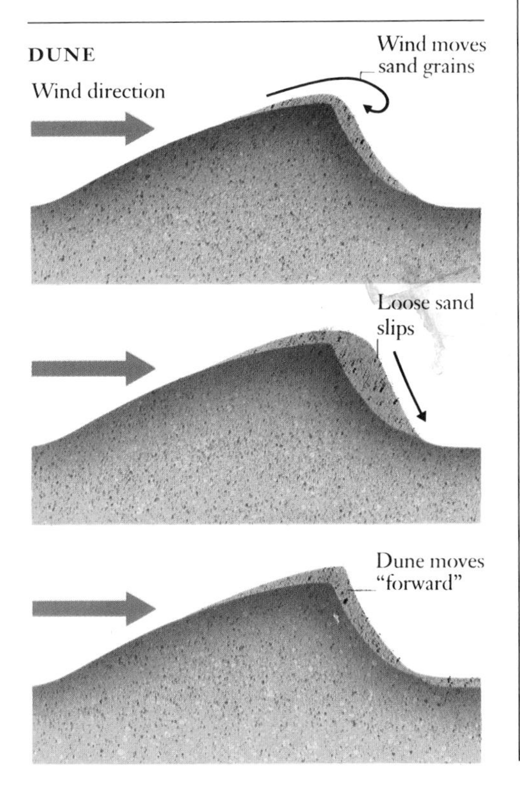

earthquake
The sudden movement of rocks along a deep-seated fault plane which sets off seismic waves generated by the abrupt release of potential energy inside the Earth. It is the resultant vibrations that are felt as earthquakes. They have been scientifically useful in allowing geophysicists to probe the Earth's interior, but can cause catastrophic damage.

CONNECTIONS
EARTHQUAKES AND SEISMIC WAVES **96**
ISLAND ARCS **116**
NATURAL CATASTROPHES **138**

eccentricity
The degree to which the orbit of an object such as a planet or moon departs from that of a circle.

echinoderm
A member of a phylum of invertebrate animals that includes brittlestars, sea cucumbers, sea urchins and starfish, some of them dating back more than 500 million years to the **Cambrian** period. Echinoderms had chalky exoskeletons, some of which have survivd as fossils.

eclipse
The partial or whole apparent disappearance of one celestial object when it passes through the shadow of another. An eclipse of the Sun occurs where the shadow cone of the Moon passes across the Earth, whereas a lunar eclipse happens when the Earth's shadow passes across the disk of the Moon. Lunar eclipses usually occur not more than twice a year, because the Moon's orbit around the Earth is tilted with respect to the ecliptic. Solar eclipses provide important opportunities for scientists to study the outer layers of the Sun. Mutual eclipses of the moons of the outer planets and of the moons by their primaries can also be observed using Earth-based telescopes.

CONNECTIONS
THE EARTH AND THE MOON **64**
THE INNER PLANETS **66**

ecliptic
The projection of the Earth's orbit onto the **celestial sphere**. It also represents the apparent annual path of the Sun and planets against the background of stars.

Einstein's equation
An equation derived by Albert Einstein as part of the theory of general relativity. It reveals that, because of the equivalence between inertial mass and energy, enormous quantities of energy are locked into an object of even quite small mass. The equation is stated as $E = mc^2$, where E = energy, m = the rest mass of an object and c = the velocity of light in a vacuum.

ejecta
The material thrown out of the cavity produced during an impact event, such as a meteorite colliding with a planet. Most ejecta consists of solid rocky fragments, but it may also include impact melt. The ejecta moves as a curtain away from the impact focus, and is deposited as the **ejecta blanket**. Lunar ejecta is located predominantly by ballistic movements of material, but the very dense atmosphere of Venus appears to have affected Venusian ejecta, which frequently shows the effects of fluid flow across the surface. On Mars, too, fluid ejecta flow has been generated by the inclusion of melted ground ice in the moving ejecta curtain.

CONNECTIONS
THE EARTH AND THE MOON 64
THE FIRST CRUSTS 90

ejecta blanket
The deposit that results after a curtain of **ejecta** moves outward from an impact point. Within it the motion of solids and gas is often turbulent, and may lead to different styles of ejecta deposition. The ejecta blanket may include smooth and featureless, dunelike or radiating materials.

electromagnetic force
The force between two electrically charged objects, one of the four fundamental forces of nature.

electron
A fundamental particle with a unit negative charge; it makes up part of every atom.

element
A substance that cannot be broken down into others by normal chemical or physical processes, characterized by the number of protons in its nucleus (atomic number).

Enceladus
The brightest and fifth largest of **Saturn**'s moons, with a diameter of 510 kilometers, Enceladus orbits at an average distance of 240,000 kilometers from its parent planet. Uncratered regions on its pitted surface suggest that it may still be geologically active.

Eocene
In the Earth's geological history, an epoch that lasted from about 55 million to 38 million years ago (during the early Tertiary period). During the Eocene, mammals were the dominant creatures, with hoofed animals (and carnivores that preyed on them) and whales evolving for the first time.

ephemeris
A set of tables for astronomers and navigators, published every few years, that give details of the future positions of the Sun, Moon and planets.

epicenter
The point on a planet's surface immediately above the focus of an **earthquake**.

equator
An imaginary line (drawn as a great circle) that encircles the Earth at zero latitude, and is therefore equidistant from the poles, dividing the Earth into the Northern and Southern Hemispheres. By analogy, other planets have an equator in the plane that is perpendicular to their axis of rotation. *See also* **latitude, longitude**.

equinox
Either of the two points where the Sun crosses the celestial equator. They occur on or about 22 March (vernal equinox) and 22 September (autumnal equinox). On these days the day and night are of equal length throughout the world, 12 hours each.

Eros
A small, irregularly shaped **asteroid** (35 by 16 by 8 kilometers) with a highly elliptical orbit that every 7 years brings it close to the Earth (sometimes within 22 million miles). It completes each orbit every 643 days.

erosion
All the surface processes that cause loose materials to move downhill or downwind. Erosion may be achieved by mechanical or chemical weathering of rocks, followed by transportation of the fragments to a depositional site.

eruption plume
The cloud of gas, solids and molten magma that rises above a volcanic vent or fissure. The height to which it rises depends on the force of the initial event, the gravity on the planet, and the density of the atmosphere.

esker
A winding ridge of glacial debris left behind by streams flowing beneath a sheet of ice.

estuary
An inlet, usually funnel-shaped, created when a relative rise in sea-level inundates the mouth of a river.

eukaryote
A living cell that has a nucleus. Eukaryotes first appeared about 1.2 billion years ago.

Europa
The smallest of the four moons of **Jupiter** discovered by Galileo in 1610, with a diameter of 3126 kilometers (not much smaller than Earth's Moon). It completes its orbit, 671,000 kilometers from its parent planet, once every 3.55 days.

exfoliation
A common weathering process in which shells of rock are removed from an outcrop. This typically produces what is known as "onion-skin weathering".

exosphere
The region at the very edge of a planet's atmosphere in which fast-moving particles may either escape completely or enter orbit around it. For the Earth, the exosphere lies between altitudes of 400 and 500 kilometers.

extinction
The abrupt demise of a group of plants or animals.

fault
A fracture in a body of rock along which some movement has occurred. Normal faults are caused by extension of the crust, and may occur in pairs to form **graben**. Individual fault planes tend to be steep, with the upper block having moved down relative to the lower. Where crustal shortening is involved, faulting may occur along thrusts, giving rise to reversed faults. In these, the block above the fault plane moves up and over the one below. In other cases, where there is lateral movement of adjacent crustal blocks with little or no vertical slippage, "strike-slip" or **transform faults** develop.

CONNECTIONS
MOUNTAINS FROM THE SEA 118
RIFT VALLEYS 120

feldspar
The general name for a group of alumino-silicate rock-forming minerals which have an atomic structure in which a framework of oxygen atoms are shared by adjacent silicon atoms. They are of considerable importance and occur in a wide variety of terrestrial and lunar crustal rocks. They can be divided into two families: the alkali feldspars, with the general formula $(K,Na)[AlSi_3O_8]$, and the plagioclases, which have the general formula $Na[AlSi_3O_8] - Ca[Al_2Si_2O_8]$.

feldspathoid
A mineral group with similar chemistry to **feldspar** but a lower silica content. Feldspathoids crystallize in igneous rocks that are undersaturated with respect to silica – they have insufficient silica in them to accommodate all of the other cations present in the "normal" rock-forming silicates. Nepheline ($Na_3\{Na,K\}[Al_4Si_4O_{16}]$) and leucite ($K[AlSi_2O_6]$) are examples.

felsic
Any rock that consists of light-colored minerals such as **feldspar** and **quartz**. Felsic minerals tend to have low density. They are characteristic of the rocks that are found in continental regions.

FAULT

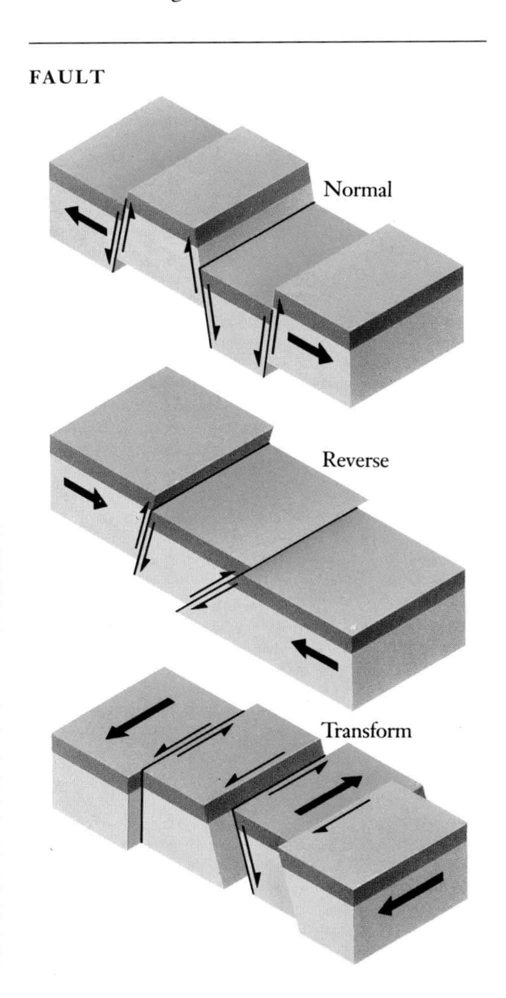

FOLD

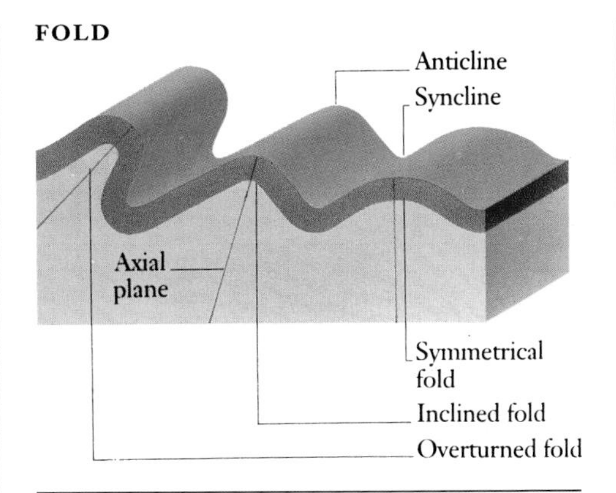

field reversal
A phenomenon in which the Earth's magnetic polarity is reversed. Currently the Earth's magnetic field affects a freely suspended magnetized needle such that its north-seeking pole points toward the north magnetic pole. Paleomagnetic data indicate that there have been numerous periods when it would have pointed toward the south magnetic pole. Studies of both continental and ocean-floor rocks show that rocks of the same age always show the same magnetic polarity. There have been at least 20 field reversals during the last 5 million years. *See also* **polarity reversal.**

CONNECTIONS
THE GEOLOGICAL JIGSAW **104**
WANDERING CONTINENTS **110**

fluviatile
Describing any processes associated with flowing rivers and streams.

fold
A phenomenon affecting sedimentary rocks. Most are deposited as horizontal strata, but during **orogenic** movements they may be warped and folded by compression. Folds may be simple open buckles with opposite limbs dipping away from each other, or tightly-packed isoclines in which the fold limbs become parallel to each other. The latter may flop over, becoming recumbent.

fold belt
A long, narrow, zone of planetary crust in which there has been strong deformation and intense development of folds. Such belts usually develop along continental margins associated with convergent plate boundaries. Fold belts have been recognized on Venus.

fold mountains
A mountain massif produced by compressional deformation that generates folding. The European Alps and the Appalachians are typical examples. *See* **fold.**

fractional crystallization
When a magma (molten rock) cools, crystallization takes place. The crystals that grow as the temperature falls do so in a sequence determined by the melting points of the various components. Thus, in a typical basaltic magma, the magnesium-rich silicate mineral **olivine** crystallizes first, thus removing magnesium from the melt. Fractional crystallization has occurred, because the remaining magma is now depleted in magnesium with respect to the other components. As other silicate species enter the solid phase (the temperature continuing to drop), so further fractionation of the melt takes place. It is this process which causes the diversity of igneous rock types found in nature.

CONNECTIONS
CORE, MANTLE AND CRUST **80**
ROCKS FROM WITHIN THE PLANET **92**

fusion
The combination of the nuclei of lighter elements to form heavier ones. It also refers to the melting of solid rocks and/or their blending together.

gabbro
A coarse-grained **igneous rock** which is chemically equivalent to basalt. Essential minerals found in gabbro are the calcium-rich plagioclase **feldspar** and **pyroxene**, usually the variety augite.

galaxy
A massive rotating star system which may consist of more than a million stars, possibly as many as 100 billion. Galaxies occur in several different shapes, such as spirals and ellipses. The Milky Way Galaxy, of which the Sun and its Solar System are members, is a disk-shaped spiral galaxy, having a bulbous core from which radiate a number of spiral arms. The Milky Way contains around 100 billion stars. The Sun is located about three-fifths of the way from the center, along one of the arms.

Galileo mission
A NASA spacecraft traveling to Jupiter, expected to arrive in 1995. On its way it returned several close-up images of the asteroids Gaspra and Ida, both of which were shown to be irregular bodies with cratered surfaces. Galileo was intended to transmit several thousand pictures and send an entry probe into the top of Jupiter's atmosphere.

Ganymede
The largest of **Jupiter**'s four moons (discovered by Galileo Galilei in 1610). It is 5276 kilometers across (larger than the planet Mercury) and completes each orbit – at a distance of 1,070,000 kilometers from the parent planet – every 7.12 days.

gas
A substance which, at everyday temperatures and pressures, is neither solid nor liquid.

geochronology
The technique of deriving absolute ages for rocks through the measurement of the spontaneous decay of long-lived radioactive isotopes. *See also* **radiometric dating**.

geoid
A term introduced to describe the shape of the Earth: a level surface that lies close to that of the ocean surface. When scientists refer to the shape of the Earth they are usually referring to that of the geoid. Its form can be determined by gravity observations at the surface. Gravity varies from place to place at the Earth's surface – a reflection of the asymmetric distribution of mass inside the planet. Because the Earth rotates, it is oblate (flattened at the poles), and as a result any element of mass on the Earth is not only subject to the gravitational attraction of the rest of the planet but also to a centripetal force. If the Earth could respond as if it were a fluid, it would have the smooth shape of an oblate ellipsoid. However, it cannot. Because the internal mass distribution is asymmetric, in places the geoid lies above the ideal ellipsoid, and in other places it is below it.

geothermal gradient
The natural increase in temperature that occurs with increasing depth in the Earth's (or any other planet's) **lithosphere**. Within the Earth's upper crust the average gradient is 30°C per kilometer.

geyser
An intermittently active vent that ejects superheated steam or hot water into the air. In volcanically active areas such as Iceland, geyser activity appears to be the result of contact between magma and groundwater circulating in the upper crust. Geysers have also been identified on certain outer planet moons, where forceful ejection of icy nitrogen compounds has taken place.

Giotto mission
The successful ESA (European Space Agency) probe that visited Halley's comet during 1986. Launched in July 1985 via an *Ariane* rocket, the 960-kilogram probe encountered the comet while traveling at a velocity of 86 kilometers per second. It survived the encounter and sent back excellent images of the cratered nucleus of the comet and a variety of non-imaging data. In July 1992 Giotto flew by another comet, Grigg-Skjellerup, although its camera could not be reactivated.

glacial
A period of relative cold and consequent ice advance during an ice age. Glacial periods are interspersed with interglacials, a period in which the climate becomes warmer and there is ice retreat.

CONNECTIONS
THE ICE AGES 106
GLACIERS AND ICE 132

glaciation
An increase in the extent of the ice cover of a planet. Glaciations have punctuated Earth's history, the earliest having occurred during Proterozoic times. A glaciation may include both **glacials** and **interglacials**.

glacier
A thick mass of ice and compacted snow that moves slowly downhill and persists throughout the year for periods of up to 100 years. The largest glaciers on Earth are located in the polar regions. Glacial activity is also known to exist on Mars.

CONNECTIONS
THE ICE AGES 106
GLACIERS AND ICE 132

Glossopteris flora
Widely scattered fossilized remains of the seedfern Glossopteris and associated species, which were buried among the rocks of what are now the southern continents on Earth, during the Permo-Carboniferous period before the breakup of the supercontinent **Gondwanaland**. Since that time Gondwanaland has split apart, but the flora remain as evidence of continental drift.

gneiss
A coarse-grained **metamorphic rock** which exhibits alternate bands of light and dark minerals. Such rocks have formed at considerable depth inside the Earth, within mobile belts. Granitic gneisses are frequently associated with granites in cratonic continental core regions. *See also* **core, craton, granite** and **mobile belt**.

GONDWANALAND

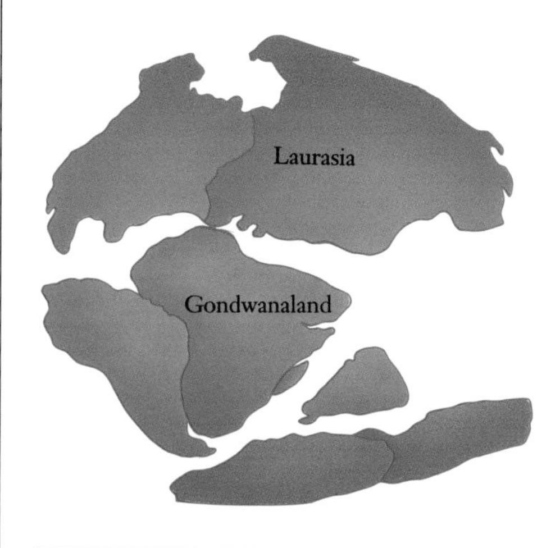

Gondwanaland
The southern part of the supercontinent of **Pangea**, comprising the ancestral continents of Antarctica, Australia, India, Africa and South America. During Proterozoic times (more than 1.1 billion years ago) Gondwanaland and North America were sometimes joined, but at other times were split apart. Furthermore, it is also apparent that one form of Pangaea existed in Late Paleozoic times, 200 million years ago, when northern Africa, the northern coast of South America and the eastern coast of North America were joined. To the east, however, the two parts of the continent were widely separated by an ancient ocean, the Tethys Ocean. Gondwanaland finally broke up during the Mesozoic era and continental drift has carried them apart ever since.

CONNECTIONS
EARTH'S FIRST CONTINENTS 102
WANDERING CONTINENTS 110

graben
A downfaulted block of crust bordered by a pair of normal faults, caused by extensional deformation. Graben formation is common on the crest and flanks of rising crustal domes, such as the Kenya Dome in East Africa, and on Venus in the region known as Beta Regio.

granite
A hard, coarse-grained **igneous rock** that consists mainly of quartz and feldspar, often with mica or other colored minerals. Most regions of granite have resulted from the crystallization of molten **magma**, although some granites may have formed through the metamorphism of other, existing rocks. The extrusive equivalent is rhyolite.

granodiorite
A coarse-grained igneous rock similar in composition to granite, but containing more plagioclase **feldspar** than the potash variety. The average composition of the Earth's continental crust is very close to this rock. The extrusive (volcanic) equivalent is dacite.

graptolite
A member of a group of extinct invertebrate animals that were abundant in the Paleozoic era (from about 590 million years ago). Graptolites lived near the surface of the oceans, like modern plankton, and are known from fossils of their simple outer skeletons. Most graptolites died out at the end of the **Silurian** period.

gravitational energy
Gravity is a fundamental, if weak, force of the Universe; however, where large masses are concerned, it is the dominant one. As a result of the mutual attraction between objects – which increases proportionally to their masses – huge amounts of potential energy are stored inside them. During the relatively rapid sinking of dense iron droplets toward the Earth's center during core formation, a massive amount of stored energy was released and converted to heat.

gravity anomaly
A discrepancy in gravity measurements between the hypothetical measurement for an ideal ellipsoid Earth and the corrected measurement taken in practice. Gravity measurements made at the surface of the Earth vary from place to place. This is a function of the Earth's oblateness, the variable distance from the center of mass and the asymmetrical distribution of mass inside the planet. The measurements for the first two can be corrected by applying what are termed free-air and Bouguer corrections. Anomalies of short wavelength or extending over small areas only reflect density anomalies that lie close to the surface. Those that persist over large areas tend to represent mantle anomalies. Interpretation of gravity data reveals that there is a correlation between gravity and topography, but that gravity variations are less than if they were due to topographic effects alone. Elevated mountainous regions are associated with a deficit of mass (negative anomalies), whereas depressed regions such as oceanic ridges show positive anomalies – a surplus of mass. The former results from the existence of a root of relatively non-dense continental crust beneath fold mountain ranges. In the latter case, the crust is thinner and dense mantle lies close to the surface. In both cases, the density anomalies partly compensate for the surface load.

gravity segregation
Settling of the early formed crystals within a magma body occurs because they are denser than their surroundings. As a result they accumulate at or near the floor of the magma chamber and become segregated from the remaining melt. This is one means whereby **fractional crystallization** occurs.

Great Bombardment
A period during the early part of Solar System history, when meteorites struck the planets, forming major impact basins and large craters. During this time, cratering was so intense that solid surfaces were saturated with craters, to the extent that existing craters were gradually destroyed by newer ones. The rate (flux) of impact appears to have declined around 4 billion years ago.

Great Red Spot
The most distinctive feature of the planet Jupiter's cloud belts, known to have persisted for at least 300 years. It is a huge weather system measuring approximately 26,000 kilometers in length and 13,800 kilometers in width, and generally has a distinctly reddish hue. The Spot is cooler than its surroundings, and is therefore probably an elevated region of high pressure. Temperature data show that the Spot has a cooler outer region above its center; this is consistent with a slow upward movement of the gases in the upper atmosphere, at a rate of a few millimeters per second.

greenhouse effect
The gradual rising of the temperature of air in the lower atmosphere, which is believed to be due to the buildup of gases such as carbon dioxide, ozone, methane, nitrous oxide and chlorofluorocarbons. Light energy from the Sun passes relatively readily through the Earth's atmosphere and is re-emitted from the surface at longer wavelengths which are reflected by carbon dioxide in the atmosphere. This contributes to heating of the atmosphere from below. As the proportion of carbon dioxide increases, as has happened on Venus, most of the incoming solar radiation is trapped and a "runaway greenhouse" develops, with a rapid and unstoppable rise in temperature. *See also* **ozone layer**.

CONNECTIONS
THE EVOLUTION OF ATMOSPHERES **84**
PLANETARY ATMOSPHERES TODAY **98**

greenstone belt
An association of ultrabasic and basic volcanic and sedimentary rocks that occurs among **Archean** terrains on most continents. The composition and textures of certain of the lavas, known as **komatiites**, within such sequences suggest that in those distant times mantle temperatures must have been higher than at present.

ground ice
A body of ice locked up in pores within a planet's **regolith**. Considerable quantities of volatiles are believed to be stored in this way on Mars, where it is too cold for water to exist at the surface.

half-life
The period it takes for half of a sample of a radioactive parent isotope to decay to its daughter product.

Halley's comet
The first "returning" **comet** to be identified (by the English astronomer Edmund Halley toward the end of the 17th century), with a period of about 76 years. Its frequently dramatic appearance in the sky has been recorded for 2000 years. At its reappearance in 1986, five spacecraft were sent to examine it; the European probe **Giotto** passed within 605 kilometers of the comet's nucleus.

GREENHOUSE EFFECT

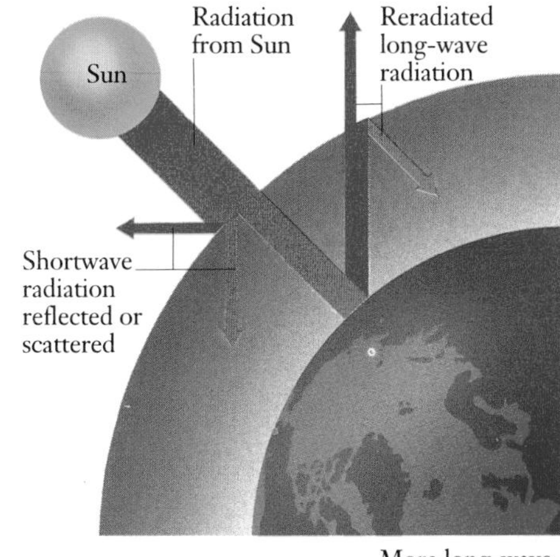

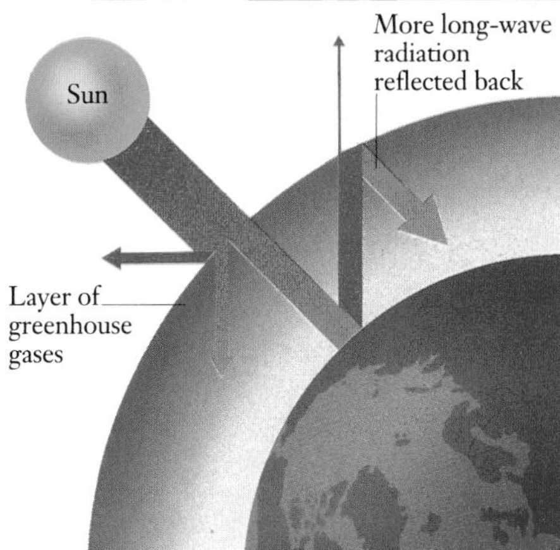

harzburgite
An ultrabasic type of rock, rich in **olivine,** considered to be very like the material of the Earth's mantle. Such rocks are occasionally brought to the surface as exotic blocks in volcanic explosion pipes.

heat flow
The rate at which internal heat is brought to the surface of a planet. It is a function of the internal rocks and the rate at which energy is supplied to the crust from below. The Earth's mean heat flow has been estimated to be 60 milliwatts per square meter.

heavy elements
Elements of atomic number higher than iron (Fe, atomic number 26). These elements are believed to have been synthesized in supernovae explosions.

Holocene
The most recent epoch of geological time, beginning about 10,000 years ago; also known simply as Recent or, sometimes, Postglacial (because it followed the end of the Pleistocene Ice Age).

hominid
A member of the zoological family which includes mankind and predecessors of the human race who had a primarily upright walking gait.

horst
An upfaulted crustal block. Horsts often occur alongside downfaulted blocks, known as **grabens**.

hot spot
A place at which a deep-seated mantle **plume** causes hot mantle material to rise toward the base of the planetary crust, producing high heat flow and volcanism at the surface. The Hawaiian islands lie above such a hot spot.

CONNECTIONS
VOLCANOES 94
GEOLOGICAL PLATE MOVEMENTS 112
BENEATH THE OCEAN FLOOR 114

hydraulic action
The destructive effect of water. Sea waves, for example, have enormous energy and, particularly during storms, exert massive pressures against obstacles such as headlands and breakwaters. Frequently the water is forced into joints and along fault planes, exerting hydraulic pressures which may become so great that the rocks are shattered.

hydrocarbon
An organic chemical compound consisting predominantly of hydrogen and carbon. Coal, oil and natural gas are three commonly occurring hydrocarbons.

hydrolysis
In geology, a weathering process in which rock-forming minerals interact with water and weakly acidic fluids, causing them to be broken down into other minerals.

hydrostatic pressure
Also known as confining pressure, the pressure exerted upon crustal rocks by the overlying load; it does not include a component of directed stress.

hydrothermal
Any process involving hot fluids. Hydrothermal solutions emanate from most magmas and are effective at altering rocks through which they penetrate. Much mineralization is achieved through hydrothermal processes.

ice age
A protracted period of cold climate during which there is a general increase in the surface area of glaciers and ice caps. *See also* **glacial; glaciation** and **interglacial**.

CONNECTIONS
THE OUTER PLANETS 68
THE ICE AGES 106
GLACIERS AND ICE 132

igneous intrusion
A body of igneous rock that has risen and forced its way into the crustal rocks. Such intrusions range from enormous **batholiths** to tiny **dykes** and sills.

igneous rock
Any rock formed by the crystallization of a **magma**. The classification of such rocks subdivides them into plutonic (intrusive or deep-seated) and volcanic (extrusive) types. Further subdivision is based on chemistry and mineralogy. Igneous rocks with a silica (SiO_2) content above 66 percent are termed acid or silicic; those with 66–52 percent are termed intermediate, those with 52–45 percent are known as basic, and those with less than 45 percent are ultrabasic.

imbricate structure
A formation resulting from the stacking up of elongated pebbles by a stream, giving rise to an overlapping series in which the long axes point upstream. The term also applies to deformed rocks in which a series of thrust slices are stacked up like a leaning deck of playing cards.

impact basin
An impact crater with a diameter in excess of 200 kilometers, formed when an object such as a meteorite strikes a moon or planet. Impact basins range from what are in effect large craters, with terraced walls and sunken floors that may have a central peak, to structures with a multiring form which may measure 1000 or more kilometers across. The youngest and best preserved lunar impact basin is Mare Orientale (largely on the far side of the Moon), in which the outer ring of mountains is as high as 4000 meters in places; it has a diameter of 900 kilometers. Associated with such basins are extensive regions of **ejecta**, which may blanket smaller craters completely. Some areas were emplaced ballistically, although elsewhere there is evidence for flow across the surface. Large impact basins occur also on Mars and Mercury. Most were formed during the **Great Bombardment** period, probably before 4 billion years ago.

impact breccia
Large quantities of broken and shattered rocks produced by major impacts, particularly when a meteorite strikes the surface of a planet. Some fragments fall back into the crater cavity, and tend to consist of a relatively restricted number of rock types. They may include droplets of impact melt that were formed by the intense heating of the cavity wall rocks and by the impacting body itself. Other fragments were ballistically ejected or, mixed with hot gases, flowed out across the surface, later to be deposited as "polymict breccias" with a more varied mixture of constituents. Microscopic features of the clasts within such rocks are shock lamellae and deformed twin planes; high temperature and pressure polymorphs of silica, such as coesite and stishovite, are also present. Around the cavity itself, the country rocks will be highly fractured and may include **shatter cones.**

index mineral
During the regional **metamorphism** of crustal rocks during mountain building, existing rocks recrystallize and new minerals grow which are stable under the changed conditions of temperature and pressure. Some of these may be used as an index of the "grade" of metamorphism that has affected the original rocks. Index minerals include chlorite, biotite, garnet, kyanite and sillimanite. The presence of such minerals indicates zones of metamorphism.

IONOSPHERE

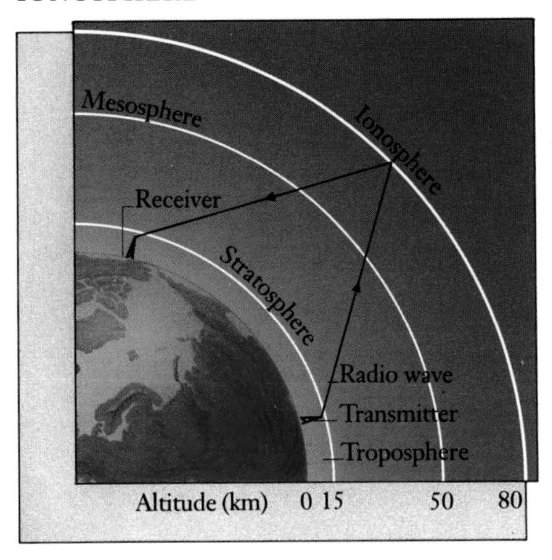

inert gas
Any of a group of nonreactive gases which take part in hardly any chemical reactions, also known as rare, or noble, gases. Five of them are present in the Earth's atmosphere: helium (He), neon (Ne), argon (Ar), krypton (Kr) and xenon (Xe).

insolation
Solar radiation that reaches the Earth's surface. It varies with latitude, the Sun's height in the sky and the length of daylight.

intercrater plain
A volcanic plain laid down between or over craters on Mercury, the Moon and Mars. This was part of a process of resurfacing after the period of early bombardment in the Solar System left the planet surfaces deeply pockmarked by basins and craters.

interglacial
A period of relatively warm climatic during an **ice age**, when glaciers and ice sheets diminish in area.

interstellar dust
See **dust**.

invertebrate
Any animal that has no backbone.

Io
The innermost of **Jupiter**'s four Galilean moons (discovered by Galileo in 1610). At 3632 kilometers across, it is the second smallest. It orbits extremely rapidly – once every 1.77 days – at an average distance of 421,600 kilometers from Jupiter. Io has highly active volcanoes on its surface.

ionization
The process of stripping an electron from the outer orbit of an atom. For example, if the one electron from a hydrogen atom is removed, a H^+ ion results.

ionosphere
The region of a planet's atmosphere in which the gases are ionized by X rays and ultraviolet radiation from the Sun. Variations in solar activity alter the concentration of ions in the ionosphere and possibly influence the frequency or intensity of thunderstorms. For the Earth, the ionosphere is above a height of about 150 kilometers.

ion tail
A tail generated by subliming or vaporizing material within the nucleus of a **comet** as it approaches the Sun and heats up. The gaseous material released gives rise to a tail which is usually almost straight, in contrast to the **dust tail** which is gently curved.

iridium anomaly
See **K/T boundary event**.

iron meteorite
A dense **meteorite** composed essentially of metallic iron and nickel, also known as a siderite.

island arc
An arced chain of oceanic islands characterized by intense seismic and volcanic activity. An arc lies parallel to a nearby deep trench in which active subduction is taking place (*see* **subduction zone**). Island arcs thus develop along destructive plate margins. Two examples are the Indonesian archipelago and the Hellenic Arc which includes the Aegean islands between mainland Greece, Santorini and Nisyros.

CONNECTIONS
EARTH'S FIRST CONTINENTS 102
ISLAND ARCS 116

isochron
A curve on a graph that illustrates the decay of one radioactive isotope pair to another and connects points of equal age.

isostasy
The principle that the rocks of the Earth's crust "float" on those of the underlying mantle. The oceanic crust is made from dense basaltic rocks, whereas the upper part of the continental crust is built largely from felsic rocks of relatively low density, with oceanic-type mafic material below. To compensate for the relative lack of density, the continents have developed deeper "roots" of less dense materials, which is why the continents stand higher than the ocean floors. This results from the greater buoyancy of the felsic materials compared with the oceanic rocks. In so doing they achieve a measure of "isostatic equilibrium". As a result, the Earth's surface is at two distinct levels.

isotope
One of a number of forms of an element which has the same number of protons in the nucleus but a different tally of neutrons (and therefore a different relative atomic mass).

CONNECTIONS
COSMIC INGREDIENTS 52
DATING THE ROCKS 86

isotopic ratio
The ratio of the amount of one isotope to another in the same rock. It can be an important indicator of origin or of genetic links between rocks.

jet stream
High-velocity winds that persist within the atmosphere of a planet, usually above the zone of the prevailing winds.

joint
A plane of fracture within a rock not associated with relative movement on either side.

Jupiter
The largest and most massive planet in the Solar System, with an equatorial diameter 143,000 kilometers; its mass 317.8 times the mass of the Earth. The fifth planet out from the Sun, which it orbits once every 11.86 Earth years, Jupiter is larger than all the other planets put together, and rotates rapidly on its axis once every 9.92 hours, causing marked flattening at the poles. It has 16 moons, largest of which are **Callisto**, **Europa**, **Ganymede** and **Io**.

CONNECTIONS
THE PLANETS FORM 56
LARGE AND SMALL PLANETS 58
THE PLANETS AND THEIR ORBITS 62
THE OUTER PLANETS 68
PLANETARY MAGNETIC FIELDS 82

Jurassic
The period in Earth's geological history that lasted from about 213 million to 144 million years ago. Best known as the age of the dinosaurs, it also saw the evolution of flying reptiles and primitive birds which made their homes in forests of conifers and tree ferns.

Ammonites, bivalves and brachiopods remained abundant in shallow seas.

Kepler's laws of planetary motion
Three laws formulated by the German astronomer Johannes Kepler (1571–1630). The first law states that planets move around the Sun in an elliptical orbit, with the Sun located at one focus of the ellipse. The second law states that the radius vector (a line drawn between the planet and the Sun) sweeps out equal areas of space in equal times; and the third law that the square of the sidereal period (the time it takes to complete one orbit of the Sun) is directly proportional to the cube of the mean distance between the planet and the Sun.

kinetic energy
The energy of a moving object. Kinetic energy is equivalent to half the mass of the object times the square of its velocity.

komatiite
An extrusive **igneous rock** that has a composition of **peridotite**. Such rocks usually contain peculiar branching crystals which suggest that they cooled rapidly. Komatiites are believed to have been widespread in **Archean** times. They are the precursors of basaltic crustal rocks.

K/T boundary event
A controversial explanation of the widespread animal extinctions that took place at the close of the **Cretaceous** period, about 66.3 million years ago. At that time, previously flourishing forms such as the dinosaurs, ichthyosaurs, pterosaurs, ammonites and belemnites were wiped out completely, while bony fishes, planktonic foraminifera, reef corals and several kinds of shallow-water shell creatures suffered badly but survived into the **Cenozoic** (Tertiary) period. In contrast, amphibians, mammals, sharks, calcareous algae and plants and insects all survived and continue to flourish.

The groups that became extinct shared one of two characteristics: either they could not live in high temperatures (or their eggs could not), or they had to come ashore to lay their eggs. The one climatic factor which could therefore result in their demise would have been a rise in world temperatures. This would not really have affected the survivors, because they did not live in regions of high temperature, were small and did not have a problem with getting rid of body heat, or simply were heat-tolerant. If the rise in temperature had been gradual, living forms would have adapted or moved elsewhere. However, it appears that it was very quick; evidence from **stratigraphy** (the study of rock strata near the surface of the planet) indicates almost instantaneous extinction.

LATITUDE, LONGITUDE

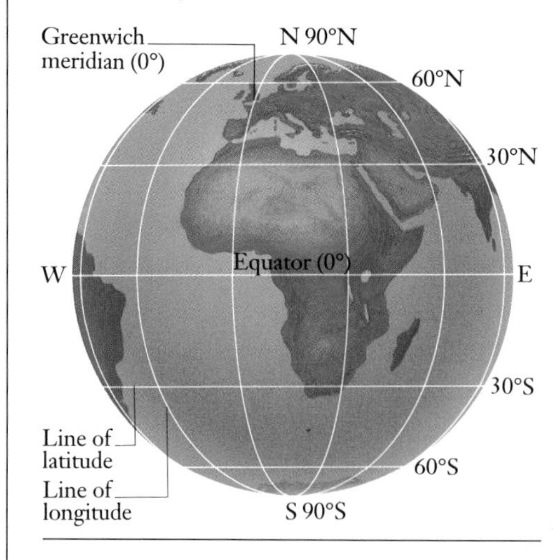

In 1978, analysis of concentrations of the element iridium in a certain rock **horizon** in central Italy showed it to be 150 times higher than in the strata above and below. This element is now rare in the outer layers of the Earth, most having sunk toward the core; however, it would have been relatively abundant in the original cosmic material and therefore in meteorites and asteroids. This iridium anomaly has been traced elsewhere within the stratigraphical succession and always occurs at the Cretaceous/Tertiary (K/T) boundary. It could have been produced by the impact of a 10-kilometer meteorite onto the Earth. Scientists now believe they have traced the meteorite to two possible sites, one in Iowa (where it formed the 32-kilometer Manson Crater) and another fragment in the Gulf of Mexico (which dug out the 180-kilometer Chicxulub Crater). The object must therefore have broken into two pieces before hitting the ground, instantly vaporizing and sending huge clouds of debris into the stratosphere. This would not only have caused darkness for several weeks but would also have raised the global temperature dramatically. Eventually the debris settled to the ground and was deposited among the normal sedimentary deposits, alongside meteoritic iridium.

lahar
A mudflow, a dangerous mixture of hot or cold water, mud and rocks which may cause widespread damage and loss of life during volcanic eruptions.

Landsat
A series of satellites originally called ERTS (Earth Resources Technology Satellite) but changed two years after the first of the series of five craft was launched, during July 1972. Landsats have proved to be highly successful for remote sensing of the Earth, and have provided a great deal of data on land use, vegetation, pollution and so on.

landslide
A type of mass movement in which rocks slump down either curved shear planes, bedding planes or faults.

lateral moraine
A glacial deposit of material scoured from the sides of a valley **glacier**. *See also* **glaciation** and **moraine**.

latitude
An imaginary line that runs round the Earth parallel to the **Equator** (which is 0° latitude). There are 90 degrees of latitude between the equator and each of the poles, so that latitude indicates a position north or south of the equator. Together with **longitude**, it defines the precise position of any place. Lines of latitude are also called parallels.

Celestial latitude indicates the angular distance of a heavenly object north (positive) or south (negative) of the **ecliptic**. Together with **celestial longitude** it can be used to indicate the location of any object on the **celestial sphere**.

lava
Molten rock or **magma** which reaches a planet's surface, often from an erupting **volcano**. Most lavas, such as basalt, are basic and flow for considerable distances. *See* **aa**, **basalt**, **pahoehoe**, **pillow lava**.

LANDSAT

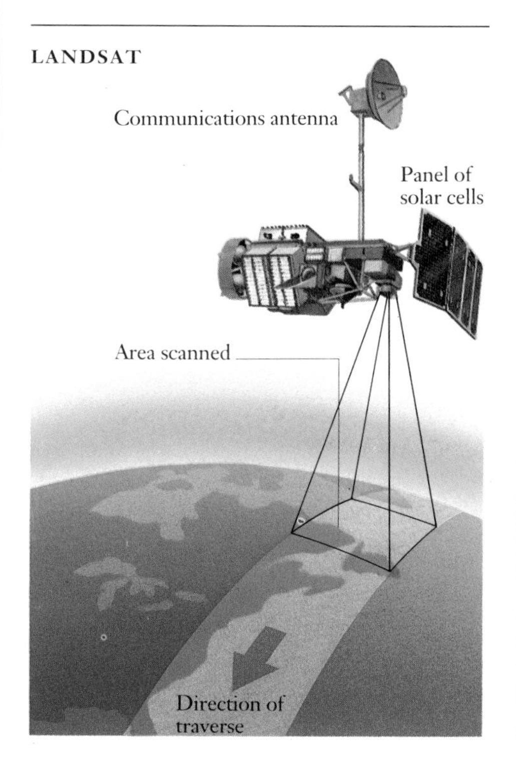

layered intrusion
During the **fractional crystallization** of basic magmas it is common for crystals of different minerals to sink because of their greater density, to form what are called cumulate rocks. Thus the intrusion solidifies largely from the base upward, as a series of layers which resemble sedimentary stratification. Such layered intrusions exist beneath modern mid-oceanic ridges.

light year
A unit of distance in astronomy, equal to the distance light travels in one year (equal to 9.465×10^{12} kilometers). For example, the nearest star (after the Sun) is about four light years distant.

limestone
A sedimentary rock composed largely or entirely of calcium carbonate ($CaCO^3$). Limestones may be made from carbonate grains and shells laid down in much the same way as quartz sand, chemically precipitated from sea water, or formed by the growth of colonial organisms such as corals and bryozoans.

lithophile element
An element that has an affinity for the silicate phase, for example aluminum, sodium and calcium.

lithosphere
The rigid outermost layer of a planet. On the Earth it lies above the **asthenosphere** and includes the **crust** and upper part of the **mantle**, and is approximately 100 kilometers in depth. The Earth's lithosphere is segmented into plates.

CONNECTIONS

THE INNER PLANETS **66**
EARTHQUAKES AND SEISMIC WAVES **96**
GEOLOGICAL PLATE MOVEMENTS **112**

lithospheric plate
A large or small segment of the **lithosphere** of Earth or any other terrestrial planet.

CONNECTIONS

THE GEOLOGICAL JIGSAW **104**
MOBILE AND STABLE ZONES **108**
BENEATH THE OCEAN FLOOR **114**

load
The amount of sedimentary material that a flowing river can carry, in suspension or by dragging along the bed. The term also applies to material carried by a glacier.

longitude
An imaginary line, part of a great circle, that joins the Earth's poles. The line of longitude that passes through London (called the prime, or Greenwich, meridian) is designated longitude 0°. All other locations have longitudes of up to 180° east or west of this meridian. Together with **latitude**, longitude can specify the position of any place on Earth. Celestial longitude indicates the angular distance of a heavenly object eastward along the **ecliptic** from the position of the vernal **equinox**. Together with the **celestial latitude**, it can indicate the location of any object on the **celestial sphere**.

long-lived radionuclide
A radioactive **isotope** of an element which decays with a very long half-life; for example, rubidium-87 and uranium-235.

longshore drift
A type of movement of beach material caused by the swash and backwash of waves that meet the shoreline obliquely. It causes sediment to migrate along the shore.

low-velocity zone
A zone inside a planet in which seismic waves travel more slowly than in the zone above or below. This slower velocity implies a lesser density than the surrounding regions. The Earth's **asthenosphere** is a low-density zone.

L waves
Seismic waves that travel around the periphery of the Earth. They propagate multiple reflections between the surface and the top of the layer below.

Magellan missions
A highly successful United States radar mapping mission to Venus which first orbited the planet during August 1990. The spacecraft carries a parabolic radar mapping antenna and a horn-shaped altimetric antenna. It has completed four mapping cycles, each taking 243 days – that is, one Venusian year. At the close of the third cycle the craft was made to orbit at a lower altitude to permit detailed gravity measurements to be made. The imagery and altimetric information has been combined to generate a series of new high-resolution maps and a wealth of spectacular images.

magma
Molten rock or other mineral material, usually of silicate composition, that is formed in the lower **crust** or **mantle** of the Earth. When it solidifies, it is known as **igneous rock**; when it emerges through the crust, as in a volcanic eruption, it is known as **lava**. See also **magma ocean** and **volcano**.

CONNECTIONS

CORE, MANTLE AND CRUST **80**
ROCKS FROM WITHIN THE PLANETS **92**

magma ocean
According to one theory of lunar history, the outer layers of the Moon became completely molten for a protracted period as a magma ocean, allowing the accumulation of calcium-bearing **feldspar** and the generation of the calcium-enriched lunar highlands crust.

magnetic dipole
Within the Earth's magnetic field, the north-seeking pole of a compass needle points to the north magnetic pole, whereas the reverse is true of the south-seeking pole. Such behavior is due to the magnetic dipole nature of the Earth's field.

magnetic field
The force field surrounding a magnetic object. Planetary fields are essentially dipole (exhibiting a north and south pole) in nature, although subsidiary fields also are present. The magnetic properties of atoms and molecules, indeed any mass, come from the movements of the constituent particles.

CONNECTIONS

PLANETARY MAGNETIC FIELDS **82**
WANDERING CONTINENTS **110**

magnetic inclination
A freely-suspended magnetized needle points vertically downward at the pole but lies horizontal above the equator. In between these two extremes is its angle of inclination, which depends on the latitude. This property is useful in determining the original latitude of rocks containing magnetic minerals.

magnetic striping
When extremely sensitive magnetometers were first towed above the sea floor at **divergent plate margins**, it was found that "stripes" of the basaltic crust had recorded successive normal and reversed magnetic polarities. On each side of a spreading axis, the pattern of stripes was mirrored, indicating that new crust was being generated and then moving away from the source zone on each side of a median rift. The discovery of this magnetic striping provided geophysicists with proof that **sea-floor spreading** occurs.

MAGNETOSPHERE

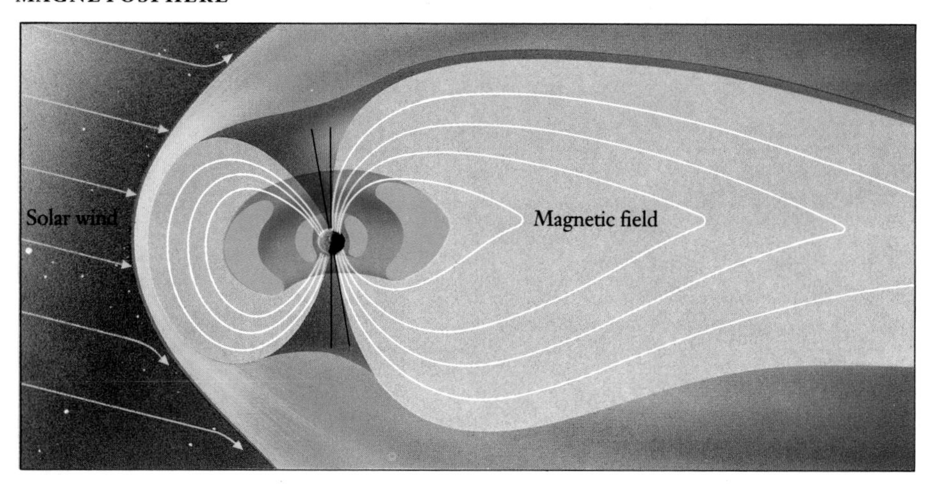

magnetopause
The boundary around the **magnetosphere** across which the **solar wind** cannot flow.

magnetosheath
A turbulent zone of confused magnetic field lines between the **magnetopause** and the **bow shock**.

magnetosphere
A region in space within which a planet's magnetic field is the predominant magnetic influence. For the Earth it extends for about 10 Earth radii on the sunward side and for 1000 radii in the opposite direction, that of the extended "magnetotail". As a result, the magnetosphere resembles a very extended horseshoe shape in cross-section.

main sequence
The part of a star's life when it is in a stable state, with energy being produced in the core and no contraction taking place. The Sun has been in this condition for at least 5 billion years and is expected to remain so for a similar period. The term applies to a zone delineated on the Hertzsprung-Russell diagram, which plots star luminosity against temperature.

major element
An element that occurs in significant amounts in most igneous rocks. For terrestrial crustal rocks the major elements are silicon, oxygen, aluminum, titanium, iron, magnesium, manganese, calcium, sodium, potassium and phosphorus.

manganese nodule
A small banded growth, rich in manganese, found on the deep ocean floor. It may contain as much as 20 percent manganese, together with smaller amounts of copper, iron and nickel oxides and hydroxides.

mantle
The layer within a planet that lies between the thin outer **crust** and the **core**. The mantles of terrestrial planets are composed of dense silicates of iron and magnesium. The Earth's mantle is nearly 2900 kilometers thick. Mantle materials of the gaseous outer planets are believed to be dominated by hydrogen compounds.

CONNECTIONS

THE EARTH AND THE MOON 64
THE INNER PLANETS 66
THE FORMATION OF THE CORE 78
CORE, MANTLE AND CRUST 80
ROCKS FROM WITHIN THE PLANET 92
EARTHQUAKES AND SEISMIC WAVES 96

mantle nodule
A block or small fragment of mantle-derived rock which may be brought to the surface during explosive volcanic activity.

margin
The edge of a tectonic plate, one of the huge slabs of the Earth's **lithosphere** that support the continents as they slowly drift around. *See* **continental drift** and **plate tectonics**.

maria
Dark-colored lava plains found on the Moon in a number of the larger impact basins, mainly on the near side. Each mare is composed of basaltic rocks and was emplaced in more than one extrusive episode. Individual maria may be of different ages, ranging from 3.85 billion to about 3.16 billion years old.

Mariner-9
A successful NASA mission to Mars, launched in May 1971. It reached the planet in November of the same year, as a planet-wide dust storm was raging. This eventually cleared and Mariner returned 7329 images of the Red Planet which revolutionized the study of Martian geology.

Mariner-10
Launched on 3 November 1973, Mariner-10 bypassed Venus at a distance of 5800 kilometers, taking photographs of its cloud tops as it did so. Then, swung by Venusian gravity, it arrived at Mercury in late March 1974. Because of the configuration of Mercury, the Sun and the spacecraft, less than half of the planet's surface was mapped, but the mission was highly successful and has provided all current geological information about the innermost planet.

Mars
The fourth planet out from the Sun, which it orbits once every 687 days at an average distance of 228 million kilometers. The Martian day (24.62 hours) is similar to Earth's but the planet is much colder – the highest temperature at the equator is only about 10°C. At its closest approach to Earth, it is 56 million kilometers away. Mars has two moons, Deimos and Phobos. Mars has been mapped by the Mariner space probes, and two Viking spacecraft landed on its surface in 1976.

CONNECTIONS

THE PLANETS FORM 56
THE PLANETS AND THEIR ORBITS 62
THE INNER PLANETS 66
THE PLANETS HEAT UP 76
PLANETARY ATMOSPHERES TODAY 98
RIFT VALLEYS 120

Mars Observer
A NASA spacecraft planned to enter Mars orbit during October 1993. However, radio contact with the craft was lost just before it was set to go into orbit.

mass
The amount of material in an object and a measure of its inertia: the extent to which it resists acceleration when a force is applied.

massif
A single mass of rock or a series of connected masses that together form the mountains in a range. The term also describes a large part of the Earth's crust, surrounded by faults, that may move as a result of **plate tectonics**.

mass wasting
The downhill mass movement of debris under the influence of gravity.

matrix
The finer material set between the larger clasts in a sedimentary rock.

mélange
A highly deformed and brecciated mass of rock usually associated with the convergence of two lithospheric plates. Sedimentary and volcanic rocks that have accumulated at the toe of an advancing plate margin, particularly one containing an **island arc**, become "scraped" up into an imbricate wedge of mélange as collision progresses. Ancient mélange deposits have been identified within the geological succession and enable geologists to establish positions of ancient plate margins. *See also* **lithosphere, impact breccia** and **polymict breccia**.

melting point
The temperature at which a substance, such as a crystalline rock, begins to melt.

Mercury
The smallest of the terrestrial planets (4880 kilometers across) and the closest planet to the Sun, around which it completes one orbit every 88 days at an average distance of 58 million kilometers. It has no moons and little or no atmosphere.

CONNECTIONS

LARGE AND SMALL PLANETS **58**
THE PLANETS AND THEIR ORBITS **62**
THE INNER PLANETS **66**
THE PLANETS HEAT UP **76**
THE FORMATION OF THE CORE **78**
THE EVOLUTION OF ATMOSPHERES **84**

Mesozoic
In the geological history of the Earth, an era that lasted from about 248 million to 65 million years ago. It was the age of reptiles, but ended with the extinction of the dinosaurs and ammonites etc. at the end of the **Cretaceous** period. The climate remained warm across the Earth's land mass, which for most of the era was joined together as the single supercontinent **Pangea**.

metal phase
All metals in the solid state are crystalline; they are also good conductors of electricity. Metal phases are ionic, with each atom being bonded covalently and very difficult to break down except by chemical decomposition.

metamorphic rock
A type of rock produced (usually from a **sedimentary rock**) by the mineralogical change known as **metamorphism**.

metamorphism
The process that changes existing rocks mineralogically, but without them entering the liquid phase – that is, without melting. Regional metamorphism affects rocks over large areas which have undergone rises in both temperature and pressure in the root regions of mountain belts. Contact, or thermal, metamorphism is achieved by a rise in temperature brought about by proximity to an igneous body, and is more localized.

CONNECTIONS

DATING THE ROCKS **86**
EARTH'S FIRST CONTINENTS **102**
MOBILE AND STABLE ZONES **108**
MOUNTAINS FROM THE SEA **118**

metasomatism
A change produced in a rock by reaction with a liquid. Metasomatic activity is typically seen near igneous intrusions and adjacent to oceanic ridges, where seawater penetrates newly emplaced oceanic crustal melts.

meteor
A "shooting star"; a small meteroid which burns up in the Earth's atmosphere.

meteorite
A rock fragment of extraterrestrial origin. Meteorites include metallic and rocky particles of a range of sizes, many of which have originated in asteroids. Stony meteorites, which make up 94 percent of all samples that fall to Earth, are subdivided into chondrites (86 percent) and achondrites (8 percent). These types are dominated by silicate minerals, the former containing spherical clasts called chondrules (*see* **chondrite**); they may also contain carbon and volatiles. Stony-irons which account for a mere 1 percent, show signs of having been differentiated and are admixtures of silicate and metal. Irons, which account for the remaining 5 percent, are also differentiated and typically contain iron-nickel alloys. Meteorites are very old, with radiometric ages of more than 4.5 billion years being typical. They are free samples of primitive Solar System materials.

meteoroid
A small rocky object that orbits the Sun. One that enters the Earth's atmosphere is seen as a meteor.

methane
A gas formed by the combination of carbon and hydrogen, with the formula CH_4. It is relatively abundant in the atmospheres and in ices in the outer Solar System.

mica
A member of a group of hydrated aluminum silicate minerals with the hydroxyl molecule (OH) in their atomic lattices. They have a crystal structure that results in a sheetlike form, with individual plates separated by cleavage planes.

mid-oceanic ridge
See **oceanic ridge**.

Milankovitch theory
A theory of climatic change which was first suggested by the British astronomer John Herschel in the late 18th century, soon after the first evidence for the recurrence of **glaciation** had been collected. It invokes changes in eccentricity of the Earth's orbit, precession of the rotational axis, and obliquity in an effort to explain the series of Quaternary ice ages. Always controversial, the theory was revived in the early 20th century by Milutin Milankovitch. *See also* **Quaternary** and **ice age**.

millet-seed grain
A very rounded sand grain formed by the action of the wind in a desert environment.

Mimas
The second of **Saturn**'s major moons (in terms of its distance from the planet) , which orbits at an average distance of only 188,200 kilometers. Mimas measures 390 kilometers across, and completes each orbit in just less than one terrestrial day.

METEORITE

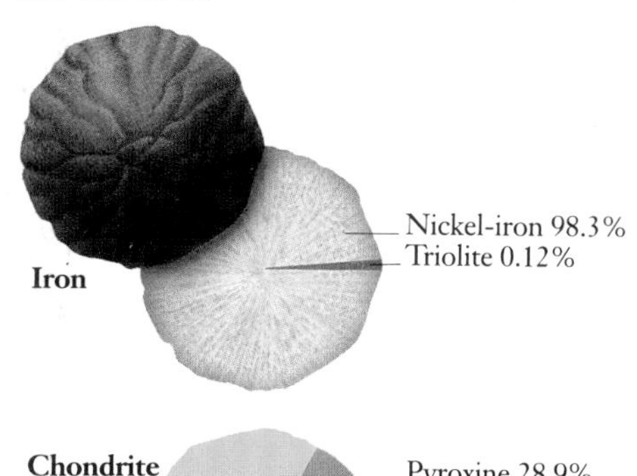

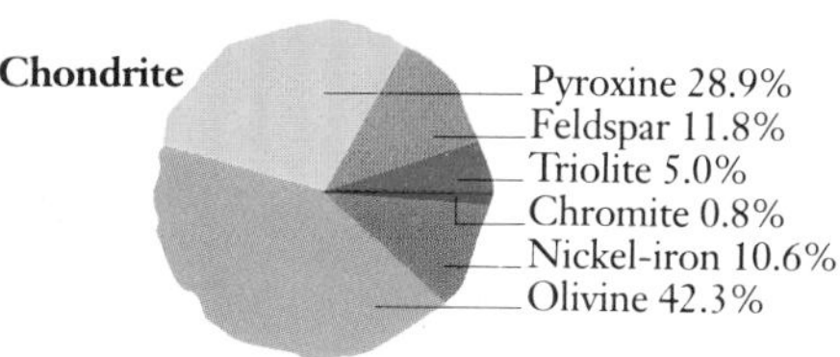

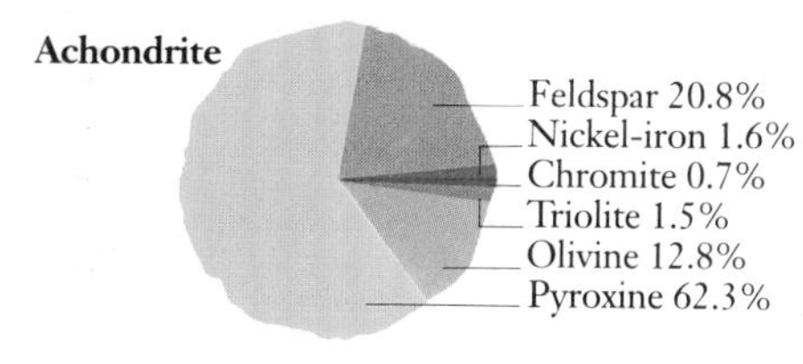

mineral
A naturally occurring crystalline substance with an ordered atomic structure and definite chemical composition.

Miocene
An epoch in the geological history of the earth that lasted from 25 million to 5 million years ago. During this time, which ended the **Tertiary** period, mammals continued to expand in number and grasslands became predominant.

mobile belt (mobile zone)
A region of intense geological activity located along a plate margin. Mobile belts are characterized by phenomena such as volcanism, seismic activity and orogenesis.

Mohorovicic discontinuity
A **seismic discontinuity** located at the interface between the Earth's **crust** and **the mantle**. Beneath the continents it is encountered at depths of about 35 kilometers and below the ocean floor at 10 kilometers; however, beneath some mountain belts it may plunge to 70 kilometers. Seismic velocities increase sharply across it.

molecular cloud
A site of active star formation within our Galaxy. With densities of 10^{10}–10^{12} atoms per cubic meter and temperatures of between 30–100 K, they contain much heated dust and are sources of infrared radiation.

molecule
The smallest particle of a substance that can exist independently.

moon
A natural satellite of a major planetary body. The planetary moons range from huge bodies the size of Earth's **Moon**, to tiny chunks of rock known as **shepherd moons** which rotate within Saturn's ring system. Inner planet moons are generally composed of silicate rocks, whereas outer planet moons are a mixture of rock and ices. The only moon with an atmosphere is **Titan**, a satellite of Saturn.

Moon, the
The only natural satellite of the Earth, which it orbits once every 27.32 days at an average distance of 384,000 kilometers. The Moon is so large (measuring 3476 kilometers across) that the Earth–Moon system is sometimes considered to be a double planet. Because the Moon rotates once on its axis during the same time that it completes one orbit of the Earth, it always presents the same face toward Earth. It reflects light from the Sun, and so exhibits monthly phases as it orbits. The Moon has no atmosphere.

CONNECTIONS
THE SUN'S FAMILY **60**
THE EARTH AND THE MOON **64**
THE FORMATION OF THE CORE **78**
CORE, MANTLE AND CRUST **80**
DYNAMIC PLANETS **88**
THE FIRST CRUSTS **90**

moraine
Rock debris scoured by a valley glacier from surrounding countryside. It may collect at margins (lateral moraine), beneath (ground moraine) or at its toe (terminal moraine).

nebula
A largely uncondensed cloud of cosmic dust and gas from which a star may form. The Sun formed out of the solar nebula.

CONNECTIONS
THE FORMATION OF THE SOLAR SYSTEM **48**
BEFORE THE SOLAR SYSTEM **50**
THE FORMATION OF THE SUN **54**
THE PLANETS FORM **56**
LARGE AND SMALL PLANETS **58**

Neptune
The fourth largest planet in the Solar System and the eighth from the Sun (at an average distance of 4.497 billion kilometers). It is slightly smaller than its neighbor Uranus, but more massive. It rotates on its axis once every 18 hours, but takes 164.8 years to complete one orbit of the Sun. It has two major moons, **Nereid** and **Triton**.

CONNECTIONS
THE PLANETS FORM **56**
LARGE AND SMALL PLANETS **58**
THE SUN'S FAMILY **60**
THE OUTER PLANETS **68**

Nereid
One of the smaller moons of **Neptune** (340 kilometers across), which it orbits once every 365.2 days. It has the most elliptical orbit of any moon, with an average distance of 5,513,400 kilometers from Neptune.

neutron
An electrically neutral particle in the atomic nucleus. The number of neutrons in an element may vary, giving rise to **isotopes**.

nickel-iron alloys
A mixture of iron and nickel, typically found in iron **meteorites**. When cut, polished, and etched with acid, they reveal intergrowth patterns called Widmanstatten figures, which indicate that they were once molten.

noble gas
Another term for **inert gas**.

non-sequence
A gap within the succession of geological strata, which arises because deposits from a particular time were never laid down, or because subsequent erosion has removed these deposits. The existence of such gaps is usually proved on the basis of other paleontological evidence.

nuclear transformation
A high-energy reaction that takes place in the interiors of stars and converts one kind of atomic structure to another. For instance, it is believed that early in the history of the Universe, some neutrons joined with protons to form the isotope deuterium, also known as "heavy hydrogen". Sometimes they joined with a further neutron, with the result that tritium (another hydrogen isotope) was produced. When a further neutron became involved, then a different element, helium, emerged. This sequence is just one kind of nuclear transformation. Further into the life cycle of a star, when temperatures and pressure become very high, hydrogen nuclei are stripped of their electrons, crushed together and fused into helium nuclei. This process (nuclear fusion) releases huge amounts of energy, which physicists are trying to duplicate in laboratories on Earth so that the energy can eventually be used for commercial and industrial purposes.

nucleus
The core of an atom, consisting of neutrons and protons. Differing numbers of neutrons give rise to different **isotopes** of the same element. The center of a **galaxy** is also known as a nucleus.

nuée ardente
A "glowing avalanche" – consisting of hot volcanic ash, fine dust, molten lava fragments and hot gases – associated with a volcanic eruption. It may move downhill at very high speeds. Such flows are extremely dangerous and, after deflation following gas escape, give rise to non-welded tuffs and welded deposits called ignimbrites. The destruction of the city of St Pierre on the Caribbean island of Martinique in 1902 was brought about by such a flow, killing 40,000 people almost instantly.

OBDUCTION

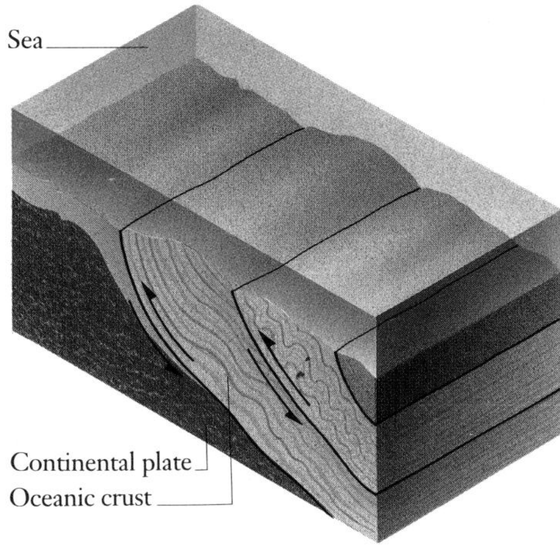

obduction
The process by which oceanic crust is thrust over the leading edge of a continental plate as huge slabs, in contrast with the more usual process in which the crust is subducted beneath the edge (*see* **subduction zone**). When such overthrust slabs of oceanic crust are found within the geological succession, they frequently comprise distinctive pelagic sediments and basaltic volcanic rocks, called **ophiolites**. *See also* **ocean**.

Oberon
The second largest and outermost moon of **Uranus**. It is 1526 kilometers across and orbits at an average distance of 583,400 kilometers. Like the other three major moons, it has a **retrograde rotation**, taking 13.46 days to complete one rotation on its axis.

obliquity
The angle that one plane makes with another. For instance, the **ecliptic** makes an angle of 23° 26' with the celestial equator (because the Earth's equator is inclined at this angle to the orbital plane); this is termed the obliquity of the ecliptic.

ocean
A depressed, tectonically defined basin, floored by oceanic crust, and filled with seawater. The deepest parts of the oceans are located in subduction trenches, such as the Marianas Trench.

CONNECTIONS

EARTH'S FIRST OCEANS **100**
BENEATH THE OCEAN FLOOR **114**

oceanic crust
The relatively dense basaltic material, on average 8 kilometers thick, that forms the ocean floor. It has an average density of around 3.0 grams per cubic centimeter. A typical sequence through it comprises a surface veneer of ocean floor sediments up to 1 kilometer thick, with an underlay by 2–3 kilometers of basaltic lava intruded by dykes. Moving down through the layers, the basaltic lava passes into coarser gabbro which, at a depth of around 8 kilometers merges into the peridotitic rocks at the **Mohorovicic discontinuity**.

oceanic ridge
A global network of ridges crosses the floors of the oceans. They rise several thousand meters above the abyssal plain, have a linear form, and crests which are dissected by rifting. The ridges are frequently offset by **transform faults** which accommodate the attempts the mantle makes at "bending" the relatively brittle, non-deformable, plate margins. In many places the ridge flanks are inflated by volcanic domes of basaltic materials. The ridges may have developed above rising convection cells in the Earth's mantle.

Oligocene
An epoch in the geological history of the Earth that lasted from about 38 million to 25 million years ago. Mammals continued to expand during this time, and the first two-toed ungulates or hoofed animals emerged.

olivine
A silicate of magnesium and iron which is one of the first minerals to crystallize from basaltic magma. It chemical composition is $[Mg,Fe]SiO_4$ and it has a greenish hue. Olivine is the principal constituent of mantle **peridotite**, and it occurs widely in stony meteorites and lunar mare lavas.

Oort cloud
A distant region of the Solar System, which is believed to extend from about 4500 to 15,000 × 10^9 kilometers from the Sun; that is, one-third of the distance to the nearest star. It is considered by many astronomers to be the source of most comets, and is named for the Dutch astronomer James Oort.

ooze
A fine-grained precipitate of carbonate or silica which is distributed over the ocean floors down to a depth of about 3.5 kilometers. Oozes are frequently associated with oceanic ridges. They have their source in the remains of living organisms such as diatoms, radiolaria and foraminifera.

ophiolite
A member of a group of rocks that include basic and ultrabasic igneous rocks, together with deep-sea sediments, which are sometimes found as thrust slices within the stratigraphic succession. Many appear to be slabs of ancient oceanic crust which have been obducted onto the continental margins (*see* **obduction**).

orbit
The path of an object revolving around another with which it is associated. Planetary orbits are elliptical with respect to the Sun (which lies at one focus of the eclipse). *See* **Kepler's laws of planetary motion.**

Ordovician
A geological period (part of the Paleozoic era) that spanned the 67 million years from about 505 million to 438 million years ago. It was a time when reef-building algae and graptolites were active, and the first jawless fish appeared. As well as graptolites, fossils from the period include those of trilobites and the first corals and land plants.

organic molecule
Any molecule containing carbon, except carbonates and the oxides of carbon.

OCEANIC CRUST

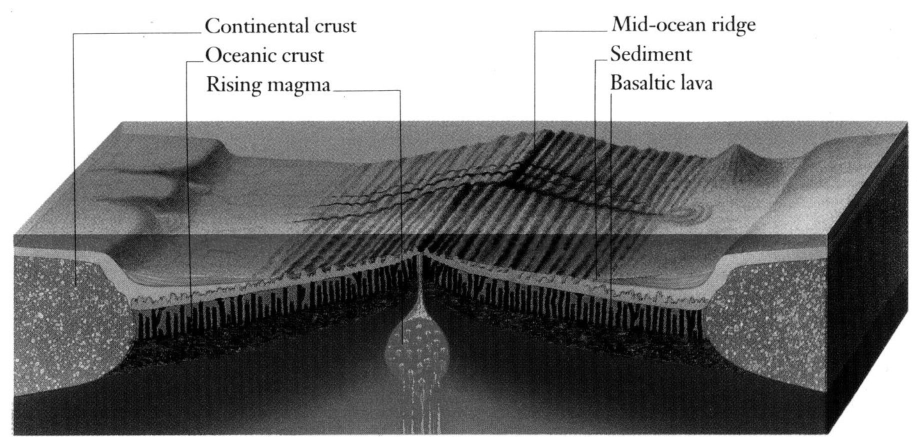

orogenesis
The tectonic process that produces **fold belts**, **metamorphism** and **magmatism**, typically from a sequence of sedimentary rocks that have been involved in plate convergence. The culmination of an orogeny is the uprise of new belts of fold mountains.

CONNECTIONS
EARTH'S FIRST CONTINENTS 102
MOBILE AND STABLE ZONES 108
MOUNTAINS FROM THE SEA 118

orogenic belt
A linear zone characterized by the effects of **orogenesis**, that is, crustal shortening with associated development of fold mountains, metamorphism and granite intrusion.

outflow channel
A kind of large-scale flood channel found in certain regions of Mars. Outflow channels have no tributaries and their floors show effects, such as scouring, which indicate rapid movement of large volumes of fluid. They are believed to have formed by the abrupt release of frozen groundwater.

outgassing
The process by which volatiles from within a planet gradually escape to the surface, there to enter either the atmosphere or hydrosphere, or escape into space.

ozone layer
Sometimes called the ozonosphere, a layer within the Earth's atmosphere between heights of 12 and 50 kilometers, which has a high concentration of the highly reactive gas ozone (O_3). Although the concentration of the gas is a mere one part per 30 million, it is an efficient absorber of harmful ultraviolet radiation. Today there is considerable concern about the detrimental effects of chlorine from the gases called CFCs (chlorofluorocarbons), used in some aerosol sprays, which are believed to be breaking down some parts of this protective layer.

pahoehoe
A Hawaiian word used for a form of basaltic lava surface that resembles strands of rope. Blocky lava is known by the term **aa**.

Paleocene
In the geological history of the Earth, the first Tertiary epoch, which ran from about 65 million to 55 million years ago. Following the extinction of most large reptiles, mammals began to dominate the land, and rodents and primates evolved.

paleomagnetism
The scientific discipline that deals with the fossil or remanent magnetization of rocks of all ages, and allows scientists to reconstruct the Earth's past magnetic field and the position of crustal features with respect to the magnetic poles. Paleomagnetism has established that the terrestrial continents have moved in relation to one another and to the **paleopoles**. Paleomagnetic data were vital in convincing the scientific community that **continental drift** is a reality.

Sensitive measuring devices called magnetometers provide most geomagnetic data; when towed behind research ships, they reveal magnetic anomalies in rocks on the sea floor. These indicate numerous reversals of magnetic polarity throughout geological time (*see* **field reversal**). The rocks that reveal them can be dated radiometrically, allowing the sequence but also the timing of such changes to be established. Paleomagnetism also allows geologists to calculate the speed at which oceanic crust is being produced, and at which the Earth's lithospheric plates move around.

The widespread presence of iron in certain of the minerals commonly occurring within rocks is the vital ingredient. Metallic iron (Fe) is ferromagnetic; that is, the tiny internal atomic magnets in it are all aligned in the same direction – that of the Earth's field. Magnetite, on the other hand, a very common oxide of iron (Fe_3O_4), is said to be ferrimagnetic because some of the tiny magnets point in the other direction. If a ferrimagnetic mineral is heated above a particular temperature, known as the Curie point (around 500°C), its magnetization is destroyed. It is then termed paramagnetic and its atomic magnets have a random orientation. If it subsequently solidifies by cooling below the Curie point, it takes on the magnetic field it finds itself in.

When igneous rocks cool below the Curie point, they fossilize a remanent magnetization known as thermoremanent magnetization (TRM). Although sedimentary rocks never become very hot, the grains of magnetic minerals which settle through water and accumulate in them tend to become aligned in the direction of the prevailing field. They take on detrital remanent magnetization (DRM). Samples of all types of rock can be used to establish paleopole positions and the latitude of the rocks at the time they encapsulated their magnetization.

CONNECTIONS
PLANETARY MAGNETIC FIELDS 82
WANDERING CONTINENTS 110

PARSEC

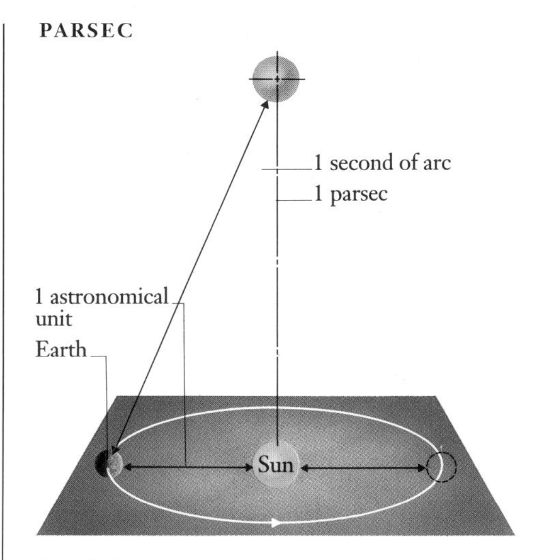

paleontology
The study of past life forms through fossil remains. Although fossils have been known and described since pre-Christian times, modern paleontology developed only during the middle of the 19th century.

CONNECTIONS
WANDERING CONTINENTS 110
CHANGING WORLDS 122

paleopole
The past position of the Earth's magnetic poles as determined in rocks from their remanent magnetization. *See also* **paleomagnetism**.

Paleozoic
A geological era that lasted from about 590 million to 248 million years ago. During the Paleozoic, the first animals with hard shells and skeletons emerged, although most marine animals died out toward the end of the era.

Pangea
The vast supercontinent that came together on Earth during the latter part of Paleozoic time. It comprised all of the present-day continents. It finally fragmented in Mesozoic times. *See also* **Gondwanaland**.

CONNECTIONS
EARTH'S FIRST CONTINENTS 102
WANDERING CONTINENTS 110

parent isotope
The original radioactive **isotope** of an unstable element that decays to another (daughter) as part of a decay sequence.

parsec
A unit of distance in astronomy, equal to 3.26 light years (3.086×10^{13} kilometers). The name derives from the terms *par*allax *sec*ond, because at a distance of 1 parsec a length of 1 **astronomical unit** subtends an angle of 1 second of arc.

partial melting
If a sedimentary rock is heated sufficiently a temperature is eventually reached when it begins to melt. The first liquid droplets to form do not, however, have the average composition of the rock but are probably close to that of sodium-rich **feldspar**. This is a reflection of the fact that rock-forming silicates have different melting points. Thus, in the above example, the first melted fraction has a composition close to that of the mineral with the lowest melting point: probably sodium-rich feldspar. This is the first "partial melt" to form. If melting continues as more heat is supplied, the feldspar-like melt changes its chemistry gradually as the chemical components of silicates with progressively higher melting points join it. Furthermore, because the silicates also show the property of a solid solution, the composition of individual mineral groups also show changes with falling temperature.

This process is exactly what happens within the upper part of the Earth's **mantle**, at depths where pressures and temperatures are sufficiently high to cause fusion of the peridotitic mantle material. Thus the pockets of melt which collect in the melting region contain only the lower temperature components of the mantle: mainly these comprise the chemical constituents of basalt (such as calcium–plagioclase feldspar, calcium–magnesium–iron-pyroxene and some magnesium-rich olivine and iron oxides). The precise composition of such partial melts is determined by the temperature, hydrostatic pressure, and amount of volatiles in the source region. The latter, in particular, lower the melting points of silicates and may have an important modifiying influence on the nature of partial melts. As it rises toward the Earth's surface, further modification of the melt occurs because a point may be reached when it begins to crystallize again. Because crystallization takes place in order of decreasing melting temperature of the rock-forming silicates, residual melts slowly become enriched in silica and volatiles. This process – **fractional crystallization** – is approximately the reverse of partial melting.

particle-particle collisions
Within the **solar nebula**, collisions between dust and gas particles must have been frequent. At some stage mutual collisions occurred in which particles began to come together. For this to happen, some force must have acted on them which exceeded that of the rebound after particle-particle collision. The most likely candidate is some kind of electrical dipole effect. Once larger particles developed a **regolith**, adhesion became considerably easier.

pediment
A sloping surface cut in the rocks at the base of mountains being actively eroded, usually in arid regions. It is common for it to be covered by alluvial sediments.

pelagic
Describing organisms that drift or swim in the sea (or a lake), or material that derives from them. *See also* **benthic**.

peridotite
An ultrabasic igneous rock consisting primarily of magnesium-olivine, with subsidiary amounts of pyroxene, spinel and amphibole. The Earth's mantle is believed to be peridotitic. Peridotite generally forms by the accumulation of dense olivine crystals under the influence of gravity.

perihelion
The point of closest approach to the Sun of an object in orbit around it. The comparable term for a spacecraft in orbit around a planet is periapsis.

permafrost
In arctic and subarctic regions of a planet, the permanently frozen soil and subsoil (or **regolith** on a planet without organic constituents). It is believed to be particularly widespread at most latitudes on Mars.

permeable rock
Any rock that can absorb and transmit a fluid (gas or liquid).

Permian
In the geological history of the Earth, the period at the end of the Paleozoic era that lasted from about 286 million to 248 million years ago. It was a time when deserts were widespread and reptiles began their great expansion; trilobites and many marine organisms became extinct. Cone-bearing plants replaced ferns as the dominant plants.

Phanerozoic
The era of advanced life, corresponding to the period from the beginning of Cambrian times to the present day, spanning approximately 590 million years. The original meaning of this term was the "era of life". But since it was introduced, organic remains have been found in older and older rocks so that now they extend back well into the Proterozoic.

phase change
In physics and geophysics, the transformation of a substance from one phase to another, such as gas to liquid, liquid to solid, etc. **In chemistry,** the change of a solid such as a silicate mineral to another form, of the same chemistry, but with different density or internal structure. Such phase changes are believed to occur within the Earth's **lithosphere** and are marked by changes in seismic velocities.

Phobos
The inner moon of **Mars**, which it orbits once every 7.65 hours at an average distance of 9270 kilometers. It has an ellipsoidal shape, measuring approximately 20 by 23 by 28 kilometers.

phonolite
An extrusive igneous rock intermediate in composition and strongly alkaline. Phonolite contains **feldspathoids** rather than **feldspars** and is widespread as lavas in rift valleys.

photodissociation
The breakdown of molecules, usually in the upper levels of planetary atmospheres, by sunlight (particularly the shortwave ultraviolet wavelengths).

photosphere
The bright visible "surface" of the Sun. It is from the photosphere that nearly all of the Sun's radiation is emitted. It has a generally mottled appearance which is caused by **convection** within its outer regions. A common feature of the photosphere is the development of **sunspots**.

PHOTOSPHERE

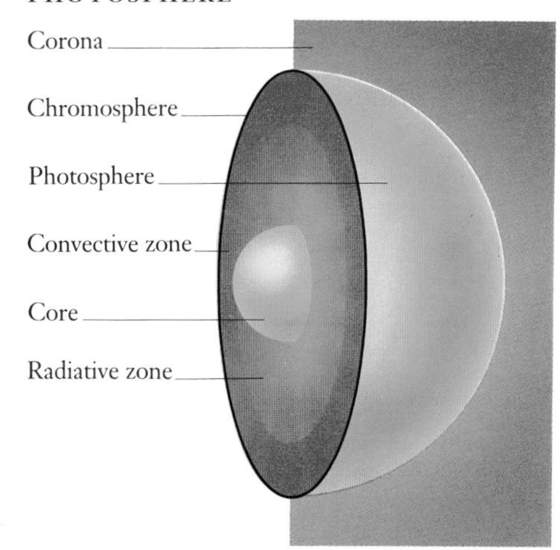

photosynthesis
The process by which water, carbon dioxide and solar energy are converted to oxygen and organic compounds by plants containing chlorophyll. Without this phenomenon the Earth's atmosphere would not have been able to support advanced life.

CONNECTIONS

THE EVOLUTION OF ATMOSPHERES **84**
PLANETARY ATMOSPHERES TODAY **98**
BEGINNINGS AND ENDINGS **134**

phreatomagmatic eruption
Sometimes groundwater may enter a magma chamber or the conduit of a volcano in which magma is rising. Its rapid conversion to superheated steam then causes a highly explosive eruption of a type which often forms maar craters. Much of the early volcanism on Mars may have been of this type.

pillow lava
Extruded basaltic lava formed into pillow shapes when it enters the sea or is erupted on the sea floor. When the hot lava meets the cold water, it chills rapidly on the outside, forming a solid shell. The pressure of the flowing lava, however, frequently breaks through this skin, forcing molten lava out, rather like toothpaste from a tube, forming pillow shapes. Flow proceeds by the continual growth of similar forms as successive pillows burst.

plagioclase feldspar
See **feldspar**.

planet
Any nonstellar differentiated solid or solid and gas body in space. The Solar System is known to have nine planets orbiting the Sun; other nearby stars may also have planetary systems.

CONNECTIONS

THE PLANETS FORM **56**
LARGE AND SMALL PLANETS **58**
THE SUN'S FAMILY **60**
THE PLANETS AND THEIR ORBITS **62**
THE INNER PLANETS **66**
THE OUTER PLANETS **68**

planetesimal
The building block of planets. The term applies to any large or small solid body (as long as it is significantly smaller than a true planet) which may be attracted by gravity toward others, eventually accreting into a planet. Planetesimals are thought to have condensed out of the original solar nebula, then accreted to form the planets.

CONNECTIONS

THE PLANETS FORM **56**
THE EARTH AND THE MOON **64**
ASTEROIDS AND METEORITES **70**

plasma
Ionized particles, such as protons, electrons and atomic nuclei, that move freely around. A great deal of the matter within the Universe is formed of plasma.

plate tectonics
The theory of the Earth that invokes the movement of lithospheric plates as an explanation of processes such as volcanism, seismicity and orogenesis. *See also* **continental drift** and **lithosphere**.

CONNECTIONS

DYNAMIC PLANETS **88**
THE GEOLOGICAL JIGSAW **104**
MOBILE AND STABLE ZONES **108**
GEOLOGICAL PLATE MOVEMENTS **112**
ISLAND ARCS **116**

playa
An enclosed flat basin in a desert, usually occupied in part by an ephemeral lake or lakes.

Pleistocene
The previous epoch in geological time, at the beginning of the **Quaternary** period, about 2.5 million years ago. The Pleistocene ended about 10,000 years ago and was marked by extensive **glaciation**.

plume
Any upwelling lobe or jet of hot and partly molten material, usually rising out of a planet's mantle. Such plumes are believed to be the source of volcanism within lithospheric plates and are numerous on Venus, where smaller ones give rise to coronae and larger ones to volcanic rises. Plume activity was probably responsible for the huge shield volcanoes of Mars and, because plate movements do not appear to have moved the lithosphere away from plume sources, the volcanoes grew to immense sizes.

CONNECTIONS

GEOLOGICAL PLATE MOVEMENTS **112**
BENEATH THE OCEAN FLOOR **114**

Pluto
The outermost planet of the Solar System, which orbits the Sun at an average distance of 5.9 billion kilometers once every 247.7 years. It has a highly eccentric orbit, and at closest approach to the Sun (perihelion) it is nearer the Sun than its neighbor Neptune. Pluto is 2284 kilometers across (smaller than Earth's Moon) and takes 6.4 days to rotate once on its axis. It has one moon, **Charon**.

CONNECTIONS

THE SUN'S FAMILY **60**
THE PLANETS AND THEIR ORBITS **62**
THE OUTER PLANETS **68**

point bar
An accumulation of crescent-shaped sand and gravel deposits on the inner bends of meanders in a slow-flowing river.

polarity reversal
The Earth's magnetic field periodically shows reversals in its polarity, giving rise to magnetic epochs of normal and reversed polarity. Exactly why it occurs is unknown but evidently the intensity of the field changes with time and it is possible that when it decreases to a very low figure (a reflection presumably of the slowing down of electrical currents in the molten outer core), it suddenly "flips over". Reversals occur at intervals of between 10,000 and 25 million years.

polar wandering
The magnetic poles show positional changes over time, though only of a modest order, and always they are focused close to the rotational poles. The apparent "polar wandering" revealed in rock samples with remanent magnetization is due largely to continental movement driven by plate tectonics.

polymorph
One of two or more forms of the same chemical compound, such as a silicate mineral. For instance, aragonite and calcite are polymorphs of calcium carbonate ($CaCO_3$); andalusite, kyanite and sillimanite are polymorphs of the alumino-silicate (Al_2SiO_5).

potential energy
The latent energy that resides in any mass by virtue of its position or deformation, and which can be translated into another form, such as kinetic energy.

Precambrian
The whole of geological time from the formation of Earth to the **Phanerozoic** era (4500–570 million years ago).

precession
The apparent slow motion of the celestial poles. It is largely due to the wobbling motion of the Earth's rotational axis induced by gravitational attraction of the Moon on Earth's equatorial bulge. Other planets behave in the same way. It is also known that planetary orbits precess. Precessional effects may have been responsible for climatic changes such as have occurred on Mars.

principal stress
Because of the activity of lithospheric plates the Earth's crust is in a constant state of stress. These stresses can be resolved into three principal components, usually arranged at right angles to each other. Because of the way in which gravity works, in rocks subject to simple hydrostatic pressure, one of the principal stresses generally lies vertically, with the other two (the minimum and intermediate principal stresses) arranged in a horizontal plane. During the compressional movements that accompany **orogenesis**, this arrangement may alter.

projection
The mapping of a spherical object, such as a planet, onto a two-dimensional surface. Three main types of projection are used: cylindrical (in which the surface of the sphere is projected onto a cylinder which is then opened out); conical (in which the sphere is projected onto a cone); and azimuthal or zenithal (in which the sphere is projected onto a plane tangential to a point on its surface, frequently the pole).

Proterozoic
One of the two eons of Precambrian time, extending from about 2.5 billion to 590 million years ago. It was during this time that the early single-celled organisms developed into more complex creatures.

proton
A subatomic particle with a positive charge, which resides in the nucleus. In the neutral atom of any element, the number of protons is matched by the number of electrons.

protostar
The large nebular clouds of dust and gas distributed throughout the Galaxy eventually collapse under the influence of gravity, whereupon star formation begins. At this stage the coalesced central region of such a cloud constitutes a protostar.

CONNECTIONS

THE FORMATION OF THE SOLAR SYSTEM 48
BEFORE THE SOLAR SYSTEM 50
COSMIC INGREDIENTS 52
THE FORMATION OF THE SUN 54

pumice
A variety of glassy volcanic rock, usually felsic in character, with a spongy texture. It forms through strong **vesiculation** of the original magmatic gases, with the result that the rock is full of the evacuated gas cavities, termed vesicles. Pumice will float in water.

P waves
Known also as primary or "push-pull" waves, the seismic waves that travel fastest through the Earth. They consist of a series of compressions and expansions of the material through which they pass. Unlike **S waves**, P waves can travel through liquid material.

pyroclast
Material that includes large volumes of fine dust, ash and blocks, ejected from the vent of an exploding volcano. Pyroclasts may enter into gaseous suspension in **nuées ardentes**, producing pyroclastic flow deposits.

pyroxene
One of the common rock-forming silicates with the general formula [Ca,Mg,Fe] Si_2O_6. Pyroxenes are generally mafic and are characterized by high density and the presence of intersecting sets of cleavage planes. The pyroxene augite is one of the essential minerals in basalt.

quartz
One of the most widespread silicate minerals of the Earth's continental regions and the chief component in clastic sedimentary rocks. It is mostly silicon dioxide (SiO_2).

Quaternary
The most recent geological period, which began 2 million years ago and extends to the present. It began with ice ages and the evolution of wooly mammoths. Humans then became the dominant species of land animal.

radar
A method of planetary imaging using reflected electromagnetic radiation at microwave wavelengths. Radar imaging systems have been used to photograph planets (including the Earth itself) from space, and have the advantage over visual methods of being able to penetrate clouds and darkness. They have been of vital importance in the mapping of the cloud-shrouded surface of Venus. Ground-based and satellite-borne radar have been used for meteorology, for studying other aspects of the Earth's atmosphere, and for tracking meteors.

radiogenic isotope
Some chemical elements have a number of **isotopes**, one or more of which (the parent isotope) is radioactive and decays at a fixed rate to another (daughter isotope). Such radiogenic isotopes, although present in only small quantities within crustal rocks, are

PROJECTION

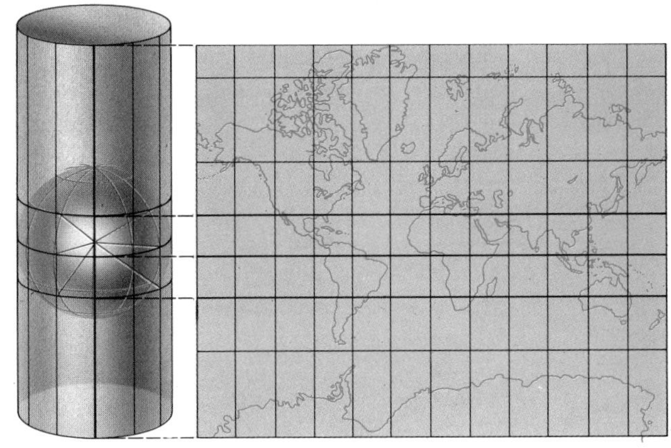
Cylindrical

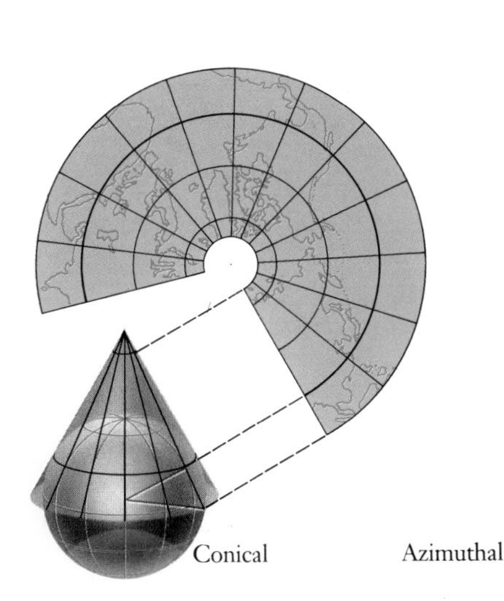
Conical

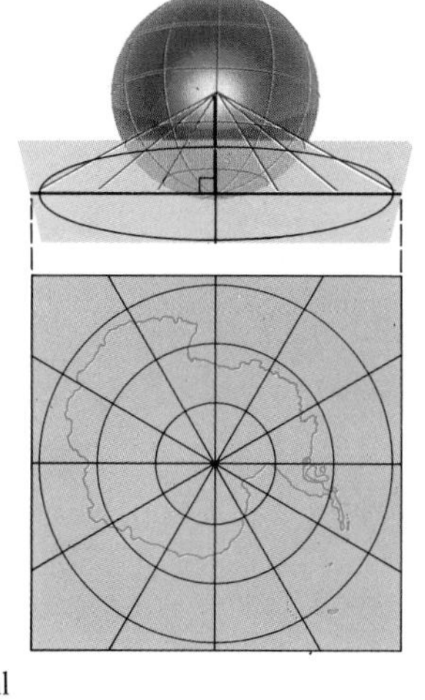
Azimuthal

distributed widely and are of great value in deriving absolute ages of silicate rocks and meteorites.

CONNECTIONS

LARGE AND SMALL PLANETS **58**
THE EARTH AND THE MOON **64**
PLANETARY HEAT ENGINES **74**
THE PLANETS HEAT UP **76**

radiometric dating
The science of deriving **absolute ages** for silicate rocks, using **radiogenic isotopes**.

raised beach
An ancient beach left stranded above sea level by a relative rise in the level of the land in relation to the sea. Raised beaches at various levels were left in northern latitudes as the Pleistocene ice sheet shrank and the land level rose isostatically. *See* **isostasy**.

rarefaction wave
A decompression wave that follows a shock wave. Following the impact of a meteorite or asteroid on a planetary surface, a compressional shock wave passes through the target rocks. This very short phase is followed by a return to more normal conditions as the rocks decompress, and the rarefaction wave moves upward and outward. Ultimately it is responsible for the ballistic ejection of a cone-shaped curtain of material which once resided beneath the cavity and which comes to rest as the **ejecta blanket**.

red giant
A star in the late stages of evolution, when it is relatively cool but highly luminous and with a very low density. The star's red coloration is an indication of its comparatively lower temperature (4000–3000 K).

refractory element
Any element with a high melting point. The order in which solid material condensed from the **solar nebula** was a function of temperature. The first to condense were the refractory elements, including the trace elements osmium (Os), rhenium (Re) and zirconium (Zr), and the major element aluminum (above 1400°C). At about 1250°C, they were joined by titanium (Ti), calcium (Ca) and some rare earth elements.

CONNECTIONS

LARGE AND SMALL PLANETS **58**
ASTEROIDS AND METEORITES **70**
CORE, MANTLE AND CRUST **80**

refractory inclusion
An irregular grain found in some carbonaceous chondrite **meteorites** along with chondrules. Usually such fragments are between 1 and 2 millimeters in diameter and are composed of high-temperature anhydrous minerals rich in refractory elements, such as titanium, aluminum, calcium and magnesium.

regolith
A layer of loose rock and mineral grains that is found at the surface of planetary crusts. A regolith becomes a **soil** with the addition of any organic material.

CONNECTIONS

THE PLANETS FORM **56**
THE FIRST CRUSTS **90**

rejuvenation
The process by which relative levels of land and sea are constantly changing on the Earth. Some of the more major changes occur during glaciation and in its aftermath. After the removal of an ice sheet the unencumbered land rises isostatically, with the result that streams begin cutting down afresh into the elevated landscape.

remote sensing
The science of obtaining information about a planet without actually landing any instruments on it. Remote sensing of the Earth can be done from an aircraft or satellite; planetary methods include imaging (at a variety of wavelengths) from orbiting satellites, and radar imaging via Earth-based telescopes.

retrograde rotation
The motion of a planet or moon around its rotational axis in such a way that, if it were viewed from north of the orbital plane, it would be observed to rotate in an east-to-west (clockwise) direction.

Rhea
The second largest of **Saturn**'s many moons. It is 1530 kilometers across, taking 4.52 days to rotate once on its axis as it orbits at an average distance of 571,000 kilometers from its parent planet (inside the orbit of the giant moon Titan).

Richter scale
A scale used by geophysicists to measure the magnitude of an earthquake in terms of the frequency and amplitude of the surface waves. It is determined by taking the common logarithm (to base 10) of the greatest ground motion observed during the tremor, and applying a standard correction involving the distance to the epicentre. The scale runs from 0 to 10.

ridge belt
A deformational zone on the surface of Venus. Each consists of a zone of parallel to sub-parallel low ridges separated by troughs which may have widths of 200 kilometers or more and extend across the surface for more than 1000 kilometers.

rift valley
A fault-bounded valley formed in a region of crustal extension (divergence). Rifts are bounded by normal **faults** usually in pairs, forming **graben**. The East African Rift Valley is 5000 kilometers long and its floor lies 2–3 kilometers below its rim. Rifting is also widespread on both Mars and Venus.

CONNECTIONS

THE GEOLOGICAL JIGSAW **104**
RIFT VALLEYS **120**

ring system
Each of the four large outer planets (Jupiter, Saturn, Uranus and Neptune) has a stream of mainly icy particles orbiting it. These constitute ring systems.

roche moutonnée
Asymmetric land forms produced where an ice sheet grinds its way over a rock outcrop, smoothing the facing slope but pluck and steepening the other.

rock
Any consolidated collection of minerals.

rock-forming silicates
Any of the minerals that make up the Earth's crust and those of the other inner planets and their moons. They are compounds of **silica** (SiO_2) and metallic cations, such as sodium, potassium, calcium, magnesium and iron. The main rock-forming silicates include: quartz, feldspars, pyroxenes, olivines, amphiboles, micas and feldspathoids.

rotational energy
The potential energy resulting from an object's rotation around its axis.

salinity
The amount of salts dissolved in a solution, such as sea water.

sandstone
A sedimentary rock consisting of sand particles cemented together with calcium

carbonate, clay or iron oxide, or consolidated by pressure alone. The sand was originally deposited by water in shallow seas or accumulated by the action of wind in deserts.

saprolite
A deposit of clay, silt or soil formed by the breakdown of rocks that remains at its original site.

Saturn
The sixth planet from the Sun and the second largest in the Solar System. It orbits the Sun at an average distance of 1427 million kilometers, and takes 29.46 years to complete each orbit. It rotates very rapidly on its axis (once every 10.23 hours at the equator, rather slower at the poles, resulting in an equatorial bulge). It is composed mainly of hydrogen, giving the planet an overall density less than that of water. It is best known for its prominent ring system (probably comprised of material left over after **accretion**); it also has 20 or more moons.

CONNECTIONS

THE PLANETS FORM **56**
LARGE AND SMALL PLANETS **58**
THE PLANETS AND THEIR ORBITS **62**
THE OUTER PLANETS **68**

scablands
A highly eroded landscape that results from sudden extensive flooding. One of the best examples is in the northwestern United States. Flooding during the Pleistocene age scoured and channeled through loess and basalt of the Columbia River Plateau when the immense Lake Missoula burst its natural dam wall. Such scablands are terrestrial analogs of Martian outflow channels.

schist
A **metamorphic rock** (such as mica) that tends to split in layers. Deep burial of sedimentary rocks causes them to recrystallize, which they do under conditions of raised temperature and stress. As a result, the existing minerals become unstable and change to others, which are in equilibrium with the new conditions.

With increasing intensity of **metamorphism**, termed the grade, there tends to be a separation of felsic and mafic constituents. Recrystallization takes place under pressure, often giving metamorphic rocks a distinctive banded appearance, in which alternate layers of felsic and mafic platy mineral crystals (such as micas) tend to be preferentially arranged in planes perpendicular to the maximum applied stress.

sea-floor spreading
A theory describing anomalies found in the sea floor. It became apparent during geophysical experiments in the 1950s that the patterns of magnetic anomalies along parallel transects of the Pacific ocean floor off the coast of California were remarkably similar. Such patterns could be traced west to east but not north to south, in which direction abrupt changes were often observed. This was found to be caused by the presence of east-west **transform faults**. Further research showed that similar anomalies formed a mirror image on each side of **oceanic ridges**. It was also discovered that there was an absence of ocean-floor sediments along ridge crests but an increase in sediment thickness with increasing distance from the crest. Subsequent **radiometric datin**g indicated that oceanic sediments also became older with increasing distance from the ridge crest, as did the age of magnetic anomalies. To explain these observations, the theory of sea-floor spreading was proposed. New oceanic crust is generated by the upward movement of mantle-derived magmas into median rift zones (caused when lithospheric plates move apart), which then spread away on each side of the rift.

CONNECTIONS

PLANETARY MAGNETIC FIELDS **82**
EARTH'S FIRST OCEANS **100**

seamount
An isolated hill that rises from the **abyssal plain** of the ocean floor. Seamounts may be as high as 1000 meters or more. Their conical shape suggests that they have a volcanic origin.

sediment
Any loose debris that accumulates at a planet's surface, formed by physical processes such as wind, water, ice and impact processes.

sedimentary rock
A rock formed from the sediment produced by weathering of continental rocks, or chemically or biologically precipitated from sea water. Chemically precipitated sediments form in a way similar to the formation of igneous rocks, although clastic sediments (accumulations of grains) usually become welded together by a cement formed after burial through **diagenesis**. Finer-grained sediments convert to rocks by dewatering and compaction. Siliciclastic rocks (those formed from mainly silicate minerals) are classified according to grain size and composition, with particular importance being attached to the proportion of muddy sediment among the larger grains.

CONNECTIONS

EARTH'S FIRST CONTINENTS **102**
EROSION AND SEDIMENTATION **124**

seismic discontinuity
An abrupt increase in the velocity of transmission of seismic waves where they pass from a less dense to a more dense medium. The **Mohorovicic discontinuity**, for example, separates the Earth's crust from the denser mantle below.

seismic reflection profiling
A geophysical profiling technique much used to explore the subsurface structure of the Earth's crust. It involves sending sound waves downward from a sonic "gun". The waves are reflected and refracted by layers of different density inside the Earth in the same way as earthquake waves. The returned signals are registered by geophones and sent to a drum recorder for later analysis.

seismometer
A sensitive instrument for recording seismic waves. The original instruments used a very sensitive spring to record vibrations caused by seismic events.

self-exciting dynamo
The accepted theory of the origin of the Earth's magnetic field. The outer core is composed of molten iron and nickel, which conducts electrical currents. Because it is in motion – because of the rotation of the Earth – it can also interact with any external magnetic field, and the field itself can influence movements within the fluid. There are weak magnetic fields all over the Galaxy, and it is believed that the Earth originally had no field but interacted with the weak Solar System field at some distant time, generating its own field. The rapid spin of the planet had profound effects on motions within the core, with the result that it behaved like a huge cosmic dynamo. It is termed a "self-exciting dynamo", because once it begins to operate it gathers momentum, regenerates the weak extraterrestrial field, and produces the relatively strong field that exists today.

shadow zone
A region that earthquake waves do not penetrate, consisting of interior rocks in a zone between 105° and 140° of a quake epicenter. It results from the refraction of waves by the Earth's molten outer core.

SHEPHERD MOON

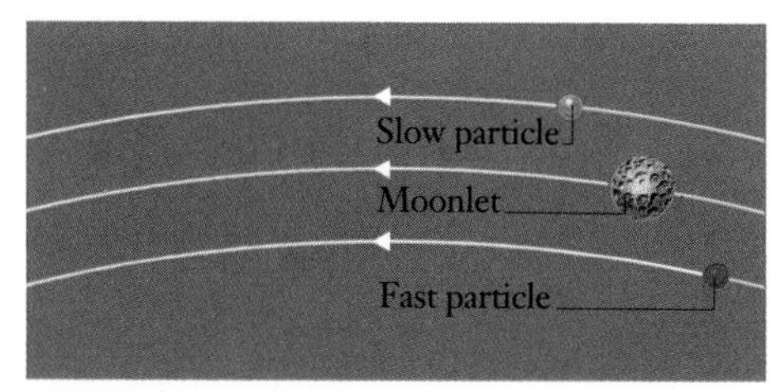

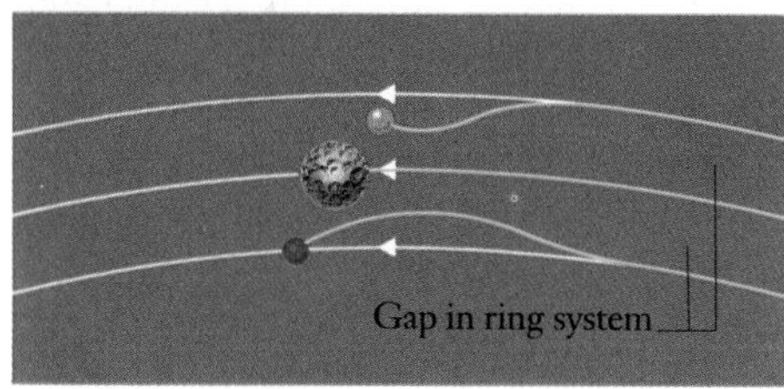

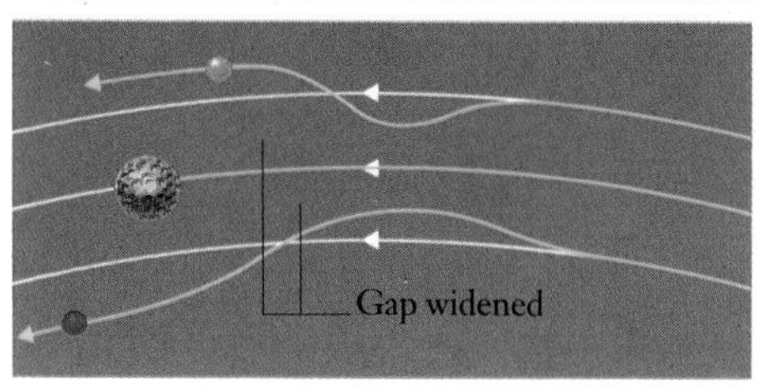

shatter cone
A striated cone-shaped feature, from less than a centimeter to several meters across, that is formed by a **meteorite** impact. Such structures are created by shock pressures of between 20 and 250 kilobars, and are arranged with the tops of the cones directed toward the focus of the impact. Mapping shatter cones may reveal the depth and force of an impact, and the degree of post-impact deformation.

shepherd moon
A small rock or rock-and-ice moonlet that forms a part of a planetary **ring system** and, through the effect of gravity and orbital resonances, acts to sweep clean certain parts of the ring, producing more or less permanent gaps. Saturn's rings are accompanied by several small shepherd moons, or guardian satellites.

shield
Also known as a **craton**, the ancient **core** regions of the Earth's continents, generally with relatively low relief and a rather flat surface. Volcanic shields are the result of effusive-type volcanoes. *See also* **volcano**.

CONNECTIONS

THE INNER PLANETS **66**
VOLCANOES **94**
EARTH'S FIRST CONTINENTS **102**
RIFT VALLEYS **120**

shocked quartz
A mineral such as quartz or feldspar with closely-spaced microscopic layers (lamellae) within it, caused by the intensely high pressures produced by impact shock.

sialic crust
The Earth's continental crust; that is, the part of relatively low density composed primarily of *si*licates of *al*uminum.

sidereal period
The time taken for a planet to complete one orbit around the Sun, or a moon to go once around its parent planet.

siderophile element
A chemical element that has an affinity for the metallic phase; for example, iron or nickel. During formation of the Earth the siderophile elements sank toward the core.

silicate
A mineral that is a compound of one or more metallic elements with silicon and oxygen.

CONNECTIONS

COSMIC INGREDIENTS **52**
THE INNER PLANETS **66**
CORE, MANTLE AND CRUST **80**
THE EVOLUTION OF ATMOSPHERES **84**
THE FIRST CRUSTS **90**
EROSION AND SEDIMENTATION **124**

Silurian
The period in the geological history of the Earth that lasted from about 438 million years ago. It was characterized by the many jawless fish (the only vertebrates) and coral reefs in the seas – nearly all animal life was marine – and by the evolution of the first primitive land plants.

simatic crust
The lower, denser, regions of the Earth's continental crust and the crust that forms the floors of the oceans. It is relatively rich in the elements *si*licon and *ma*gnesium.

smoker
A vigorously active vent on the sea floor which emits high-pressure hydrothermal fluids. At some smokers, emission of a high proportion of dark sulfur compounds gives rise to **black smokers**.

SNC meteorite
Any of a small number of **meteorites**, specifically those of Shergotty, Nakhla and Chassigny, that are significantly different from the normal meteorites. Their chemistry indicates that they originated in a differentiated planet that had an atmosphere (trapped residues of which are found in the samples). Shergotty, in particular, is chemically very similar to the Martian crust analyzed at the Viking Lander sites. Strangely, most of them are cumulate rocks, that is, they show distinct **fractional crystallization** effects. They further differ from other meteorites in that their crystallization ages are far younger: 1.2–1.3 billion years (as opposed to 4.5–4.6 billion years). This fact, together with their chemistry, has been taken to indicate an origin on Mars. If so, then they must have been derived from a mantle source that melted very early in Mars' history.

SOLAR PROMINENCE

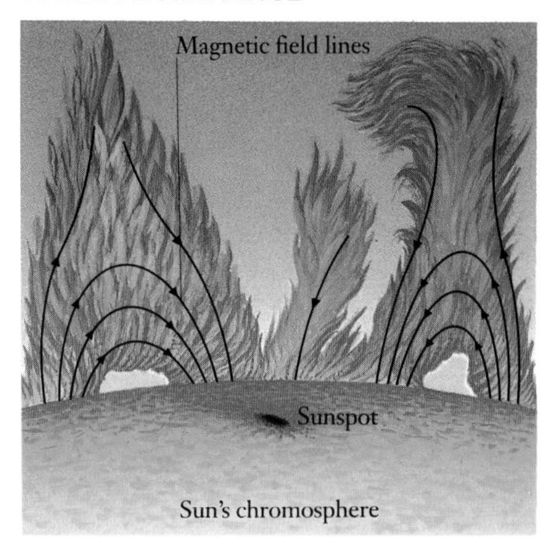

soil
The surface accumulation of rock debris and humus found on the Earth. The organic-free debris below it is known as **regolith**.

solar nebula
The cloud of dust and gas from which the Sun and its family of planets, moons and comets eventually condensed.

solar prominence
An enormous eruption of luminous gases that surge out up to 50,000 kilometers into the solar corona and occasionally off into space. Prominences may show streamers, arches, loops and other shapes, and they may be classified as active, eruptive, sunspot, tornado, quiescent and coronal. Some prominences have temperatures as high as 100,000 K; others as low as 10,000 K.

solar wind
The constant stream of protons and electrons that the Sun sends out into the Solar System. It takes several days for this solar wind to reach the Earth, which it passes at a velocity of about 600 kilometers per second.

The solar wind interacts with the Earth's **magnetosphere** and, indeed, those of all the planets that have one. The United States Mariner-2 spacecraft recorded the solar wind when it flew by Venus in 1962. Both Pioneer-11 and 12 also recorded it in the far reaches of the Solar System as they sped away from the planet Jupiter.

CONNECTIONS

THE FORMATION OF THE SUN 54
PLANETARY MAGNETIC FIELDS 82

solstice
The times at which the Sun is at its northernmost and southernmost points in the heavens. This occurs every year around 22 June (the longest day of the year in the Northern Hemisphere) and 22 December (the longest day in the Southern Hemisphere).

sorting
The process by which sedimentary grains of varying sizes and densities are sorted by moving fluids such as water, or by the wind. High-energy transportation generally is more effective at sorting than low energy, but time also plays an important part.

spit
A long, narrow tongue of sand or shingle attached to a coastline at one end. Spits are frequently formed by currents flowing across the entrance of a bay or inlet.

spring line
The interface between permeable and non-permeable rock strata, at which continuous or intermittent flow of groundwater from the subsurface occurs. As a consequence of this, springs often are found at intervals along the spring line.

stable isotope
An **isotope** of a chemical element that does not spontaneously decay to another.

stable zone
A region of the Earth's crust that is not subject to orogeny or other deformational process. Stable zones are typically found within the continental interiors, away from plate margins.

stalactite
A mineral formation, consisting of calcium carbonate or limestone, that hangs down (and slowly grows) from the roof of a cave as the mineral comes out of solution from dripping groundwater. *See also* **stalagmite**.

stalagmite
A mineral formation, consisting of calcium carbonate, that rises (and slowly grows) up from the floor of a cave as the mineral comes out of solution from dripping groundwater. *See also* **stalactite**.

star
A large incandescent ball of gases held together by its own gravity, heated by nuclear processes at its core, and giving off radiation at various wavelengths. Stars are believed to originate as condensations "protostars" out of interstellar material; as the protostar grows in mass but contracts under its own gravity, the core gradually becomes hot enough for fusion to begin.

CONNECTIONS

THE FORMATION OF THE SOLAR SYSTEM 48
BEFORE THE SOLAR SYSTEM 50
COSMIC INGREDIENTS 52
THE FORMATION OF THE SUN 54
BEGINNINGS AND ENDINGS 134

stone polygon
A polygonal pattern of coarse pebbles that may develop in regions of permafrost. Alternate freezing and thawing gradually causes the coarser particles to rise to the surface and make roughly polygonal patterns.

stratigraphic succession
The succession of the Earth's rock strata – their age and development over time. The study of the strata has enabled geologists to build up an understanding of the geological history of the Earth. In any one place the complete sequence of strata is seldom present, erosion and tectonic activity having removed a part or parts of the sequence. Thus, although the total maximum thickness of strata present within the stratigraphical succession is about 40 kilometers, the average thickness present at any locality is closer to 2 kilometers. The complete succession is divided into a small number of major time units called eons which are subdivided into eras and periods, the latter being further subdivided into epochs. Strata also contain fossil remains which show evolutionary changes with time. Certain fossils (called index fossils) are typical of specific zones and can be used to assign relative ages to parts of the sequence. *See also* **paleontology**.

stratigraphy
The study of the geological strata that lie at or near to the surface of a planet. A stratigrapher has to piece together the paleogeography of an area from fragments of information gleaned by mapping rock outcrops. A complete sequence of strata is rarely found in one area and it becomes necessary to compare one region with another where a more complete sequence is preserved, and to take account of paleontological evidence. In the case of the Earth, detailed mapping is possible and fossil remains aid the understanding of the **stratigraphic succession**. On other planets it is generally possible only to study transection/superposition relations between rock units and to assign them relative ages. Relative ages may be assigned by counting the number of impact craters on a surface; such crater dating methods were developed on the Apollo program of lunar missions.

CONNECTIONS

CHANGING WORLDS 122
BEGINNINGS AND ENDINGS 134
THE EMERGENCE OF LIFE 136

stratosphere
The layer of the Earth's atmosphere in which temperature increases steadily with height and in which most of the ozone is found. It lies between 14 and 50 kilometers from the surface, immediately above the layer in which the weather occurs.

stratovolcano
A major volcanic edifice built from a mixed sequence of lavas and pyroclastic rocks. It is also known as a composite volcano.

stratum
Any layer of sedimentary rock. The study of the sequence of sedimentary strata is the science of **stratigraphy**.

stromatolite
A layered sedimentary fossil formed from layers of blue-green algae and carbonate sediment. Algal mats are composed of sticky interwoven layers of blue-green algae. The gluey surfaces trap fine-grained calcareous sediment, with the result that the algae and carbonate sediment grow as a sequence of interleaving layers. Such growths have been recorded in shallow tidal sediments since the Proterozoic. *See* **blue-green alga**.

subduction zone
The inclined zone at the boundary of two converging **lithospheric plates**, along which consumption of one plate occurs. Subduction zones dip away from parallel **trenches** and are characterized by a high degree of seismic activity. *See* **Benioff zone** and **earthquake**.

SUPERNOVA

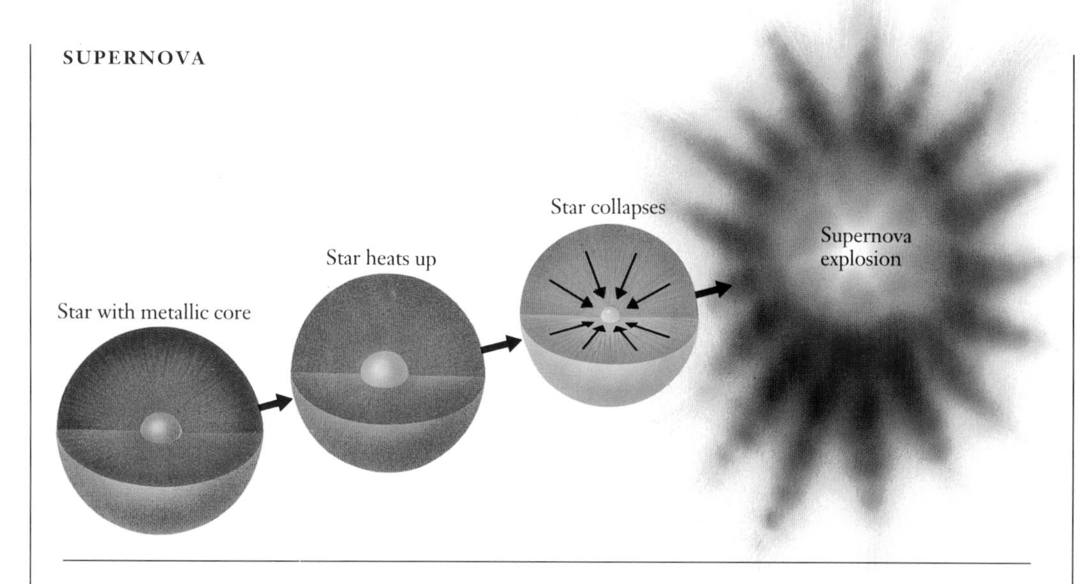

subsolidus
Convection within the **mantle** which is believed to occur in a completely crystallized material. Any movement is called subsolidus convection. Igneous rocks are chemical systems with crystallized from rock melts. The temperature at which such melts begin to crystallize is called the liquidus temperature. At some lower temperature crystallization will be complete; this is called the solidus.

Sun
The **star** at the center of the Solar System, around which all planets (and comets and asteroids) orbit. It is a ball of gas, 1,393,000 kilometers across, rotating on its axis once every 24.65 days at the equator (though rather slower, every 34 days, at the poles).

CONNECTIONS

THE FORMATION OF THE SOLAR SYSTEM **48**
BEFORE THE SOLAR SYSTEM **50**
COSMIC INGREDIENTS **52**
THE FORMATION OF THE SUN **54**
THE SUN'S FAMILY **60**

sunspot
A relatively dark patch on the surface of the Sun. Large spots have a dark umbral region surrounded by a lighter penumbra. They are regions of the **photosphere** cooler than their surroundings, which therefore emit less energy and are darker. Sunspots are associated with strong magnetic fields.

supernova
A star that suddenly increases in luminosity to the extent that it may become as luminous as an entire galaxy. It then declines in brightness. Such an event is thought to be due to a cataclysmic explosion brought about by the collapse of a very massive star, resulting in the intense heating of the surrounding layers, which are then ejected into space. The intense "burning" of the core regions generates explosive activity in which it is believed the very heavy elements in the Universe are synthesized. The generation of neutrinos removes significant amounts of the energy produced during such events, resulting in a collapse of the star, rapid core contraction, increase in temperature, and eventual onset of iron-burning reactions. The core then implodes and the outer layers are ejected.

suspended load
Solid material carried by a river. A flowing river is able to carry in suspension only a certain amount of sediment, the volume being largely dependent on the cross-sectional area of the channel, the rate of flow, the gradient and the volume of water present. In times of flood, the suspended load has a greater volume, and the river can pick up larger fragments. When the river reaches the sea, its energy is dissipated, and the load deposited.

S waves
Seismic secondary waves. They are shear waves and consist of elastic vibrations at right angles to the direction of transmission. They travel less quickly than **P waves**.

syncline
A downward fold in deformed rock strata, with the youngest beds at the center.

talus
A deposit of generally large angular fragments that accumulates at the foot of a cliff or steep slope. Its is composed predominantly of unweathered bedrock.

tectonics, or tectonism
The deformation that affects the **lithospheres** of planets. The term includes folding, thrusting, shearing and faulting, and any features associated with epeirogenic movements (that is, mainly up-and-down movements of the crust). This may be global or regional in scale.

CONNECTIONS

THE GEOLOGICAL JIGSAW **104**
MOBILE AND STABLE ZONES **108**
WANDERING CONTINENTS **110**
GEOLOGICAL PLATE MOVEMENTS **112**
MOUNTAINS FROM THE SEA **118**

tektite
A small glassy object, often black or green, composed of silica compounds unlike any in the neighboring rocks. Tektites are found in groups on the Earth's surface and probably result from meteorite impacts.

terminal moraine
A deposit of poorly-sorted sand, gravel, boulders and mud left by a glacier at its farthest point of advance. The moraine usually has a somewhat flowing form.

Tertiary
A period in the Earth's geological history that lasted from about 65 million to 2 million years ago. During this time, modern mammals came to dominate the land, and flowering plants such as shrubs and grasses evolved.

tessera
On Venus, highly deformed blocks of what appear to be relatively brittle rocks that have the appearance of tiled flooring. These tessera regions, located in the highlands, are also known as complex ridged terrain (CRT), because of the sets of intersecting ridges and grooves that characterize them.

Tethys
The smallest of Saturn's five major moons (1060 kilometers across), Tethys rotates on its axis every 1.89 days as it orbits at an average distance of 294,700 kilometers from its parent planet.

thermoremanent magnetization
Also known as TRM, the permanent magnetization that igneous rocks acquire after they have cooled to a temperature below the **Curie point** under the influence of the Earth's **magnetic field**.

thermosphere
A layer of the Earth's atmosphere characterized by large fluctuations in temperature (in excess of 300°C). It lies between 85 and 500

kilometers above the surface. At these heights the number of molecules present is low, therefore heat retention is low; the average temperature is high because of their inability to reemit at longer wavelengths radiation incoming at shorter ones.

threshold wind speed
The minimum wind velocity that allows air to pick up sand grains. It varies from among planets, depending largely on the density of the atmosphere and the surface gravity.

thrust fault
A low-angle reversed **fault** in which younger rocks are forced over younger ones. Major thrusts may transport slices of crustal rocks for many kilometers along a thrust plane, as has happened along the Moine Thrust in the northwestern highlands of Scotland. Large folded slices of rock may be moved far from their "roots" along such fault planes, giving rise to structures called nappes.

tide
Any periodic deformation of the crust of a planetary object or the atmosphere of a gaseous one by the gravitational attraction of another object. Tidal fluctuations experienced by the terrestrial oceans are a function of the gravitation effects of the Moon and, to a lesser extent, the Sun.

tillite
A consolidated sedimentary rock composed of glacial material, or till. It was the realization that glacial rocks of the same age occurred on each of the southern continents that helped geologists piece together the details of continental drift.

Titan
The largest moon of **Saturn**, which it orbits at an average distance of 1,221,000 kilometers. It is 5150 kilometers across (larger than the planet Mercury) and has an atmosphere. Its axial rotation is slow: nearly 16 days.

Titania
The largest moon of **Uranus**, Titania is 1580 kilometers across and orbits at an average distance of 583,400 kilometers. It has **retrograde rotation**, taking 8.71 days to turn once on its axis.

Titius-Bode rule
A relationship between the orbital distances of the planets noticed by J.D. Titius, published in 1772 and popularized by J.E. Bode. If one takes the numbers 0, 3, 6, 12, 24, 48, 96, 192 and 384 (each, apart from zero, being double the previous one) and adds 4 to each, then, assuming the distance from Earth to the Sun to be 10.0, the other orbital distances of the planets are given by the Titius-Bode rule as follows:

Planet	T-B Rule	Actual
Mercury	4	3.9
Venus	7	7.2
Earth	10	10.0
Mars	16	15.2
Jupiter	52	52.0
Saturn	100	95.4

When Uranus was discovered in 1781, it was found to have a distance of 191.8 (the rule predicted 198). Later, when the first asteroid (Ceres) was found and asteroidal orbits were calculated, the gap equivalent to 28 was filled. However, Neptune – which was discovered in 1846 – does not fit into the scheme (the rule predicts 388, but the actual measure is 300.7): the rule figure more closely fits the distance of Pluto/Charon (actual 394.6). The "rule" is considered to be merely coincidental.

TRACE FOSSIL

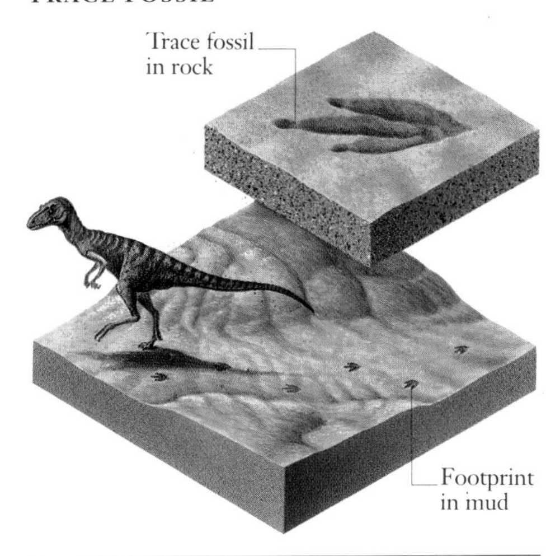

trace element
An element present in crustal rocks but only in trace amounts, less than 1 percent (often less than 0.001 percent). Trace elements are nevertheless important because the partitioning of them into different rocks is governed by factors that allow geochemists to establish genetic relationships between rock types and their parent magmas.

trace fossil
An imprint, such as a footprint, preserved as a fossil. Such fossils include not only footprints of larger mammals but also trails left by burrowing shell creatures, crustaceans, trilobites and the casts of worm burrows.

traction load
The part of a river's sedimentary load that can only be dragged along the bed.

transcurrent fault
A movement of crustal blocks of rock that involves lateral movement in the horizontal plane along each side of the fault plane. The Californian San Andreas Fault is of this type.

transform fault
A strike-stop **fault** that occurs when blocks of crustal rock, at tectonic margins, slide sideways past each other.

transient cavity
A temporary crater formed briefly during a **meteorite** impact. During the initial stages of crater excavation, the incoming object plunges into the target rocks and deforms them. During the very first seconds the object itself is thought to mold itself into a horseshoe-shape lining the transient cavity that forms. This cavity exists only during the compressional phase of the event, and is modified when the **rarefaction wave** passes through the target materials and decompression occurs. The final cavity is considerably less deep than the transient one.

trench
A long, narrow sea floor trough that forms along convergence zones as a part of plate tectonic activity. Such trenches mark regions of subduction, and they plunge to depths of the order of 10 kilometers. They are a feature of the seaward side of **island arcs**.

CONNECTIONS
BENEATH THE OCEAN FLOOR 114
ISLAND ARCS 116

Triassic
In the geological history of the Earth, the period that lasted from about 248 million to 213 million years ago. At this time, mammals began to evolve but reptiles were the dominant land animals; cone-bearing trees were abundant in the non-desert areas.

trilobite
A type of extinct arthropod that scavenged on the bottom of shallow seas in the Precambrian and paleozoic areas. Trilobites had plated, segmented bodies with many Y-shaped legs, somewhat resembling modern wood lice. They died out at the end of the Permian period, about 248 million years ago, but are abundant as fossils in rocks dating from that time.

triple junction
A point at which three lithospheric plates meet. It occurs, for instance, off the coast of Oman, where the African, Somali and Indian

plates come into contact. Another occurs off the western shore of Central America, where the Nazca, Cocos and Pacific plates are in contact. In the latter case the margins involve subduction, spreading and transform movements along individual plate boundaries. *See* **lithospheric plate, tectonics** and **subduction zone**.

Triton
By far the largest moon of **Neptune**, which orbits with retrograde motion at an average distance of 354,800 kilometers from its parent planet. It is 2705 kilometers across, and rotates (again with **retrograde rotation**) once every 5.88 days.

tropopause
The interface between the Earth's **stratosphere** and **troposphere**. It lies at a height of about 13 kilometers. The maximum horizontal pressure gradient lies just below this boundary.

troposphere
The lowest layer of the Earth's atmosphere; the zone of weather. It is the densest of the atmospheric layers and shows a general decrease in temperature with increasing height. About 80 percent of the atmospheric mass is found within it.

tsunami
A large and often highly destructive wave which is triggered by seismic events, often associated with violent volcanic eruptions. An enormous tsunami was set off during the explosion of the volcano Krakatoa in Indonesia in 1883.

T Tauri activity
Irregular variation in brightness exhibited by some very young stars that have not yet reached the **main sequence**. They appear to be surrounded by clouds of material which may well be similar in some respects to the stuff from which the **solar nebula** was made. It is possible that planet formation is taking place within T Tauri-type stars.

tufa
A type of rock that consists of smooth, rippled formations of calcium carbonate. It forms as dissolved minerals precipitate out of spring water and on the floors of caves.

tundra
The treeless plains of northern America and Eurasia, lying approximately along the Arctic Circle where – for most of the year – the average temperature is below freezing point. The tundra regions are characterized by short summers and long hard winters. Below a depth of about 20 centimeters the ground is permanently frozen. *See also* **permafrost**.

turbidite
A sedimentary rock laid down by a **turbidity current**. A typical turbidite layer shows graded bedding and sedimentary "bottom structures", which are infills of grooves and hollows eroded into the sea bed by swiftly moving sediment-laden slurries.

turbidity current
A current that occurs when sediment which has accumulated toward the edge of the continental shelf is shaken down the **continental slope**, usually by seismic tremors. The dense sediment suspensions move swiftly and bring about considerable erosion of the sea floor during their descent to the **abyssal plain**. Once they reach the deeper, flatter, parts of the ocean floor, their energy dissipates and they deposit their load. The mixture of sand and mud-grade sediment then separates, often giving rise to graded bedding, and, because of the rapid rate of transportation and deposition, to poorly sorted rocks called **turbidites**.

ultraviolet radiation (UV)
Short-wavelength electromagnetic radiation in the wavelength range 3 nanometers to 0.4 micrometers. Incoming UV radiation at wavelengths greater than 0.3 micrometers is completely absorbed by ozone in the Earth's upper atmosphere. Photographic UV radiation (wavelengths 0.3 to 0.4 micrometers) is transmitted through the atmosphere and is detectable on photographic film.

Umbriel
The third largest moon of **Uranus**, which it orbits at an average distance of 266,000 kilometers. It is 1190 kilometers across and rotates, with **retrograde rotation**, once every 4.14 days.

unconformity
Within the part of the **stratigraphic succession** of the Earth's rock that is exposed at any one location, there may be sections of the complete sequence that are missing. (The missing strata, or layers, may have been removed by erosion or never have been deposited; *see* **non-sequence**.) The plane that separates the strata on each side of such a gap is known as an unconformity. If orogenic (mountain-building) movements cause the older strata to be folded or tilted before erosion and the subsequent deposition of younger beds can occur, angular unconformity is the result. *See also* **orogenesis** and **stratigraphy.**

Uniformitarianism
"The present is the key to the past" is the tenet of the principle of Uniformitarianism, first proposed by James Hutton of Edinburgh in 1785. Subsequently it was popularized by Charles Lyell in his book *Principles of Geology*, published in 1830. It implies that the processes that form and mold present-day rocks and landforms are the same as those that operated in the distant past. In other words, by understanding how things work today, it is possible to understand what happened millions of years ago. What the principle does not state is that the *rate* at which processes operated may have been different hundreds of millions of years ago from the today's rates, and that the relative importance of the processes themselves may have changed.

Uranus
The third largest planet in the Solar System and the seventh out from the Sun, which it orbits at an average distance of 2870 million kilometers. It is similar to Neptune but slightly larger (52,000 kilometers across). It has **retrograde rotation** with its axis tilted over at 98° to the orbital plane, rotating once every 17 hours. It has five major moons and many smaller ones.

CONNECTIONS

LARGE AND SMALL PLANETS **58**
THE SUN'S FAMILY **60**
THE PLANETS AND THEIR ORBITS **62**
THE OUTER PLANETS **68**

Valles Marineris
An equatorial canyon system on Mars. It is four times deeper, six times wider and at least ten times longer than Arizona's Grand Canyon. It is a huge tectonized zone comparable in scale to the Earth's East African Rift Valley, made up from canyons, collapsed depressions and regions of "chaotic terrain". It begins on the eastern flank of a huge swelling in the Martian **lithosphere** – the Tharsis Bulge – and trends roughly west-east along the equator for 4500 kilometers, ending in extensive regions of collapsed terrain in the east.

Van Allen belts
Donut-shaped regions formed by the lines of force of the Earth's magnetic field, within which protons and electrons are trapped. They were identified by the United States physicist James Van Allen following experiments carried out by the spacecraft Explorer 1 in 1958. He showed that charged atomic particles formed a belt, about 3000 kilome-

ters distant. In the same year, the US probe Pioneer 3 proved the existence of a second belt at 20,000 kilometers.

Venus
The planet second out from the Sun, at an average distance of 108 million kilometers. At 12,100 kilometers across, it is about the same size as the Earth but takes 243 days to rotate once on its axis (with **retrograde rotation**). Its "year" – the time it takes to orbit once around the Sun – is 225 days long. The dense carbon dioxide atmosphere creates a permanent **greenhouse effect** that keeps the surface temperature in excess of 450°C. Venus has no moons.

CONNECTIONS
THE SUN'S FAMILY 60
THE PLANETS AND THEIR ORBITS 62
THE INNER PLANETS 66
THE EVOLUTION OF ATMOSPHERES 84
VOLCANOES 94

vesiculation
The process by which gas forms from **magma** and builds up pressure below the Earth's surface, finally erupting – often from a **volcano**. Magma contains volatiles which, deep inside a planet, are held in solution by high confining pressures. As the magma rises toward the surface, the pressure lessens, with the result that the gases begin to leave solution. Once this occurs, bubbles form, and the process of vesiculation has begun. The existence of bubbles in the melt helps it to rise toward the surface, with the result that more bubbles form as more gas is produced. Eventually the gas pressure may become so great that an explosive eruption is triggered. The evacuated gas cavities often found in volcanic rocks are called vesicles.

VAN ALLEN BELTS

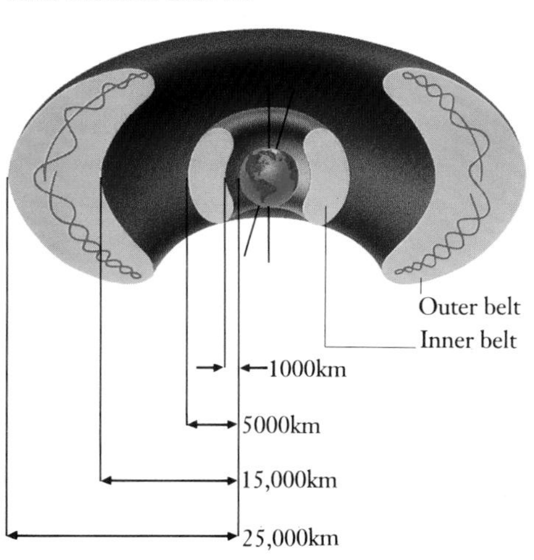

Viking missions
Two United States exploratory missions to Mars. The spacecraft Viking 1 reached Mars on 19 June 1976, and Viking 2 arrived on 3 September the same year. Each spacecraft consisted of an orbiter and a lander probe. A gas chromatography experiment sought to establish whether organic molecules had proliferated on Mars; despite initially encouraging results, they eventually recorded that they had not. The landing sites were in the plains of Chryse and Utopia.

volatile element
Any of the elements that show an affinity for the atmosphere, also termed volatiles or atmophiles. They include hydrogen, nitrogen, carbon, oxygen and the **inert gases** Molecules such as carbon dioxide and water are known as volatile molecules.

volcanic rise
On Venus, a prominent rise in the lithosphere characterized by major volcanism and rifting. An example is Beta Regio.

volcanism
Also known as vulcanism, the geological processes associated with the rise of **magma** to the surface of any planet and its emission from volcanoes and volcanic fissures.

CONNECTIONS
THE FIRST CRUSTS 90
VOLCANOES 94
ISLAND ARCS 116
MOUNTAINS FROM THE SEA 118
NATURAL CATASTROPHES 138

volcano
Any conduit that allows **magma** to reach the surface of a planet. In the broad sense, fissures (cracks) – such as those that fed the high-volume lavas which formed the Deccan basalt plateau of India and the Columbia River plateau of the western United States – are volcanoes, as are those that built the plains of Mars. More centralized activity gives rise to the growth of what more typically is thought of as a volcano. The gently effusive activity found in the Hawaiian islands generally builds volcanic shields, which have low flank slopes but great size. These contrast with the steep-sided, but generally smaller, stratovolcanoes, which involve more viscous lavas and a degree of explosive activity. Maars are craters produced by extremely violent explosions generated by the entry of water into a magma source, or of magma into a lake or river. They have virtually non-existent rims, in sharp contrast to a typical stratovolcano. Volcanic domes are built from extrusions of viscous magma that is unable to flow far.

Voyager missions
The United States missions to the outer Solar System, launched in 1979 with the spacecraft Voyager 1 and 2. They studied not only the giant planets but also a large number of their moons and ring systems. Voyager 1 reached Jupiter in March 1979, four months earlier than Voyager 2. Voyager 1 then swung around the giant planet and swept off to Saturn, which it reached in November 1980; Voyager 2 reached Saturn in August 1981. Voyager 1's work was then largely complete, but Voyager 2 went on to study Uranus (January 1986) and Neptune (1989). It is now passing out of the Solar System.

wadi
A steep-sided valley in a desert region temporarily occupied by an ephemeral stream.

water table
The underground level to which permeable strata are saturated with groundwater.

weather
The variation in the atmospheric conditions experienced at a given site.

weathering
The mechanical and chemical processes that break up bedrock and reduce it to its individual constituents or their derivatives. Generally, the more humid and warmer the climate, the more intense are the effects of chemical weathering.

xenolith
Fragments of wall rocks, pieces of earlier igneous rocks, or particles brought up from depth by a magma. Zones rich in xenoliths often occur along the margins of an intrusion, where pieces of wall rock are broken off during magma ascent.

X-ray fluorescence
An analytical technique used by geochemists to determine the major and trace element composition of minerals. A sample is bombarded by X rays, the secondary emission produced by the individual mineral species being indicative of the elements present.

zeolite
A group of rock-forming silicates that contain water. Typically they infill cavities in igneous rocks or grow at the expense of other silicates under the influence of volatiles in the magma. *See* **igneous rock**.

1

OUT of the Cosmos

OUR SUN AND PLANETS are by no means the oldest objects in the Universe. Before they existed, other stars had been created from atoms of simple elements such as hydrogen and helium. It appears that the Universe was created in a huge explosion which took place about 15 billion years ago. The theory that explains it in this way is known as the Big Bang.

The Big Bang theory has received strong support from the discovery of radiation apparently left over from the initial explosion. This effect, first detected in 1965, takes the form of 3.2-centimeter microwave radiation that emanates from all directions and is detectable by radio telescopes. Its existence had already been predicted, on the theory that the Universe has been cooling since its immeasurably hot birth; by the present time it must have cooled to around 3 K (3° above absolute zero). This is exactly what was found.

Galaxies are massive star systems in which new stars are forming and others are in their death throes. The Milky Way is an edge-on view of our own galaxy. New stars condense out of nebulae – vast clouds of gas and dust dispersed within galaxies. Newly formed stars – or protostars – tend to spin rapidly, and rotational effects cause the surrounding cloud of matter eventually to flatten into a rotating disk. Planets may eventually form from this disk as the material cools.

The Vela supernova remnant, 1500 light-years from Earth, is what remains of a gigantic stellar explosion. Also known as the Cygnus Loop, it is a giant expanding shell of gas and interstellar dust which formed about 50,000 years ago. A supernova explosion occurs at the close of a star's life as its core collapses inward and an ever-expanding shell of gas is sent spiraling into space. The energy output of such an event is staggering: for about one week, its light often outshines the entire light of its parent galaxy. Supernovas are rare, taking place about once every 100 years in our own galaxy.

Cosmic Ingredients

A KNOWLEDGE of what planets are made of helps in understanding how they evolved. The fundamental ingredients of everything in the Universe, including the stars and planets, are the chemical elements. Each element is composed of a single type of atom, itself built from protons (positively charged particles), electrons (negatively charged) and neutrons (electrically neutral). Neutrons exist within the nucleus of an atom but do not dictate its chemical identity. But if present in different numbers within atoms of the same element, they give rise to different isotopes. The numbers of neutrons and protons in an element determine its atomic mass.

KEYWORDS

CHALCOPHILE ELEMENT
ELECTRON
HEAVY ELEMENT
LITHOPHILE ELEMENT
NEUTRON
PROTON
SIDEROPHILE ELEMENT
SILICATE
VOLATILE ELEMENT

Hydrogen is the simplest of the elements, normally containing just one proton and one electron. It is also the lightest element, with an atomic mass of 1. If there are differing numbers of neutrons in the nucleus, isotopes with different atomic masses result. For example, deuterium is an isotope of hydrogen with one proton and one neutron in its nucleus; it has a mass of 2.

The extreme simplicity and abundance of hydrogen atoms in the Universe have led physicists to conclude that all other elements were formed from the primordial hydrogen atoms, which were created in the Big Bang. For this to happen – that is, for hydrogen atoms to undergo suitable nuclear transformations to give rise to elements with greater atomic masses – very high temperatures and pressures are required. The process by which such elements are created involves the fusion of light atomic nuclei to form heavier ones. Every time nuclear fusion occurs, a large amount of energy is released; this may also produce other particles such as electrons or hydrogen nuclei.

Such extreme conditions are known to exist deep inside certain types of stars and it is there that new elements are being created. Within a star of mass equivalent to the Sun's, the element hydrogen may "burn" to form helium, which has two protons and two electrons. This requires temperatures of the order of 10 million K. More massive stars have even higher temperatures and pressures at their cores, and in them helium may fuse to form carbon (6 protons and 6 neutrons); this in turn may combine with more helium to form oxygen (8 protons, 8 neutrons); and so on.

In this way, a variety of chemical elements may be produced. If the star is sufficiently massive it may eventually become unstable and a gigantic stellar explosion, called a supernova, may eject the elements across interstellar space. The very heavy elements such as lead are also formed inside stars, by a variety of processes not entirely understood.

Many of the chemical elements join together to form molecules and compounds. Among these are very mobile ones which are termed volatiles. Water, carbon dioxide and sulfur dioxide are three important volatiles. These tend to be stable as gases at quite low temperatures (below 300°C). Other combinations of elements produce minerals such as those which build rocks, most of which are silicates. They tend to solidify at quite high temperatures (450 –1200°C). Elements such as aluminum and calcium that combine with oxygen to form silicates are called lithophiles. Zinc, lead and silver are chalcophiles – they tend to form sulfides. Elements such as gold and nickel that tend not to form compounds are called siderophiles.

Volatiles and silicates are found in meteorites, small cosmic bodies of great age which may bear close similarities to the earliest solid objects that grew within the solar nebula. This suggests that at a very early stage in the history of the Solar System there was a wide range of matter available for planet building and that both high- and low-temperature particles were well mixed together.

△ **Supernova 1987a (top right), near the Tarantula Nebula. The original blue supergiant collapsed in a few seconds, blasting the supernova remnant out into space.**

▷ **Hydrogen burning in a star of solar mass lasts about 10 billion years. When the hydrogen is used up, the helium core contracts, gravitational potential energy is released, and the star leaves the main sequence. An expanding shell of hydrogen gas envelops the core, which collapses; it becomes a red giant. In more massive stars with hotter cores, helium fuses to carbon, silicon or oxygen, synthesizing the heavier elements. Even more massive stars may burn iron, generating a cooling effect: the core implodes and the outer layers of the star are blown away as a supernova. The most massive stars of all may go beyond this stage, such that even the dense core of neutrinos is crushed and a black hole is left.**

1 Protostellar nebula
2 Luminous phase of 1 solar mass star
3 Main sequence phase
4 Expansion phase
5 Red giant phase
6 Contracting phase
7 White dwarf phase
8 Star of 10 solar masses
9 Supergiant phase
10 Supernova
11 Neutron star
12 Star of 30 solar masses
13 Supergiant phase
14 Supernova
15 Black hole

◁ A false color image of Supernova 1987a, taken by the Hubble Space Telescope in 1990, shows an expanding gas ring (yellow) around the supernova remnant. The original blue supergiant star was 155,000 light-years away from Earth. The tightly knotted debris left by the catastrophic explosion appears as the red area in the center of the ring. Many of the elements that make up the planets were formed in such explosions.

BEFORE THE SOLAR SYSTEM

LONG before the Sun was born, generations of stars lived and died across the Universe. Their remains hung in unimaginably large clouds of dust and gas throughout space, providing the material from which new stars were made.

About 4.6 billion years ago, one of several regions in such a giant molecular cloud in the Milky Way galaxy began to collapse under its own gravity, a process that is believed to have taken about 100,000 years. Its core of material slowly began to rotate, and then to collapse while still embedded in an envelope of dust and gas. This cloud, called the solar nebula, had a diameter somewhat larger than the present orbit of Pluto (the most distant modern planet). It was the embryonic Solar System to which the Earth belongs.

KEYWORDS

ANGULAR MOMENTUM
COMET
METEORITE
PHOTOSPHERE
PROTOSTAR
REFRACTORY ELEMENT
SOLAR NEBULA
T TAURI ACTIVITY
VOLATILE ELEMENT

After about 100 million years, the central mass had grown large enough to become a protostar. It was not yet burning, but started to spin rapidly and forced the cloud to flatten into a slowly rotating disk. The disk also continued to grow; by this stage it is believed to have had about 4 percent of the eventual total solar mass (today it contains only 0.1 percent of the mass of the Sun). This disk gradually evolved into the Sun's family of planets and moons.

In the center of the nebula, close to the newly created proto-Sun, temperatures began to climb through the effects of ever-increasing collisions between cloud particles. The cloud had initially been cold, about 50 K, so that only helium and hydrogen could exist as gases; the other matter in the nebula was dust or ice. Now, as the temperature near the center rose, the volatile matter nearby was vaporized, while the outer regions of the nebula remained cold.

The temperature gradient in the nebula – about 2000 K near the proto-Sun, but 50 K in the outer reaches – affected the distribution of its molecules. The densest, such as metals and

▽ The Solar System to which the Earth and Sun belong began its life as a huge, cold cloud of interstellar dust and gas. 1 As the cloud began to shrink, its central regions collapsed faster than those nearer the edge. The core warmed up, and the cloud began to rotate. 2 After hundreds of thousands of years of contraction, the cloud rotated faster and began to flatten. A hot "proto-Sun" – emitting several times more radiation than the present Sun – formed within the central regions. 3 The bulk of the rotating disk of dust and gas became concentrated into an accretion disk. Matter continued to fall inward onto the protostar, which became so hot that it ejected matter violently through its poles, where it was easiest to do so. This swept away much of the matter surrounding the protostar. 4 The Sun began its life as a "main-sequence star", driven by fusion reactions in its core, whereby hydrogen was converted to helium. 5 From the remaining cloud of matter surrounding the protostar, the planets eventually condensed.

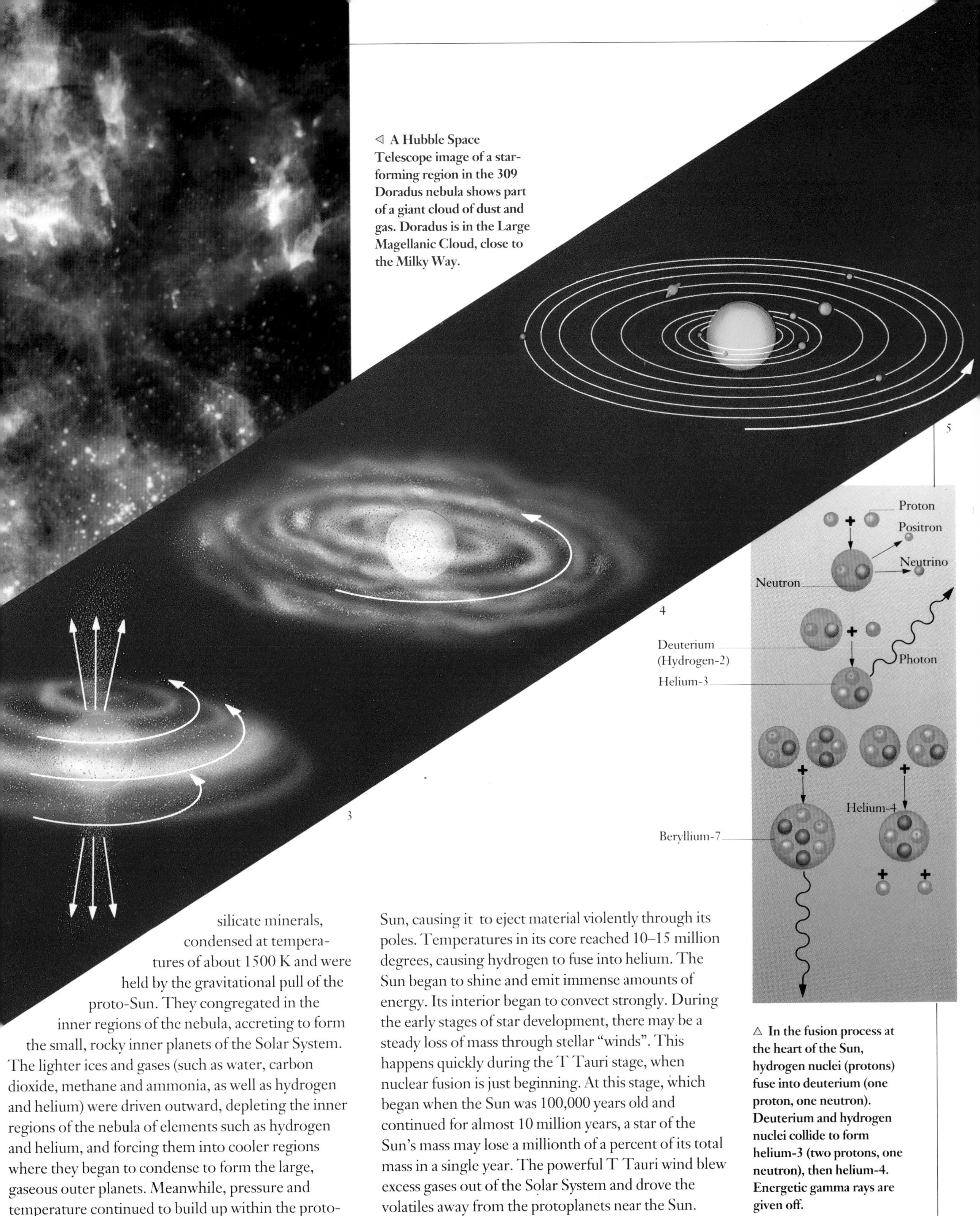

◁ **A Hubble Space Telescope image of a star-forming region in the 309 Doradus nebula shows part of a giant cloud of dust and gas. Doradus is in the Large Magellanic Cloud, close to the Milky Way.**

silicate minerals, condensed at temperatures of about 1500 K and were held by the gravitational pull of the proto-Sun. They congregated in the inner regions of the nebula, accreting to form the small, rocky inner planets of the Solar System. The lighter ices and gases (such as water, carbon dioxide, methane and ammonia, as well as hydrogen and helium) were driven outward, depleting the inner regions of the nebula of elements such as hydrogen and helium, and forcing them into cooler regions where they began to condense to form the large, gaseous outer planets. Meanwhile, pressure and temperature continued to build up within the proto-Sun, causing it to eject material violently through its poles. Temperatures in its core reached 10–15 million degrees, causing hydrogen to fuse into helium. The Sun began to shine and emit immense amounts of energy. Its interior began to convect strongly. During the early stages of star development, there may be a steady loss of mass through stellar "winds". This happens quickly during the T Tauri stage, when nuclear fusion is just beginning. At this stage, which began when the Sun was 100,000 years old and continued for almost 10 million years, a star of the Sun's mass may lose a millionth of a percent of its total mass in a single year. The powerful T Tauri wind blew excess gases out of the Solar System and drove the volatiles away from the protoplanets near the Sun.

△ **In the fusion process at the heart of the Sun, hydrogen nuclei (protons) fuse into deuterium (one proton, one neutron). Deuterium and hydrogen nuclei collide to form helium-3 (two protons, one neutron), then helium-4. Energetic gamma rays are given off.**

THE SUN

WITHOUT the star around which our Solar System is built, there would be no life. Our Sun has a diameter of 1,392,000 kilometers – more than 109 times that of the Earth. It is a modest star, and shines with a yellow light that shows it to be stable. The temperature at the core is about 15 million K, forming an environment in which atomic nuclei are stripped bare of their electrons. The outer layer of the Sun's atmosphere, called the photosphere, is the "surface" that is visible on Earth. Its temperature is about 6000 K.

KEYWORDS

CORONA
ECLIPSE
IONOSPHERE
PHOTOSPHERE
SOLAR PROMINENCE
SOLAR WIND

Viewed through suitable filters or by projecting an image of the solar disk onto a white board, the Sun's surface shows variations in brightness, referred to technically as granulation. The differences in brightness reflect differences in temperature that are caused by convection in the outer layers. It is possible that hydrogen is undergoing a change from being completely ionized in the interior of the Sun to being neutral at the surface.

About 500 kilometers above the visible surface, the atmospheric pressure drops quickly and the temperature falls by at least 2000 K. The gas here is transparent to most wavelengths of the radiation escaping from the photosphere, but absorbs radiation at wavelengths characteristic of atoms in that layer. It is here that the complex solar spectrum is produced. Analysis

Core
Radiative zone
Convective zone
Photosphere

△ The dense, hot core of the Sun extends some 175,000 kilometers from the center. This is enveloped in a radiative layer, succeeded by a convective layer that transports material to the surface. The visible layer, or photosphere, is only some 400 kilometers deep. Above this lies the chromosphere, a tenuous zone in which the absorption lines of the solar spectrum are generated. Outside the latter is the highly rarified corona, which merges imperceptibly with space.

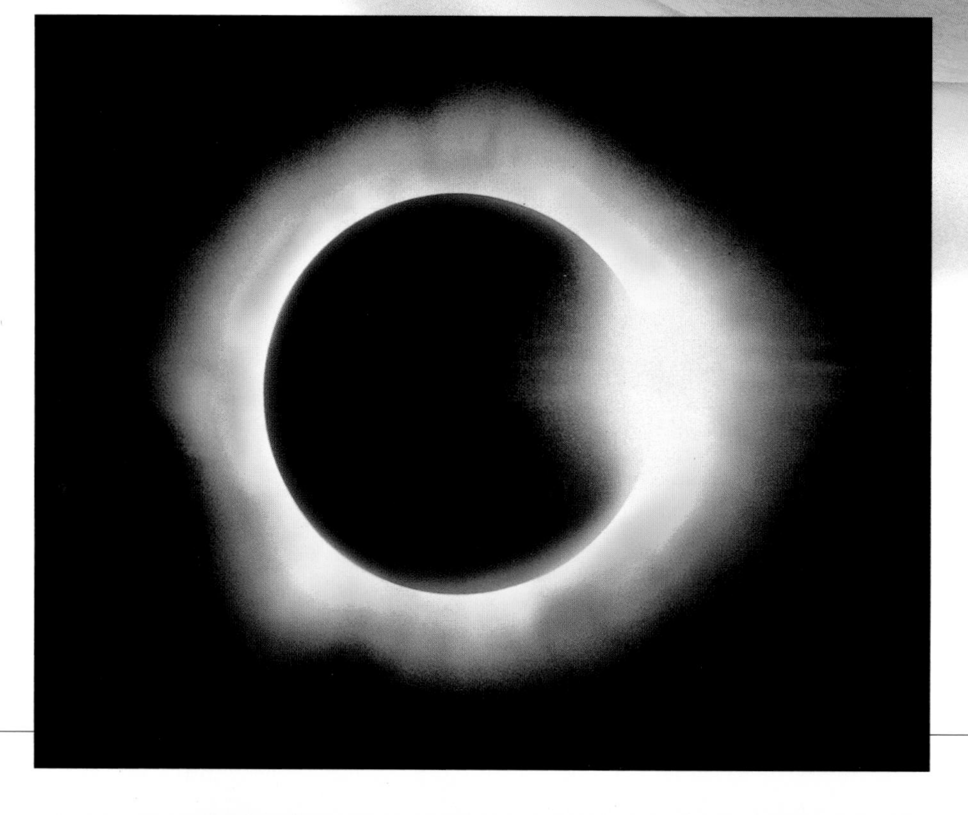

▷ In a total solar eclipse, the Moon passes directly in front of the Sun, blocking its light. It offers a rare opportunity to view the corona. In this enhanced image of the eclipse of March 7, 1970, the colors around the edge show the corona. Although very hot (2 million K) and vast (extending several solar radii into space), it is usually too faint to be seen without special instruments. The white light at the right is from the photosphere, and the eclipse will soon be over.

Prominence

Sunspots

Chromosphere

Corona

▲ **The flame-like tongue of a solar prominence extends into space from the outer layers of the Sun. Prominences are dense clouds of gas associated with the magnetic fields that link groups of sunspots. The gas is cooler but more dense than the solar material surrounding it. If the magnetic field is suddenly distorted, the gas is blown out into space. Gentle prominences may hang suspended in the corona for months or longer, while more short-lived, violent prominences may flare 100,000 km into space.**

of this has enabled astronomers to determine the abundances of the chemical elements in the Sun.

The cool zone, often called the "reversing layer", lies at the bottom of a layer known as the chromosphere, several thousand kilometers deep, which envelops the photosphere. The chromosphere passes upwards into the corona, which merges into interplanetary space and the solar wind. The chromosphere is visible optically only immediately before and after a total eclipse. It has a reddish color due to the emission of hydrogen. Studying this outer region with special instruments reveals that there are networks of jetlike spikes ascending from the chromosphere. Spectacular prominences – huge streamers of incandescent gas – rise up into the coronal regions, sometimes forming intricate arches and loops as they become involved with the Sun's magnetic lines of force.

The corona is continually in motion, activated by shock waves sent up from the photosphere into the chromosphere. The corona's expansion into space gives rise to the "solar wind" – the mixture of rapidly-moving electrons, protons, helium nuclei and other ions – which streams out through the Solar System and beyond. This may approach the Earth at velocities of around 500 kilometers per second, whereupon it interacts strongly with the magnetic field. The X-ray and ultraviolet radiation which reach the Earth ionize the upper atmosphere, producing the ionosphere.

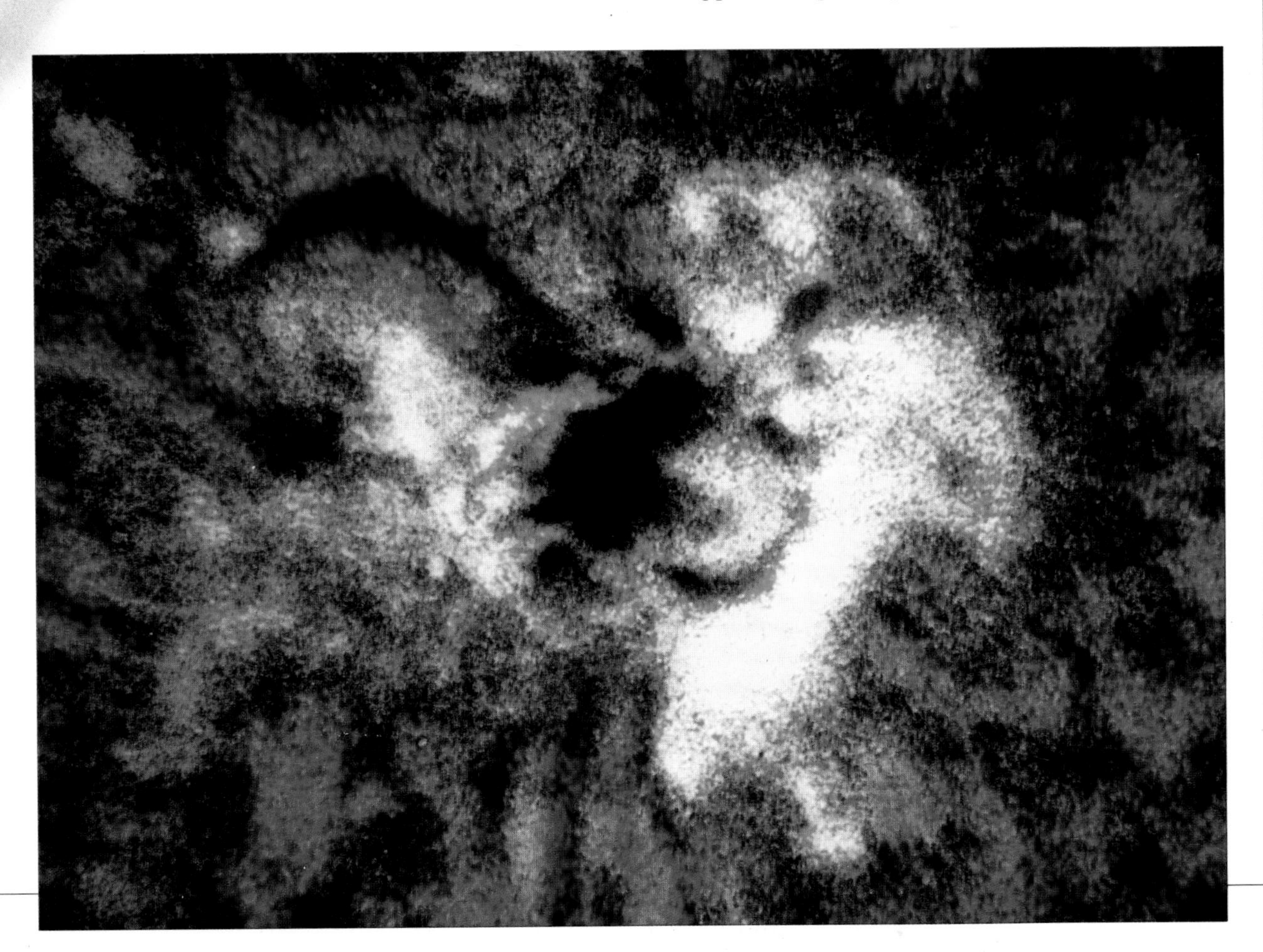

▶ **Sunspot regions on the face of the Sun appear black in this false-color optical photograph. Sunspots are up to 2000 K cooler than the surrounding photosphere. They are generally shortlived and occur in groups in a band 60° wide astride the Sun's equator. They are associated with strong magnetic fields, shown here as white areas in the chromosphere above the sunspot. Sunspots tend to appear in cycles that recur approximately every eleven years.**

Accretion of the Planets

Most accounts of the emergence of the planets from the original solar nebula describe the growth of these large bodies from much smaller ones by process of accretion, as their orbits drew them into collisions, or as the force of gravity sucked smaller bodies toward larger ones. There are two possible ways in which such growth might have begun. Some scientists have argued that dust-sized particles of hot material that had condensed out in the nebula gradually collected (accreted) into the large bodies we know as planets. Others consider that the nebula itself separated into discrete areas, which cooled to give rise to a group of relatively large objects, known as planetesimals. They believe that it was collisions between the planetesimals that caused the planets to accrete. The first theory assumes that the various materials constituting the planets condensed out of the gaseous nebula at different times and different distances from the Sun, as nebular temperatures dropped. However, if this had been so, the meteorites – small rocky bodies which formed in the early history of the Solar System, and emanate from the region now known as the asteroid belt – would have formed throughout the period of cloud cooling, and through the whole extent of the nebula. This would lead us to expect that some were made of silicates (which condense at high temperatures) and others of lower-temperature volatiles. In fact, the oldest known meteorites are mixtures of both low- and high-temperature minerals, which suggests that the process was more complicated.

KEYWORDS

ACCRETION
ASTEROID BELT
ELECTROMAGNETIC FORCE
PARTICLE-PARTICLE COLLISIONS
PLANETESIMAL
REGOLITH
SILICATE
VOLATILE ELEMENT

Most scientists now think that the planets accreted from larger original bodies, known as planetesimals or protoplanets. The building blocks of the modern planets certainly include much larger fragments than dust particles, since huge impact basins scar the oldest surfaces of most solid bodies in the Solar System. These must have been produced by the impact of bodies up to a kilometer in diameter.

Collisions between fast-moving particles, both large and small, were surely very frequent within the solar nebula. During some collisions one fragment may have been shattered completely or even vaporized; during others – particularly collisions involving one large body and another much smaller – part of the smaller body may have become embedded

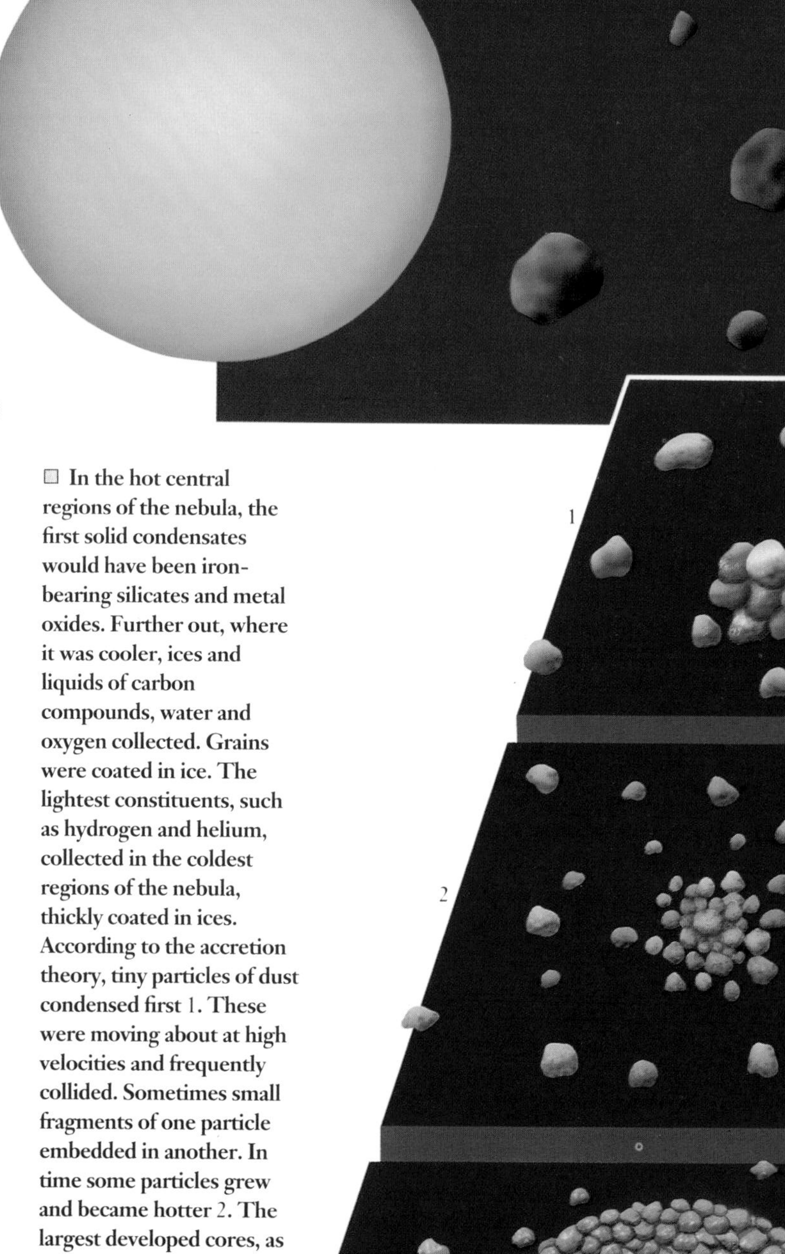

□ **In the hot central regions of the nebula, the first solid condensates would have been iron-bearing silicates and metal oxides. Further out, where it was cooler, ices and liquids of carbon compounds, water and oxygen collected. Grains were coated in ice. The lightest constituents, such as hydrogen and helium, collected in the coldest regions of the nebula, thickly coated in ices. According to the accretion theory, tiny particles of dust condensed first 1. These were moving about at high velocities and frequently collided. Sometimes small fragments of one particle embedded in another. In time some particles grew and became hotter 2. The largest developed cores, as the dense constituents such as iron sank 3.**

inside the former, increasing its mass. In this way, larger bodies grew at the expense of smaller ones. Exactly how the fragments stuck together (accreted) is not entirely understood, but the production of a dusty soil-like surface known as a regolith is thought to have assisted the process.

Each collision caused the enormous energy associated with rapid movement to be transferred instantly from one particle to the other. Some of this kinetic energy was converted into thermal energy, thus generating intense heat. When the fragments were small, this heat was soon lost to space, but for larger pieces, heat gradually accumulated deep inside. Consequently the larger planetesimals slowly heated up as they grew.

Eventually three distinct types of planet emerged, associated with the regions of the System: the small, rocky, inner planets; the gas giants; and the ice giants. All of these, together with millions of smaller bodies – asteroids, meteoroids and comets – formed within the solar nebula as it cooled and evolved into its present stable configuration. The gas giant Jupiter grew to a relatively enormous size; yet its mass remained far less than that of even the smallest star.

Each planet describes a mildly elliptical orbit around the Sun, while the planetary orbits all lie in the plane of the original solar nebula. The mutual gravitational attraction of the Sun and planets – due to their various masses – keeps the system together.

□ The core regions of the outer planets are made mainly of metallic hydrogen, which accreted very quickly. It is believed that Jupiter's core (10 to 30 times the mass of Earth's core) could have grown in only 300,000 years. The gaseous and icy components were pulled onto the cores by gravity, forming the mantle and atmosphere. The gases had been swept off the inner planets by the solar wind.

Large and Small Planets

At first the solar nebula was probably well mixed, reached a temperature approaching 2000°C, and was mainly gaseous. As it cooled, the nebula separated into materials that were physically and chemically unlike. Temperatures were highest close to the proto-Sun, and the first solid particles to form there were made from refractory elements such as tungsten, aluminum and calcium, forming oxides. As further cooling occurred, these reacted with gases in the nebula to form silicates. The inner planets, which formed close to the Sun, are largely composed of silicate minerals rich in elements such as magnesium, aluminum, calcium and iron, and are therefore relatively dense. Many meteorites and asteroids are made from these minerals.

KEYWORDS

- ACCRETION
- ASTEROID BELT
- ELECTROMAGNETIC FORCE
- PARTICLE-PARTICLE COLLISIONS
- PLANETESIMAL

Farther from the Sun, where the nebular temperatures were around 50° C, compounds rich in carbon would have crystallized out. Water ice may have existed as snowflakes and become incorporated into the solids that formed. In more distant regions, at even lower temperatures, volatile elements such as argon and compounds such as ammonia and methane would have crystallized. The lightest elements, hydrogen and helium, probably never condensed at all.

The volatile elements were far more abundant than the refractory ones, which probably accounted for a mere 0.5 percent of the total mass of the nebula. Once planets had begun to condense, growth accelerated, with very rapid accretion of carbonaceous materials,

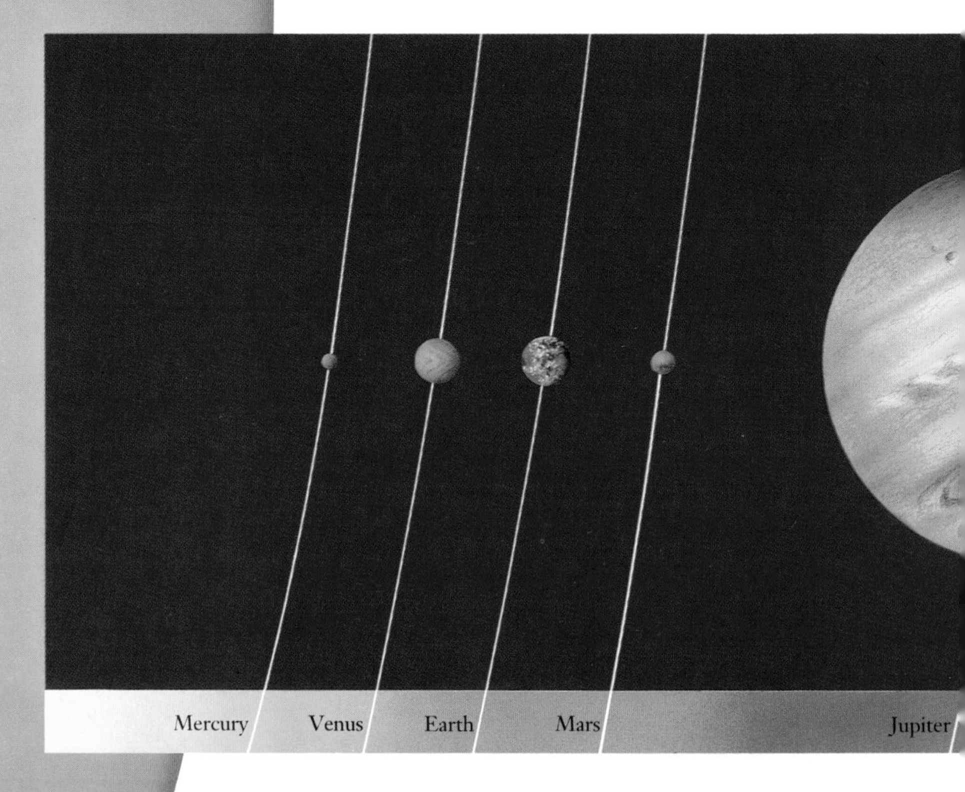

▲ **Drawn to scale, the largest of the planets, Jupiter, could swallow up over 1300 Earths. Of the outer group of large gaseous worlds, Saturn has long been known to have an attendant ring system made from small particles of ice and rock. Spacecraft have now discovered similar rings around Jupiter, Uranus and Neptune, as well as dozens of small moons. Earth's Moon is much larger – it is only slightly smaller than Mercury.**

▷ **This view of the Earth was taken by the US Apollo 11 spacecraft as it traveled to the Moon in 1969. This image, in which a part of the west coast of the United States is visible, shows the white clouds with cyclonic patterns and the bluish hue of the oceans. Land comprises about 30 percent of the planet, the oceans about 70 percent. Mountain chains and huge cracks in the surface of the planet (such as rift valley and fault lines) can also be seen from space.**

▷ **This mosaic image shows the intensely cratered surface of Mercury. This image, obtained by Mariner 10 in 1974, features the 1300-km-diameter multi-ringed Caloris Basin, surrounded by mountains rising to heights of 2 km. Mercury, closest to the Sun, is a hot, rocky planet. Other surface features may be due to wrinkling as it cooled.**

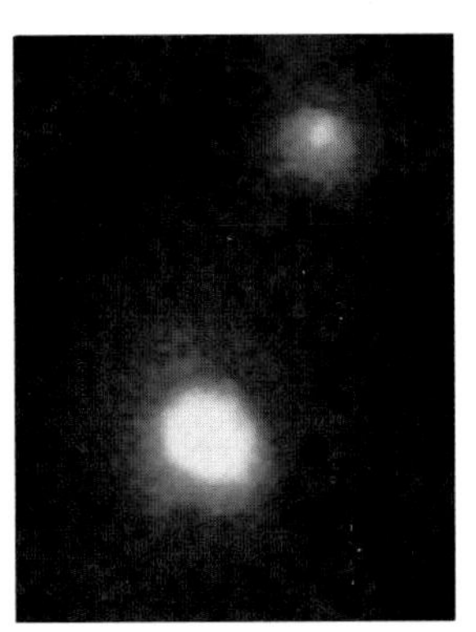

◁ **The most distant planet, Pluto, with its companion Charon (seen here in an optical telescope image), is a small, cold, rocky body. Pluto is coated in methane frost; Charon – which is about half the size of Pluto – in water ice.**

volatile-rich silicates and ices of compounds such as water, methane and ammonia. The abundance of icy material partly explains the large masses of Jupiter, Saturn, Uranus and Neptune. More of this cold volatile matter formed comets – smaller icy bodies which accumulated even farther from the Sun.

The speed of these developments can be estimated from chondritic meteorites that contain radioactive isotopes. These decay to other isotopes, giving rise to a decay sequence, at a rate referred to as the half-life of the isotope. One of the isotopes found in them is Al-26 (an isotope of aluminum of mass 26): its half-life is 720,000 years. Another is the iodine isotope I-128 (half-life 16 million years). Neither of these is native to the Solar System and must have come from distant supernova stars. They were captured by the solar nebula and locked into meteorites that formed in the early Solar System. This proves that accretion must have been quick: perhaps a few million years after the emergence of the proto-Sun. The proto-planets had formed recognizably by about 4.6 billion years ago.

2

THE *Sun's Family*

ENERGY FROM THE SUN bathes its attendant family of planets, which travel in almost circular orbits around it. The planets shine in the reflect sunlight; they do not generate light by nuclear reactions.

Earth is one of an inner group of four rocky planets, including Mercury, Venus and Mars. Farther out are four much larger bodies, two of which (Jupiter and Saturn) are composed primarily of gas, and two predominantly of ice (Uranus and Neptune). Pluto, the ninth planet, is probably made of rock and frozen methane.

Small rocky asteroids are concentrated in orbits between those of Mars and Jupiter. Many such objects crashed into the solid surfaces of the planets during the early years of the Solar System, forming impact craters and large basins. The Sun's family is completed by icy comets which originated in the far reaches of the Solar System and have mainly parabolic orbits. Many come close to the Earth and occasionally become spectacular objects in the night sky. Meteors are other small particles which often stray close to the Earth.

Many of the planets have rock and ice moons, or orbiting systems of rings. Although not the largest of the natural satellites, the Earth's Moon is a splendid object and was the first extraterrestrial world to be visited by astronauts, in 1969.

When new, the Moon is between the Earth and the Sun, and is dark on the side that faces the Earth. Usually, its orbit passes above or below the Sun, because it is slightly inclined to the ecliptic. But if the Moon is new when it approaches one of the nodes at which the orbit meets the ecliptic, it is in exact alignment with the Sun, and an eclipse of the Sun will occur. This multiple-exposure photograph shows the stages of a solar eclipse, beginning on the left and ending on the right. If the Moon is at its apogee as it crosses the Sun, the alignment may be perfect, but a sliver of the Sun will remain visible.

THE PLANETS AND THEIR ORBITS

IN 1609 the astronomer Johannes Kepler discovered that the planets circled the Sun in elliptical, rather than circular, orbits and that the Sun occupied one focus of each ellipse. This discovery was formulated into his laws of planetary motion. When closest to the Sun, a body is said to be at perihelion; at its most distant, at aphelion. The difference in the two measures is described as the eccentricity, a measure of the ellipticity of the orbit. With the exceptions of Pluto/Charon and Mercury, though, most planetary orbits are nearly circular. The minimum distance to the Sun of the innermost planet, Mercury, is 45.9 million kilometers, while the maximum distance of the outermost planet (Pluto) is 7.375 billion kilometers. The mean radius of the Earth's orbit is 149.6 million kilometers. This distance is used frequently by astronomers as a convenient unit of measurement, known as the astronomical unit (AU).

KEYWORDS

APHELION
ECCENTRICITY
ECLIPTIC
KEPLER'S LAWS OF PLANETARY MOTION
OBLIQUITY
ORBIT
PERIHELION
PRECESSION
SIDEREAL PERIOD

Except for Mercury, Pluto and Charon, the planetary orbits all lie in the same plane, due to the gravitational attraction exerted by the Sun. Because the Sun and the planets share this orbital plane, each moves against the star background along the same path, known as the ecliptic. This passes through the twelve constellations of the zodiac. Material that did not get swept up into major bodies does not always follow this pattern, and many comets have orbits highly inclined to the plane.

Individual planets rotate on axes which have varying inclinations with respect to the plane of the ecliptic – a property known as obliquity. Thus the Earth's rotational axis is inclined at 23.5°, and that of Jupiter is 3.2°. Most strange of all is Uranus, which is tilted on its side and has an axial inclination of 97.86°. The varying obliquities may be related to ancient collisions, which may also have affected the ultimate distances of the planets from the Sun. Indeed, the rotation of Venus, which is unusual in that it rotates the opposite way to the Earth, is believed to be due to such an event.

The inclination of planets' rotational axes is not fixed; long-term changes give rise to the phenomenon of precession, or wobbling of the axis. Today, for instance, the axis of Mars is inclined at 24° to the plane of its orbit. But this has not always been so. It is known that the axis "wobbles" slightly, like a child's spinning top, completing one cycle every 175,000 years. The axis of its orbit also changes slowly, precessing once every 72,000 years. Similar changes also affect the Earth. In both cases, such changes give rise to rapidly changing climatic patterns.

Seen from the Earth, the Sun moves around the sky once every year, reaching its northernmost point around 22 June (summer solstice) and its southernmost point around 22 December (winter solstice) in the northern hemisphere. During northern summer the North Pole is tilted toward the Sun, and northern latitudes enjoy summer temperatures while southern latitudes experience winter. The reverse occurs during northern winter. During the year, the Sun crosses the celestial equator twice: at the vernal equinox (around 22 March) and at the autumnal equinox (around 22 September). At these times, day and night are of equal length in both hemispheres. Because the difference between perihelion and aphelion for the Earth is only 5 million kilometers, this has only a modest influence.

■ Eccentricity is the degree to which an orbit departs from circularity. Mercury is the most eccentric (0.2056) and Venus the least (0.0067). The rotational axes may be nearly perpendicular to the orbital plane or inclined at various angles. The planets' spin gives rise to a wobble called precession. The outer planets have much longer orbital periods. Pluto's orbit is eccentric and often lies within Neptune's. Jupiter has a small axial inclination; Saturn and Neptune are slightly more inclined than the Earth; and the axis of Uranus is almost in the orbital plane. Most comets move around the Sun in highly elliptical orbits. Some, like Halley's Comet, have a period short enough to be predicted; others orbit over hundreds or thousands of years.

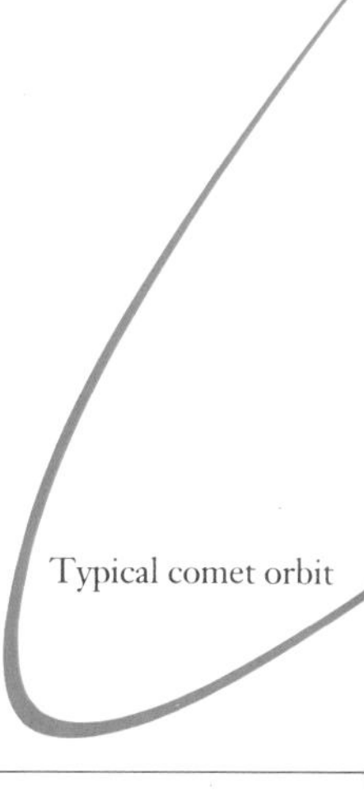

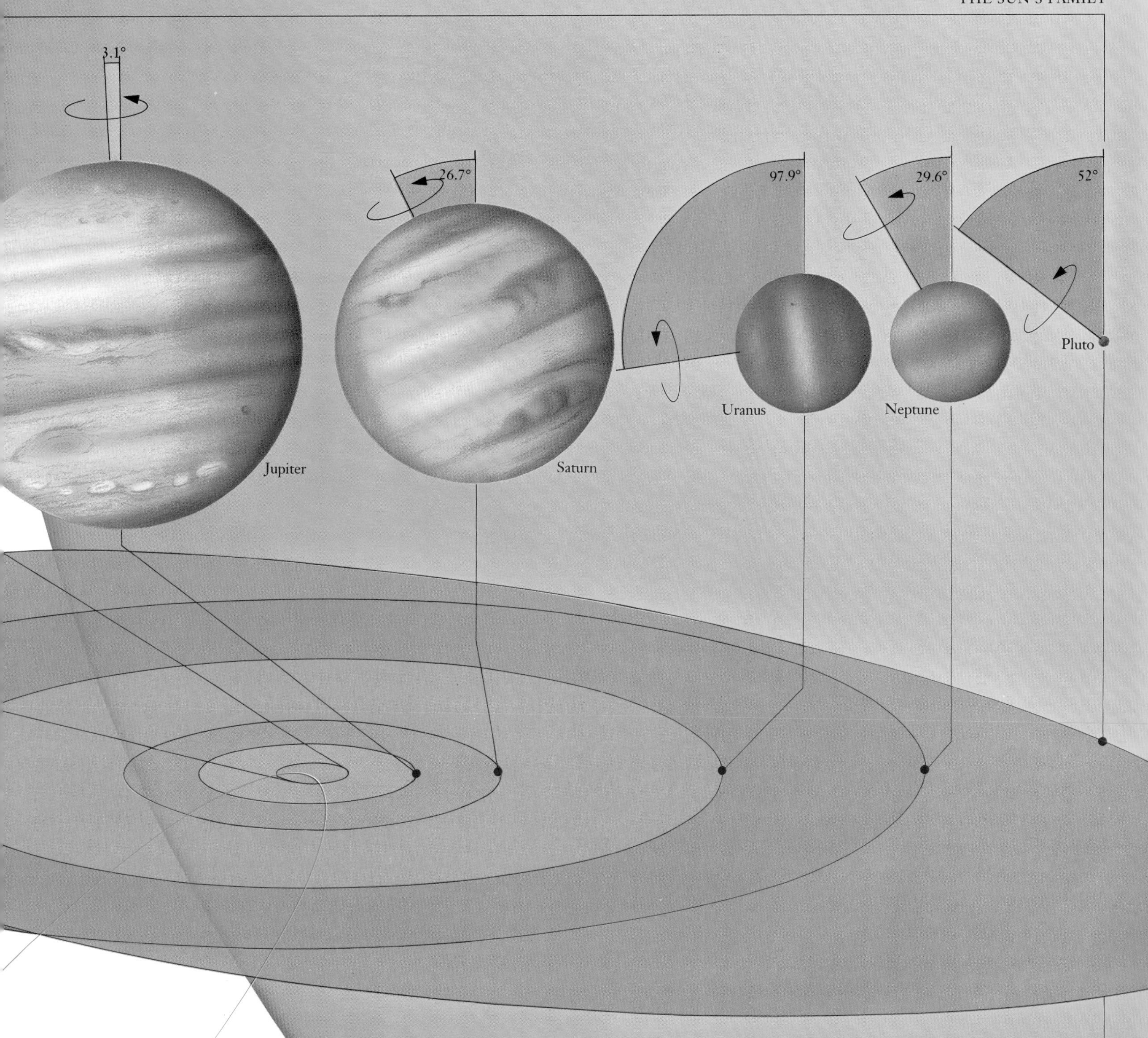

Planetary Data

Planet	Mean distance from Sun	Equatorial diameter	Sidereal period	Equatorial rotation	Mass
Mercury	57.91×10^6 km	4878 km	87.969 days	58.65 days	3.303×10^{23} kg
Venus	108.20	12,012	224.701	243 days	4.87×10^{24}
Earth	149.60	12,750	365.256	23.93 hours	5.97×10^{24}
Mars	227.94	6786	686.980	24.62 hours	6.42×10^{23}
Jupiter	778.33	142,984	4332.71	9.8 hours	1.90×10^{27}
Saturn	1426.98	120,536	10,759.50	10.6 hours	5.68×10^{26}
Uranus	2870.99	51,118	30,685	17.9 hours	8.684×10^{25}
Neptune	4497.07	49,500	60,190	19.2 hours	1.024×10^{26}
Pluto	5913.52	2300	90,800	6.4 days	1.29×10^{22}

Earth and Moon

The Moon is unusually large in relation to its planetary neighbor, with a diameter more than a quarter that of the Earth. The two are often seen as a double planet system, orbiting around a common point deep within the Earth. The strong gravitational pull of Moon upon the Earth, and its oceans in particular, gives rise to twice-daily tides.

The Moon orbits the Earth at a mean distance of 384,392 kilometers. It takes one month to complete each revolution and the same period to rotate once on its axis. As a result it directs the same face toward the Earth at all times, and the far side is never seen from Earth. The Moon's monthly phases are dependent on the angle between the Earth, Sun and Moon at different times. There are also occasionally lunar eclipses, when the Earth casts its shadow on the Moon.

KEYWORDS

ANORTHOSITE
APOLLO MISSIONS
CAPTURED ROTATION
CORE
EJECTA
IMPACT BASIN
IMPACT BRECCIA
MARIA
TIDE

The mean density of the Moon is significantly less than that of the Earth. Since much of the Earth's high average density derives from the heavy material in its core, this suggests that, unlike Earth, the Moon does not have a large dense core.

The Moon's surface shows the results of considerable cratering. Since the Moon has no atmosphere and apparently has never had one, its surface was afforded no protection from attack by planetesimals. As a result, the lunar surface is pockmarked with impact craters and basins, some parts more than others – the older a surface, the more

▷ **Geologist and astronaut Jack Schmitt inspects a huge boulder in the highland regions of the Moon in 1972. Beyond can be seen the lunar module's landing site. Samples of Moon rock brought back to Earth have revealed a great deal about the geological history of the Moon, and heatflow detectors on the spacecraft have revealed that parts of the lunar interior are hot. The energy outflow, which is about half that of Earth's, derives from the decay of radioactive isotopes deep in the Moon.**

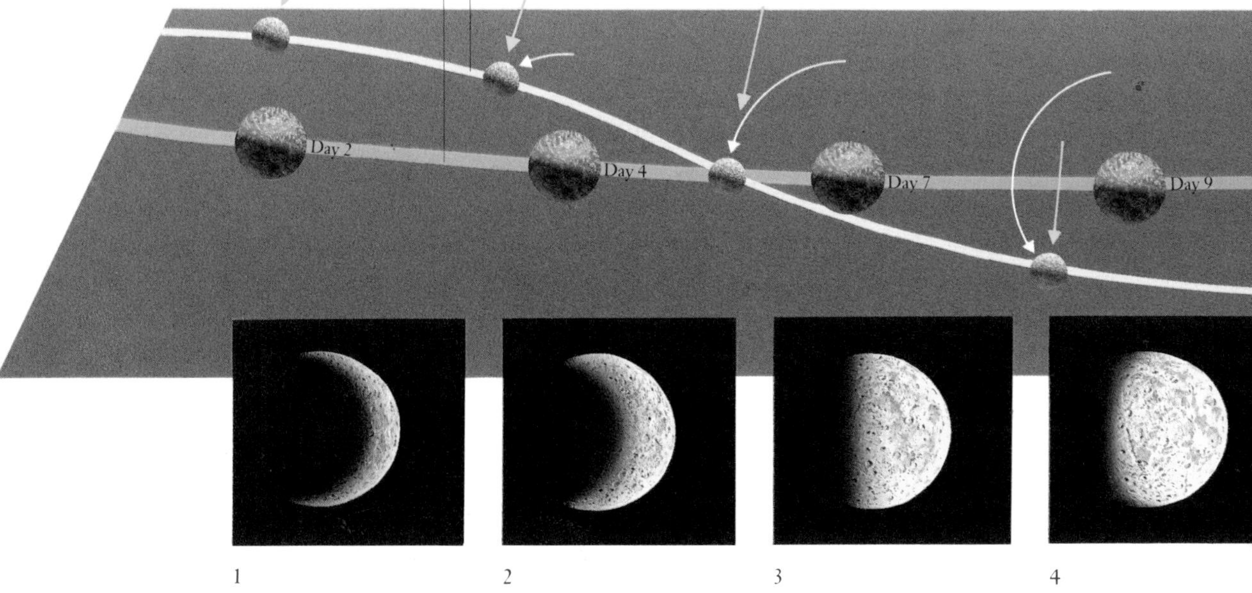

▷ **The Moon completes an orbit of the Earth once every 27.3 days although, since the Earth is itself moving around the Sun during this period, the cycle from one full moon to the next takes 29.5 days. The Moon's orbit is inclined at just over 5° to the ecliptic, which means that lunar eclipses, when the Earth's shadow passes across the face of the Moon, or solar eclipses, when the Moon obscures the Sun, are infrequent.**

craters it has. The most heavily cratered regions, the highlands, have a higher albedo (reflect light more brightly) and lie at a higher level than the darker regions known as "seas" or maria. The reflectivity of the former is due to the light-colored rock anorthosite, which is rich in the elements calcium and aluminum and forms the ancient crust. Rock samples from the highlands have been shown to be some 4.5 billion years old, older than any known rocks from the Earth's crust. Unlike terrestrial rocks of comparable type, lunar rocks are free from volatiles.

These ancient rocks suffered severe bombardment by planetesimals until 4 billion years ago. The highland rocks are impact breccias, produced as impacting bodies disrupted the crust or distributed ejecta across the lunar surface.

The maria are younger and have smoother surfaces. They are made from the volcanic rock basalt. This rock rose from inside the Moon and flowed out as lavas which filled several huge impact basins, such as Mare Imbrium. This took place in phases between 3.9 and 3 billion years ago, which implies that the interior of the Moon was hot for at least this long. Most maria are located on the near side, as the crust that lies beneath the Earth-facing hemisphere is thinner than on the far side of the Moon.

At one time it was believed that the Moon derived from the Earth, somehow ejected from the Pacific, but this idea is now discounted. Modern research suggests that soon after the Earth's core had formed, a massive celestial object gave the Earth a glancing blow. This object was vaporized, and the Moon formed largely from its mantle. This explains why the Moon and Earth are chemically dissimilar and why the Moon lost all of its volatile constituents.

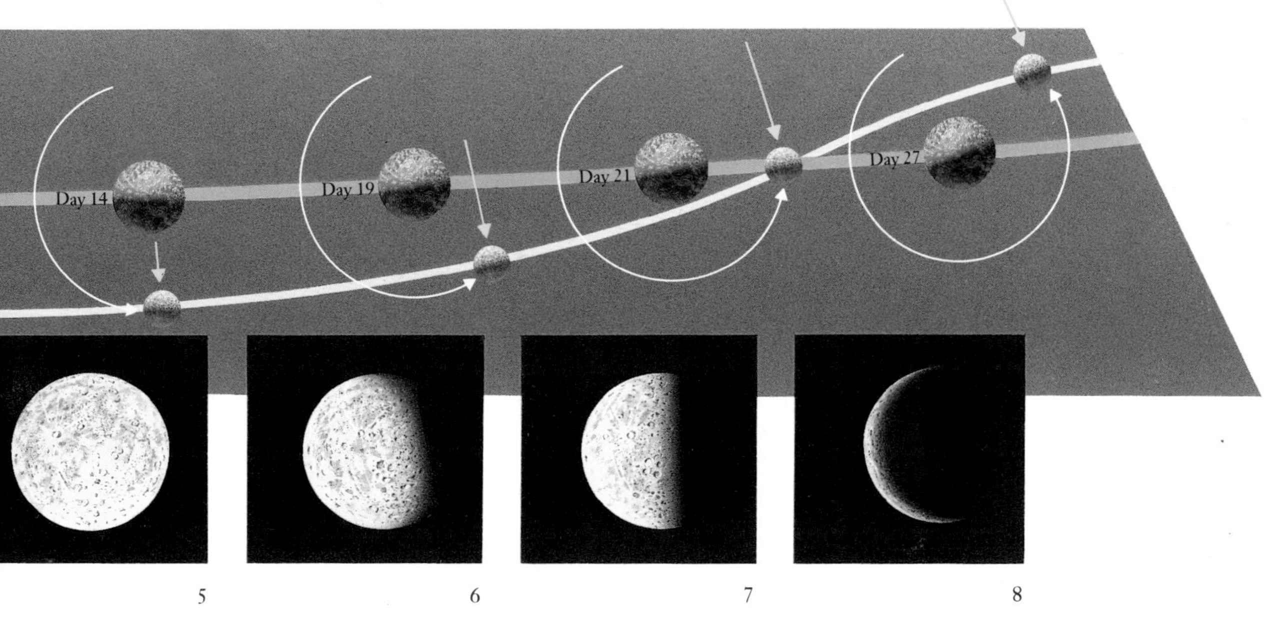

◁ The Moon is new when the illuminated face is turned away from the Earth. The crescent Moon 1 appears as the lit side turns toward the Earth. The crescent grows 2 and reaches first quarter 3 when half of the earthward face is illuminated. The Moon then becomes gibbous 4 and is full 5 when the earthward face is totally illuminated. The sequence reverses, through gibbous 6 to last quarter 7 and crescent 8 to become new.

The Inner Planets

Mercury, Venus, Earth and Mars, the four inner or "terrestrial" planets, accreted close to the proto-Sun where nebular temperatures were high and only dense silicate materials were condensing. In consequence they all have high mean densities, ranging from 3.3 grams per cubic centimeter for Mars to 5.5 grams per cubic centimeter for Earth. All have a crust (surface layers) and mantle (inner solid layers) made from silicate rocks, and a denser core which is rich in iron.

KEYWORDS

CORE
CRUST
IMPACT BASIN
LITHOSPHERE
MANTLE
REMOTE SENSING
SHIELD
TESSERA
VOLATILE ELEMENT

Mercury orbits the Sun at a mean distance of 57.91 million kilometers, but is peculiar in that it has a more elliptical orbit than most planets, while its orbit is inclined to the ecliptic plane at 7.2° – a greater inclination than any other planet except Pluto, though the significance of this is not yet understood. Mercury is particularly dense for its size, and has a large core that extends to three-quarters of the radius. It has a weak magnetic field, which suggests that at least a part of the core is still fluid.

The surface of Mercury is pockmarked with impact craters and there is one huge basin – the Caloris Basin – first photographed in detail by Mariner 10 in 1974; it is 1300 kilometers across. Between the craters are smoother plains, presumably of volcanic origin. Unique to Mercury are 1000-meter-high compressional fault scarps that cross the surface. These faults indicate that the planet underwent a contraction of about 2–4 kilometers, which probably occurred about four billion years ago.

Mercury's relatively high density and unusually large core size are generally explained by a collision that is thought to have occurred long ago with a body of about one-fifth the mass of Mercury. This would have stripped away most of the planet's mantle. Some of the lost material may eventually have accreted onto Venus, or even onto the Earth.

Venus is comparable with Earth in terms of size, mass and density, but has a dense carbon dioxide atmosphere, a slow retrograde rotation (its day is 243 Earth days), and no moon. The atmosphere prevents heat from escaping and generates a greenhouse effect, and surface temperatures rise as high as 500°C.

The Magellan space probe, which took radar images of the planet, revealed fine details of the Venusian surface. It is dominated by volcanic plains on which are found hundreds of circular structures called coronae, different kinds of volcanoes, domes and impact craters. The relatively small number of impact craters implies that the age of much of the surface is relatively young (450–500 million years).

There are also highland regions of complex structure. One of these, Ishtar Terra, has surrounding mountains apparently formed by compressional forces; at Maxwell Montes they rise more than 11 kilometers above "mean datum". East of Maxwell, and elsewhere on the planet, are intricately grooved regions called tesserae which bear witness to past movements of the crust. Similar activity gave rise to long linear ridge belts which cross the plains.

Soviet Venera space probes landed on Venus during the 1970s, photographing the blocky surface and analyzing the surface rocks. Most of these were discovered to be slabby and basaltic: volcanoes clearly played the dominant role in Venus's history.

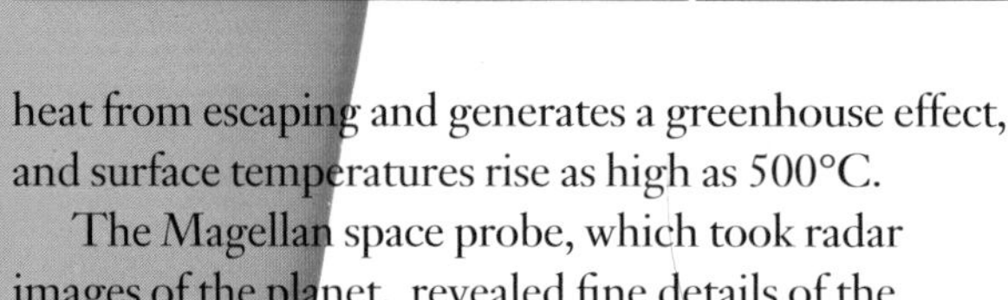

▵ **Mercury is the smallest of the inner planets, with a diameter of 4878 km. Venus is almost a twin of the Earth, having a diameter of 12,012 kilometers (Earth 12,750 kilometers). Mars is a much smaller world (diameter 6,787 km) and is the most distant of the rocky planets, orbiting at between 206.7 and 249.1 million kilometers.**

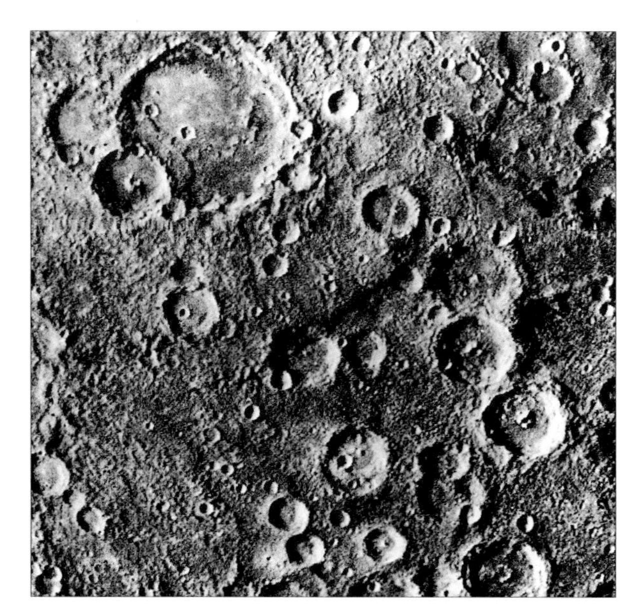

▹ **The highly-cratered surface of Mercury is visible in this mosaic compiled from images taken by Mariner 10. As on the Moon, the younger impact craters are surrounded by radiating bright rays of ejecta. The smoother, darker regions between the craters are believed to have a volcanic origin.**

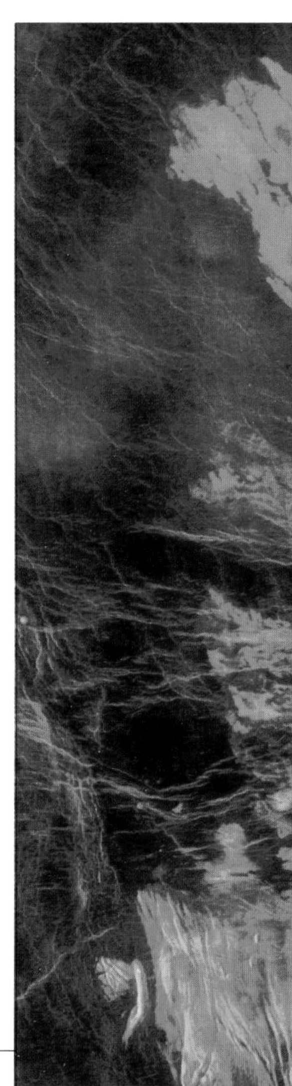

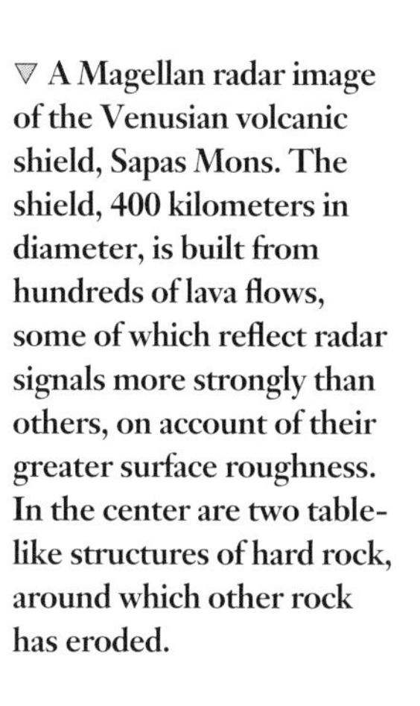

Mars

▽ A Magellan radar image of the Venusian volcanic shield, Sapas Mons. The shield, 400 kilometers in diameter, is built from hundreds of lava flows, some of which reflect radar signals more strongly than others, on account of their greater surface roughness. In the center are two table-like structures of hard rock, around which other rock has eroded.

▷ The central part of Candor Chasma, a part of the Valles Marineris rift valley on Mars. The layering visible in the walls to the left of the image, may have formed in a vast glacial lake. Although the canyon owes its origin to subsidence and faulting, the scalloped walls to the bottom of the image, are the result of the collapse of the steep scarp slopes.

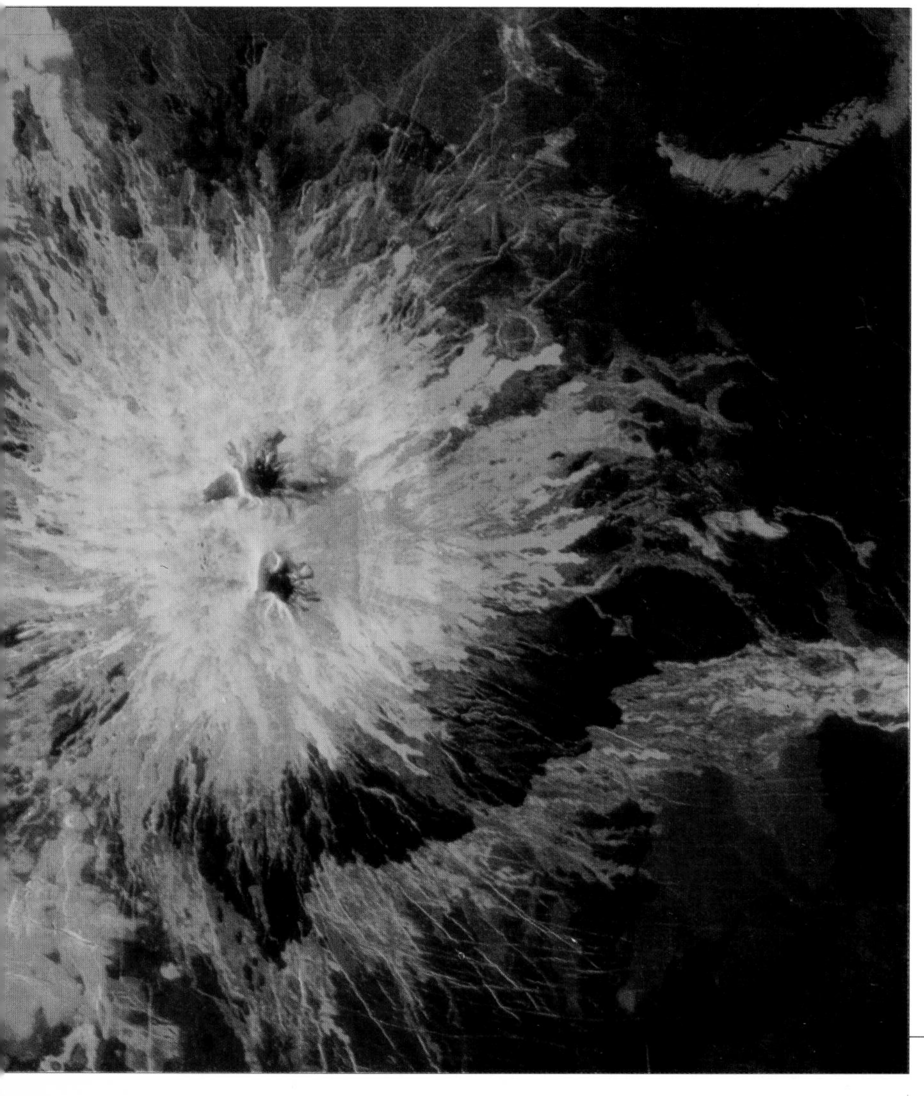

Mars has an orbit beyond the Earth's (its mean distance from the Sun is 227.9 million kilometers). It is only half as big as Earth (diameter 6787 kilometers) and considerably less dense (3.93 grams per cubic centimeter). The surface temperature is low, ranging between –140°C and +20°C; liquid water cannot exist on its surface, owing to the low temperatures and an atmospheric pressure only a hundredth that of Earth. The thin atmosphere is largely carbon dioxide. This, together with some water vapor, freezes at the poles, and gives rise to polar "ice" caps, which expand and contract with the seasons.

North of a boundary inclined at about 28° to the equator, the surface of Mars consists of volcanic plains and huge shield volcanoes, many concentrated on a massive swelling in the lithosphere (the planet's hard outermost layer) known as the Tharsis Bulge. The spectacular equatorial canyon system, Valles Marineris, splays eastward from this bulge.

The southern latitudes are heavily cratered by impacts, and are older than the northern plains. Extensive valley networks and massive outflow channels have developed there, bearing witness to the past activity of running water. Most of the original volatiles (elements and molecules that normally exist in gaseous form) probably remain, but they are locked in the porous rock of the surface layer. This implies that the planet's atmosphere has probably changed in the course of its history.

Distant Companions

Jupiter, Saturn, Uranus and Neptune are collectively called the Jovian planets because the other planets are similar in many respects to Jupiter, which dominates the group. There are, however, significant differences between the Jupiter–Saturn pair and the Uranus–Neptune pair. Pluto and its satellite, Charon, are somewhat separate.

Jupiter, the largest planet, has a mass 318 times that of Earth, yet its mean density (1.33 grams per cubic centimeter) is only one-quarter of the Earth's – similar to the Sun. This giant world is composed mainly of hydrogen and helium which, in the outer cloud layers and down to a depth of around 1000 kilometers, occur as gas. Below this, gas gives way to liquid hydrogen for a further 20,000 kilometers where the pressure is so great that the hydrogen behaves like a metal. There is believed to be a dense rocky core of about 10 to 30 Earth masses at the center.

KEYWORDS

CLOUD BELT
CRYOVOLCANISM
GREAT RED SPOT
REFRACTORY ELEMENT
RING SYSTEM
SHEPHERD MOON

Despite its huge size, Jupiter has the shortest rotation period of all the planets, spinning once every 9 hours 50 minutes, a phenomenon which causes the equatorial regions to bulge significantly. Parallel bands of light and dark clouds can be seen, as well as one semipermanent atmospheric feature, the Great Red Spot, a vast revolving weather system. The light zones comprise cold, high-altitude ammonia-ice clouds, while the darker belts represent lower-level clouds made from various hydrogen compounds.

Saturn has a diameter nine times that of Earth and a composition and structure similar to that of Jupiter. Most distinctive is its magnificent ring system, which has a diameter of 273,000 kilometers. This was explored by the Voyager spacecraft in 1979 and shown to consist of billions of small rock and ice particles, ranging in size from dust to the dimensions of a house. Small moons (called shepherd moons) orbit within the system, maintaining gaps between the components.

Uranus, with a diameter four times that of Earth, was the first planet to be discovered by telescope. Like Saturn and Jupiter, it has a hydrogen-rich atmosphere. However, it has a far greater proportion of ices (in particular, water ice and ammonia). Neptune, though somewhat larger and denser, is similar. Prominent cloud bands and a Great Dark Spot characterize its bluish disk. It is likely that beneath the outer gaseous envelopes of both worlds – which contain much methane – there is a deep layer of slushy ice surrounding a denser rocky-metallic core.

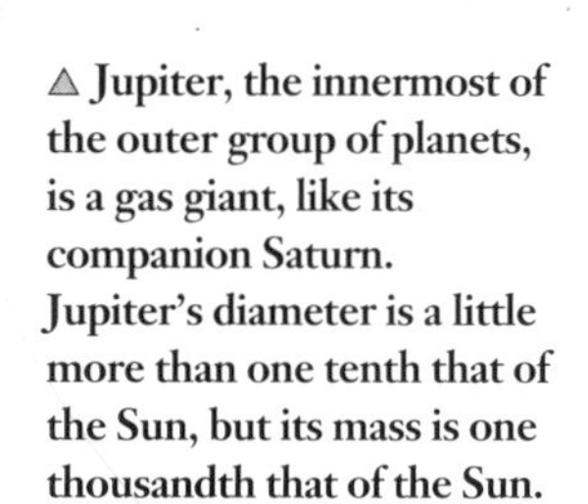

△ Jupiter, the innermost of the outer group of planets, is a gas giant, like its companion Saturn. Jupiter's diameter is a little more than one tenth that of the Sun, but its mass is one thousandth that of the Sun. The next pair of planets, Uranus and Neptune, are gas-and-ice giants, and the tiny pair, Pluto and Charon, complete the outer reaches of the Solar System. At times Neptune is the most distant from the Sun.

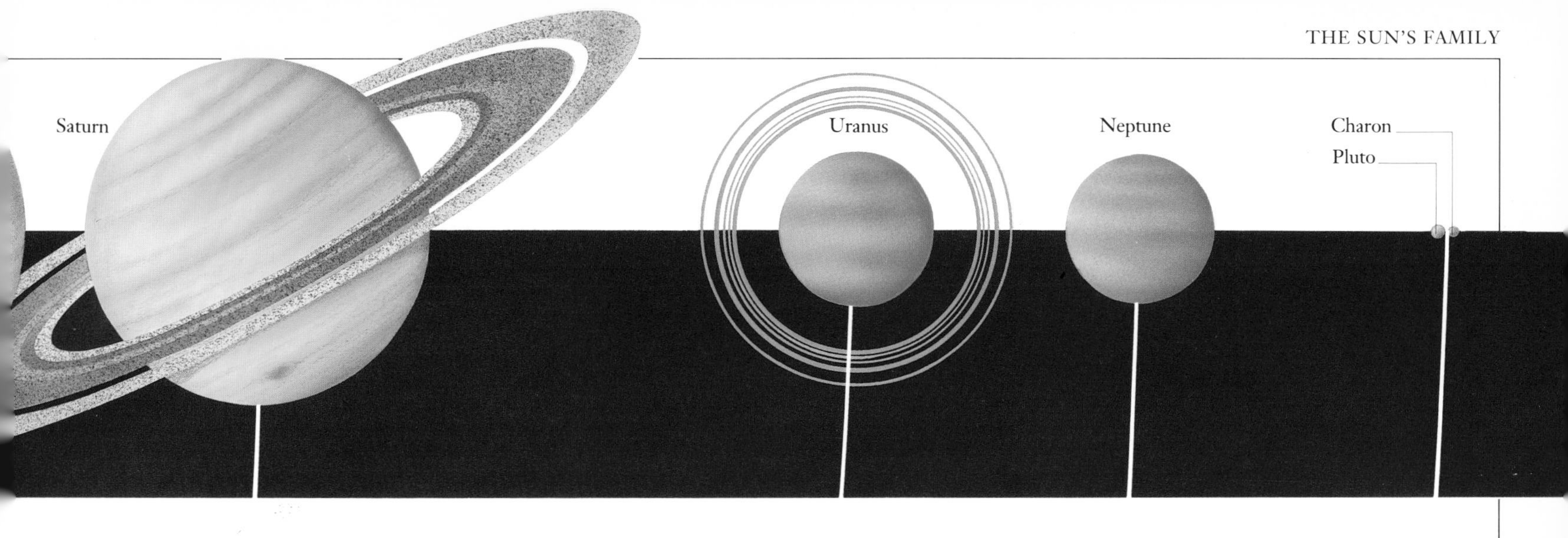

▽ **Saturn's rings are composed of a mixture of particles, including water-ice. The gaps between the rings result from gravitational resonances between them and the shepherd moons.**

△ **Uranus (shown here as a crescent) is greenish-blue and the only discernable features are faint cloud belts. A system of rings was discovered in 1979, much thinner and darker than those of of Saturn.**

▷ **Voyager 2 passed within 5000 kilometers of Neptune in August 1989. It revealed a blue disk with a number of dark spots and streams of white cloud made of methane ice, drawn out by violent winds.**

The giant planets probably formed cores of the more refractory elements very quickly and then collected their gaseous envelopes by accretion. The large mass of Jupiter may be due to its position near the "snow line", the point where water and other volatiles collected and cooled as intense solar activity cleared regions closer to the Sun of lighter elements.

Each of the giant planets has many moons: Jupiter has at least 16; Saturn 18; Uranus 15; and Neptune 8. The largest of Neptune's moons, Triton, is the only large moon to have a retrograde orbit, while Saturn's giant moon, Titan, has a deep atmosphere. The Voyager spacecraft passed close by many of these worlds, revealing active sulfur-silicate volcanism on Jupiter's moon, Io; impact cratering and tectonic deformations on many other moons; huge fault scarps, and peculiar icy volcanism known as cryovolcanism on Triton. Voyager also revealed a ring system around Jupiter, and also Uranus and Neptune.

Pluto and Charon are unlike the giant planets. They may be the brightest of a group of asteroids in the far reaches of the system. Alternatively, Pluto may be an escaped moon of Neptune.

BITS AND PIECES

MILLIONS of small bodies, as well as the nine planets of the Solar System, are in orbit around the Sun. Among these bodies, the asteroids, or "minor planets", form a prominent group. Most orbit the Sun between Jupiter and Mars, but some have paths that cross the orbit of the Earth. The largest asteroid, called Ceres, was discovered by the Italian astronomer Giuseppe Piazzi (1746–1826) in 1801; it measures 1003 kilometers across. The smallest yet found (6344P-L) is a mere 200 meters in diameter. It is believed that there are at least one million asteroids larger than one kilometer across. Some have very irregular shapes, shown by the images of Gaspra from the Galileo mission.

KEYWORDS

ACCRETION
ASTEROID
CARBONACEOUS CHONDRITE
CHONDRITE
METEOR
METEORITE
METEOROID
NEBULA

Asteroids originally resembled the other nebular objects that accreted to form planets. But before some of the asteroids could stick together to form large bodies, they were affected by the gravitational pull of the Sun and the planets, and were sent into tilted, elongated orbits. They collided into one another so that such fragmentation continued. This destruction continues to occur, though less strongly, to the present.

The powerful gravitational influence of Jupiter as it grew must have inhibited the growth of a major planet in the position where the main asteroid belt now exists. Its gravitational effects would have sent some particles flying toward the planets (colliding with them and forming craters), and others out of the Solar System completely. Those smaller particles with Earth-crossing trajectories are known as meteoroids.

When viewed through a telescope, many asteroids show changes in brightness. In large part this is due to their irregular shape, but some have different reflectivity from one side to the other, which suggests they have a variable composition. The most abundant types of asteroid are type C or carbonaceous class. These bodies are darker than coal and tend to be found in the outer regions of the belt. S-type bodies are silica-rich, of intermediate albedo (reflectivity), and dominate the mid-regions of the belt, while the M class have moderate albedos and are metallic. M-class asteroids probably represent the metal-rich cores of larger differentiated parent asteroids that broke up.

Meteorites are even more abundant than asteroids and show a similar range in chemistry. When they come under the influence of terrestrial gravity and enter the Earth's atmosphere, the resulting friction causes them to heat up, and a fireball or shooting star may be seen. Most such bodies fragment in the atmosphere, but pieces of larger ones may reach the Earth's surface, providing planetary scientists with invaluable geochemical data about the early Solar System.

Meteorites are traditionally categorized as stones, irons, or stony-irons – distinctions comparable with those of the asteroid groups. A more meaningful distinction can be made between "differentiated" and "undifferentiated" types. In the undifferentiated group are the chondrites, which contain chemical elements in similar proportions to the solar atmosphere. When these are dated, they reveal their ages to be about 4.5 billion years, representing some of the most primitive Solar System material known to modern science. The differentiated types have undergone chemical changes and are considered the products of melting and separation of more primitive planetary matter. A few younger objects, known as SNC meteorites, resemble the surface of Mars and may have come from an impact with that planet.

Within chondrites are high-temperature aluminum-rich inclusions, volatile materials and peculiar spherical particles called chondrules which are the products of primordial melting. From the presence of these ingredients, it is evident that the material that comprised the solar nebula was well mixed at the time of planet accretion.

Precise photography of some meteorites allows their orbits to be calculated and indicates that these orbits closely resemble those of Earth-crossing asteroids (such as Apollo and Icarus2). Such bodies once belonged to the main belt of minor planets but were knocked out of this belt by the strong and perturbing influence of Jupiter's gravity.

▷ **A time-exposure photograph of the Leonid meteor shower that occurs every November when the Earth's orbit crosses a lane of meteoroids. At its peak, 60,000 meteors per hour enter the atmosphere. Meteor trails are due to small rocky particles burning up in the atmosphere. A meteorite is a meteoroid that reaches theEarth's surface; the pre-impact orbits of three meteorites are shown FAR RIGHT.**

▷ **Most asteroids orbit the Sun in a belt between Mars and Jupiter. The Trojan groups move along the same orbit as Jupiter, one group 60° ahead of the planet and one group in a similar position behind. Some asteroids, such as Hidalgo, have eccentric orbits highly inclined to the plane of the Solar System.**

1 Saturn
2 Jupiter
3 Earth
4 Mars
5 Asteroid belt
6 Trojans

Asteroid orbits

7 Hidalgo
8 1983 TB
9 Apollo
10 Icarus
11 Eros

Meteorite orbits

12 Pribram
13 Lost City
14 Innisfree

▷ **The asteroid Gaspra was photographed by the Galileo spacecraft in 1991. It measures 19 kilometers long by 11 kilometers wide, and orbits the Sun at the inner edge of the main asteroid belt. The rocky surface is heavily cratered.**

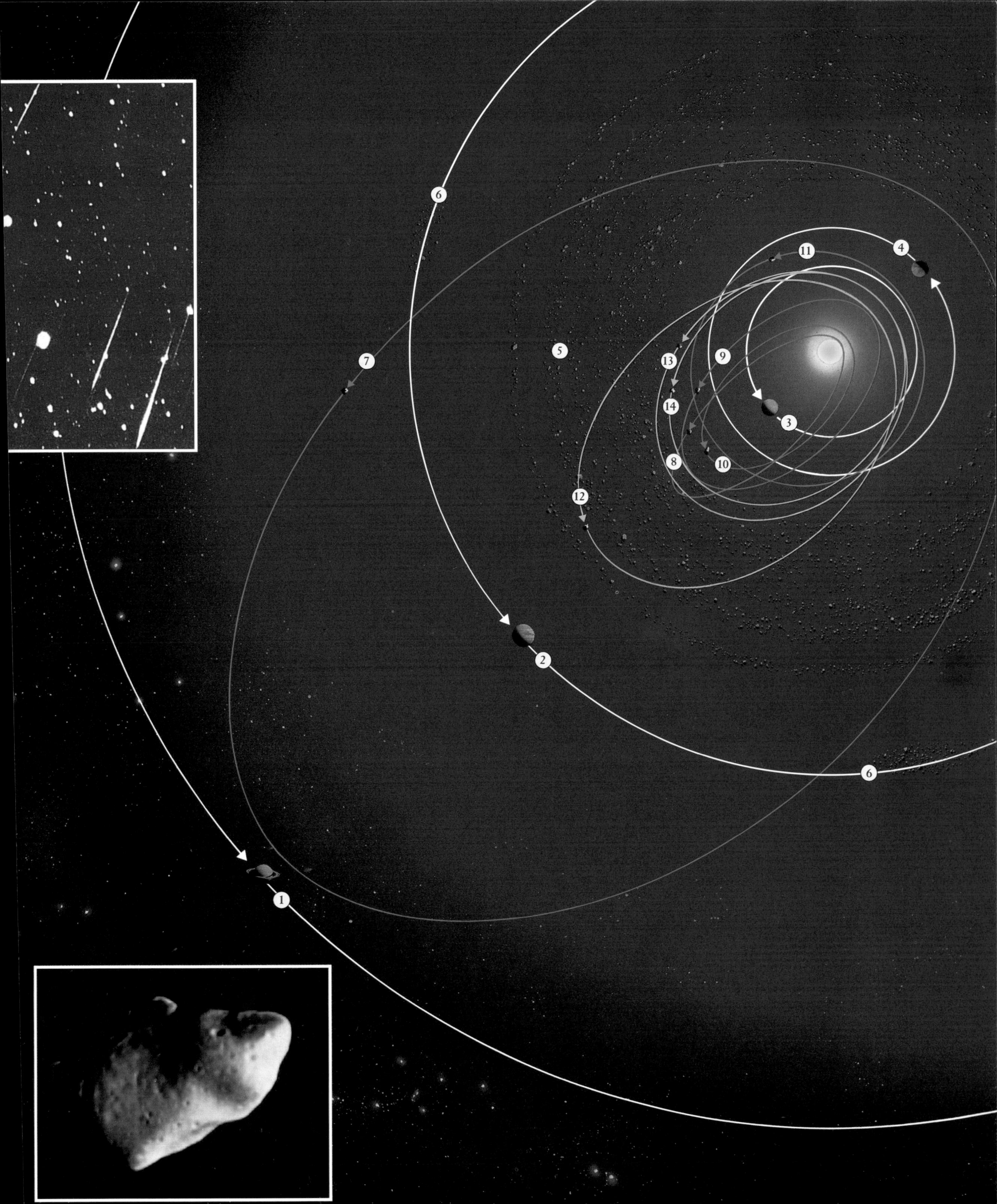

6
11
4
5
7
13
9
14
3
8
10
12
2
6
1

BEYOND THE FRINGE

ALTHOUGH they are among the smallest bodies in the Solar System, comets are also among the oldest. Their origin is intimately linked to that of the system itself because they appear to have condensed directly from the primordial solar nebula material. Although their appearance has traditionally been seen as an omen of doom, the prediction of the regular reappearance of comets was one of the first achievements of early astronomers. Today their return is celebrated by scientists as an exceptional opportunity to glean information about the early history of the Solar System .

KEYWORDS

CARBONACEOUS CHONDRITE
COMET
GIOTTO MISSION
HALLEY'S COMET
IMPACT BASIN
KEPLER'S LAWS OF PLANETARY MOTION
OORT CLOUD
PLASMA

Comets have only a very small mass, which means that they have undergone little chemical differentiation since they formed. They are therefore thought to be relics of the primitive solar nebula material that accreted to form the outer planets.

Comets are composed of ice and dust and have been graphically described as "dirty snowballs". They may have originated in the outer reaches of the Solar System, near the present positions of Uranus and Neptune, but were gravitationally perturbed at an early stage, and probably thrown out into a vast cloud, of hundreds of billions of comets, that encircles the Solar System. This cloud, known as the Oort cloud, is found about one third of the distance to the nearest star. Some of these objects may become disturbed into shorter-period, but highly elliptical, orbits which bring them into the inner Solar System, though not necessarily in the same plane as the planets.

As a comet approaches the Sun, its icy nucleus partly vaporizes, generating a diffuse bright coma, or cloud of dust and gases, and a tail of gaseous particles, ionized by the solar wind. The tail always points away from the Sun, and may be hundreds of millions of kilometers in length; it may appear very brilliant from Earth. A shorter second tail of dust particles "left behind" by the nucleus is sometimes also found.

The passage of Halley's Comet through the Solar System in 1986 was particularly informative because it allowed scientists to send five space probes (notably Giotto) directly into the heart of the comet. In doing so they learned that the nucleus was a tumbling, irregular-shaped mass 16 kilometers long and 8 kilometers wide, with a cratered surface. The surface was very dark indeed. It seems that this darkening is induced when the comet passes close to the Sun, and the interior ices melt, causing a thick residue of carbonaceous matter to build up on the surface of the nucleus. Jets of gas were observed streaming out from the nucleus, sometimes erupting material at the rate of 10 tonnes per second.

Spectroscopic studies revealed that the nucleus is composed of various volatile molecules involving hydrogen, nitrogen, carbon and sodium. Carbon monoxide is also present. When such a comet approaches the Sun, magnesium, iron, nickel and silicon have also been detected, presumably in dust particles released as the comet is bathed in the warmth of the Sun.

This research confirmed that a cometary nucleus consists of carbonaceous materials and hydrated silicates, mixed in a slushy matrix of water and other ices, such as methane, ammonia and carbon dioxide.

▽ The jagged, irregular nucleus of Halley's Comet as shown by the Giotto probe in March 1986. Two bright dust jets can be seen pouring out from the dark nucleus, on the side facing the Sun.

▷ Comet West became a bright object during 1976. The bright coma, or halo, is made from dust and gas. The tail streams out behind, forced away from the nucleus by the pressure of the solar wind.

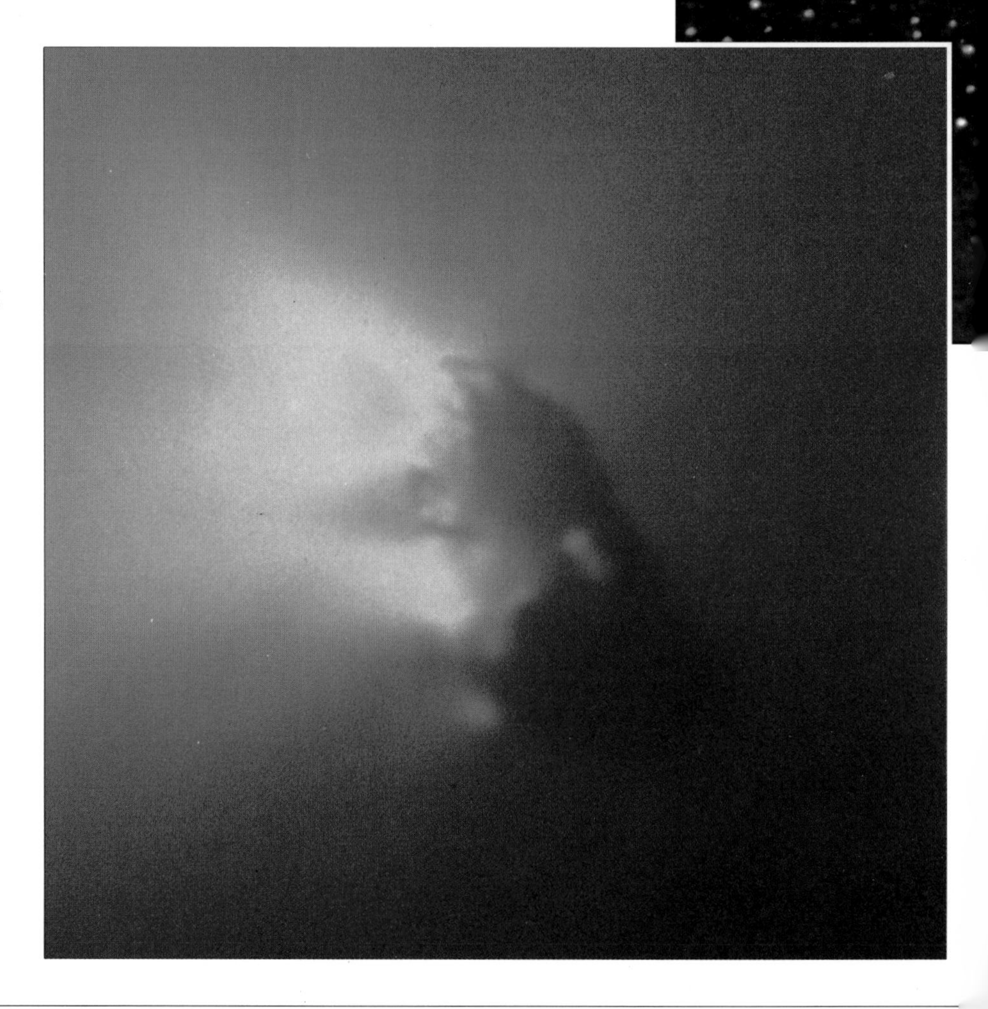

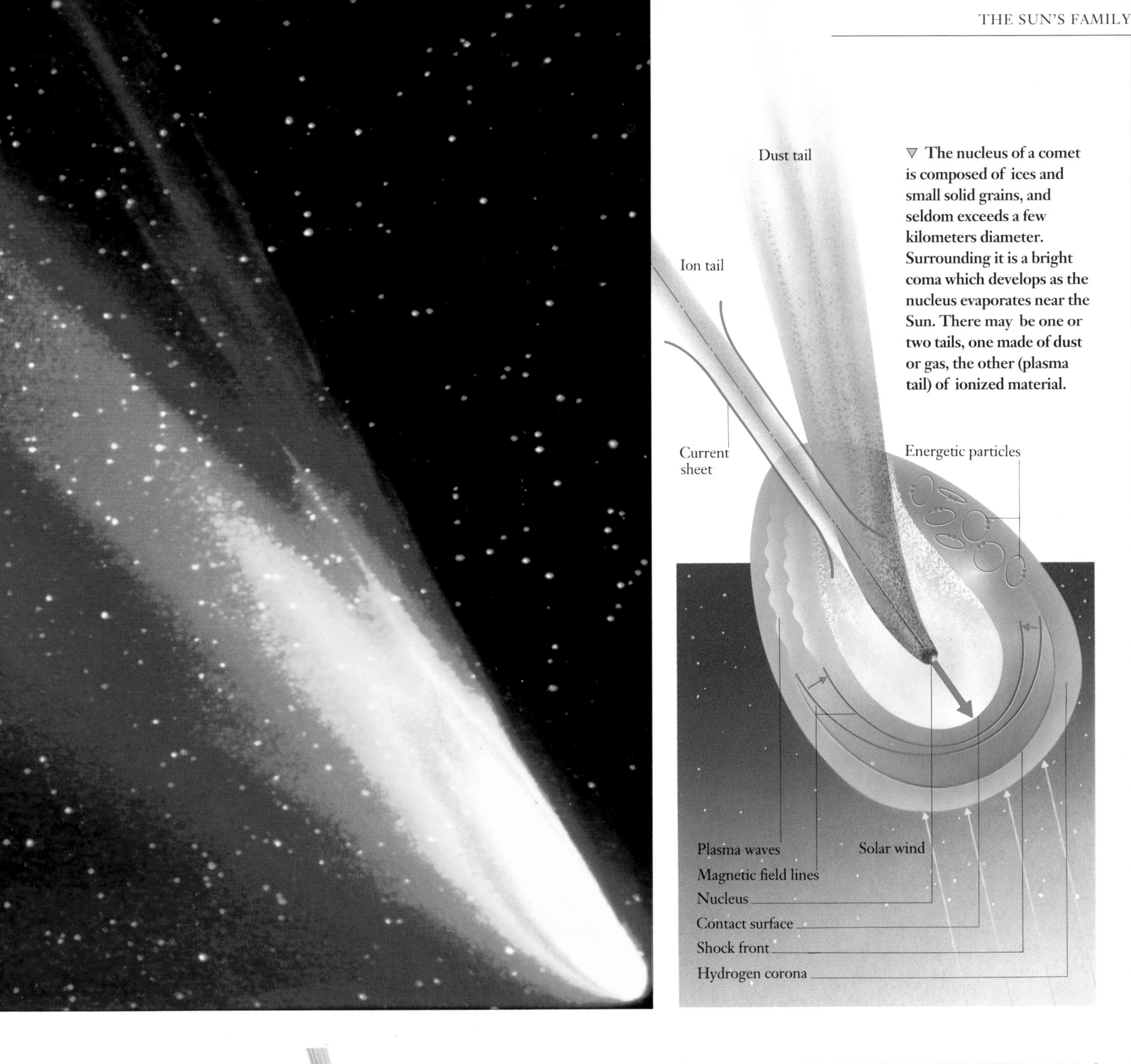

▽ The nucleus of a comet is composed of ices and small solid grains, and seldom exceeds a few kilometers diameter. Surrounding it is a bright coma which develops as the nucleus evaporates near the Sun. There may be one or two tails, one made of dust or gas, the other (plasma tail) of ionized material.

▷ When a body the size of a comet encounters a rocky planet (such as the Earth), the impact of their collision forms a crater on the planet, vaporizing the surface rocks. A shock wave in the crust forces rock to uplift in the center of the crater. This effect can be seen in the 20-kilometer-wide crater of Gosse Bluff, in the Australian desert, which is the result of a cometary impact that occurred 130 million years ago.

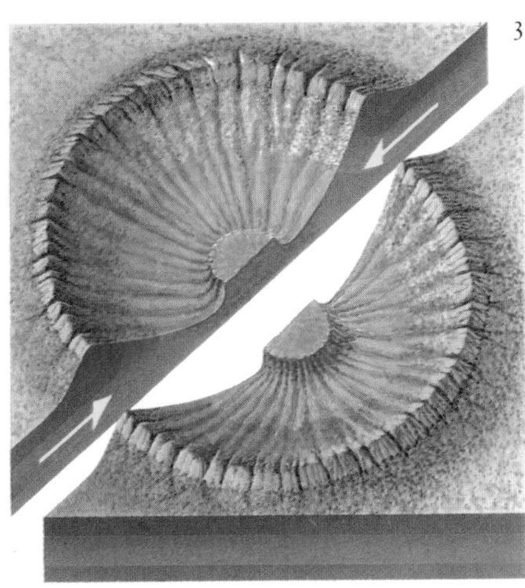

3

HEAT *Engines*

ALTHOUGH NO ONE IS SURE exactly how the planets formed, eventually they developed into dynamic, energetic worlds. The greater the original mass of the planet, the greater the amount of energy stored from the process of accretion, and the longer it has been retained. From the myriad of individual impacts which went into building a planet the size of the Earth, huge quantities of energy become locked inside. This follows the realization that vast stores of energy can be locked up within bodies of small mass. For example, the impact of a meteorite only 10 meters across onto the Earth releases energy equivalent to that of a moderately strong earthquake.

The pressure exerted by overlying rocks on those below a planet's surface is also capable of raising the temperature of the center by around several hundred degrees. The infall of dense material toward the center of a planet, forming a core, also produces heat – up to 2000°C, in the case of the Earth.

Energy released from the slow decay of long-lived radioactive isotopes contributes the greatest proportion of heat to the Earth's energy budget at the present time. Such radioactive heat has undoubtedly been escaping from all the planets since their formation and continues to be a major heat source for Venus too.

There has been a slow escape of heat from inside the Earth from very early times, as demonstrated by the extensive outcrops of very ancient volcanic rocks within the continental regions. Volcanism played an important role in the early story of all the inner planets. The outflow of volatiles from the planetary interiors also shows that heat is being generated within. This process was and still is responsible for the evolution of atmospheres and, in the Earth's case, oceans. Finally, the deformation of the rocks of some planets' crusts is caused by internal motions driven by thermal energy within.

The Planets Heat Up

During the first stage of Solar System evolution, all of the planets were bombarded by cosmic particles. This process would have contributed hugely to their "energy budgets", as the impacts of the particles alone had the potential to raise the surface temperature of Earth to 10,000° C. As each impact added more mass to the growing planets, the kinetic energy of every impacting grain was transferred and almost entirely converted to heat. As a consequence, the proto-planets became hotter as they grew.

KEYWORDS

ACCRETIONAL ENERGY
ADIABATIC HEATING
CONDUCTION
CORE
LITHOSPHERE
POTENTIAL ENERGY

If all of the infalling particles were the size of grains of dust, the accretion of the Earth would have to have been completed within 10,000 years for the temperature to become sufficiently high for a molten metallic planetary core to form. If, as seems likely, there was a proportion of larger particles, kinetic energy would have penetrated the growing planet more deeply; it would have been slower to escape, and accretion consequently could have taken longer. It is now believed that core generation took a few hundred million years.

There is strong geochemical evidence to suggest that the Earth's core grew both during and after accretion, so scientists have concluded that the energy required for core formation must have come primarily from impacts. And although much of this heat was lost through conduction, there were other sources of energy that made up for this loss. Another effect was adiabatic heating, produced by the pressure exerted on materials by an overlying mass. In other words, the deeper the matter inside a planet, the higher is its temperature (and pressure). In the Earth's case, immediately after accretion had been completed, adiabatic effects had the potential to raise the core temperature by around 900°C.

The third major source of thermal energy inside the Earth is that due to certain radioactive nuclides. Thus various isotopes of uranium, thorium and potassium – all elements present on Earth – decay slowly to daughter products, generating thermal energy in the process. Such elements would have become trapped within the atomic lattices of minerals during crystallization and immediately begun their transformations, generating heat in the process. Once each planet had cooled sufficiently to form a rigid outermost layer (lithosphere), this would act as an insulator, and cause the radioactive heat to be trapped inside. Together, these processes established conditions necessary for core formation, while the sinking of dense iron-rich droplets had the potential to raise the Earth's temperature by a further 2000°C.

If all the bodies in the Solar System were bombarded for a similar length of time by planetesimals and smaller grains, the temperatures at which the proto-planets formed must have been proportional to their present masses. Venus, which has a mass very similar to that of Earth, is thought to have a comparable thermal structure, but with the melted region closer to the planet's center. Mercury is believed to have its hottest region closer to its center than the Earth, but the temperature itself ought to be less. Mars, on the other hand, has a heating capacity only one-tenth that of Earth; therefore, if a melted region ever developed, it probably was located at about one-ninth of the planet's radius.

△ Two large impact craters are visible in this color-enhanced composite image of Mars. The Schiaparelli crater, near the center, has a diameter of 470 km; at the bottom right is the vast Hellas basin, 2000 km across, which is filled with white carbon dioxide frost. Mars is not as pockmarked as the Moon or Mercury, but it is more scarred than the Earth, on which traces of the impact of bombardments have been eroded by wind, water, ice and recycling of the crust.

The early bombardment of the planets by particles from space produced cratered surfaces, and transferred enormous quantities of kinetic energy to them in the form of heat. Although much of this escaped into space, enough was stored to provide internal heat. Over time, the different chemical elements were separated. Among these were both short- and long-lived radioactive nuclides, which decayed to form other "daughter" isotopes, releasing heat as they did so. Adiabatic heating – the increase in temperature brought about naturally by increasing pressure at depth – was another source of internal heating. When the internal temperature of the new planets had reached a critical level, the formation of molten iron droplets occurred. These sank towards the core regions under the influence of gravity.

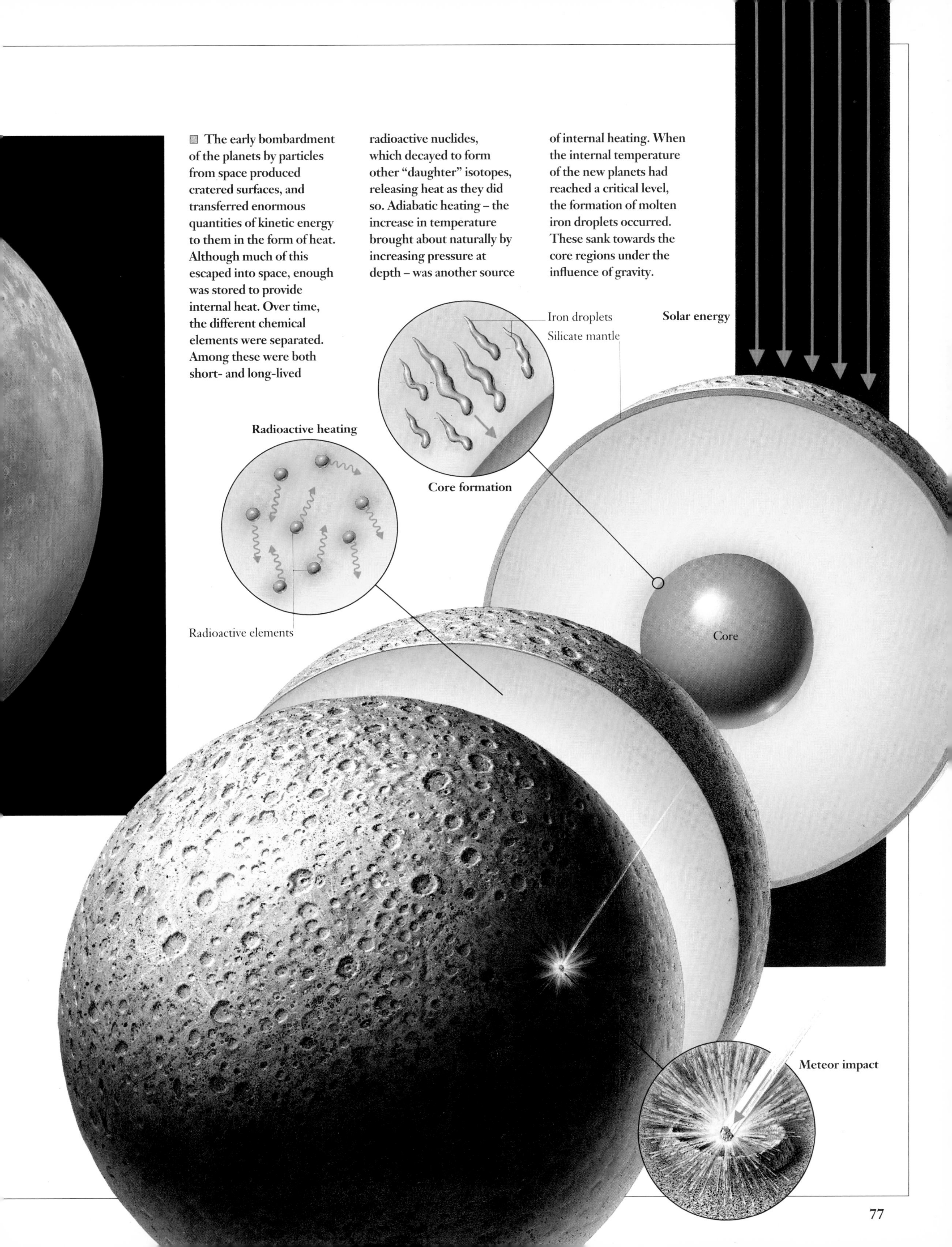

FORMING CORES

NO ONE knows exactly how the planets formed. But, by making reasonable assumptions about early temperatures and pressures inside the Earth, and the contribution made by radioactive elements, it becomes possible to make intelligent guesses about the process. From these computations of early temperatures and pressures, geologists have concluded that the accretional and radioactive energy produced in the early stages was unable to escape fast enough to prevent the heating up of our planet. Consequently, around 3.5 billion years ago, the interior temperature was sufficiently high to cause the metallic iron at the core to melt. This had far-reaching effects. A similar process occurred on the other inner planets, though the timing may have been different.

KEYWORDS

CHALCOPHILE ELEMENT
DIFFERENTIATION
GEOTHERMAL GRADIENT
GRAVITATIONAL ENERGY
LONG-LIVED RADIONUCLIDES
METAL PHASE
NICKEL-IRON ALLOYS
SIDEROPHILE ELEMENT

Because the Earth and other inner planets almost certainly began accreting after metallic and silicate grains had condensed out of the solar nebula, it is likely that metallic particles containing iron and nickel were accreted first. Such iron-rich substances have greater ductility and density than the more brittle silicates. Thus, the iron-rich particles gravitated to form the cores, while the silicates were concentrated nearer to the planets' surfaces and eventually accumulated to form the mantles. As the proto-planets continued to grow, their gravity increased. Calculations show that the sinking of metallic iron within the Earth would have begun around the point at which one-eighth of the planet's mass had accrued: it is clear that core formation had started well before the entire Earth had accreted.

Once the separation of iron and silicates had begun – a process called differentiation – the planets heated up even more quickly as gravitational energy was released. If accretion was very rapid, radiation of thermal energy into space would have slowed and internal heating accelerated. Earth's internal temperature must have risen at least 2000°C by the time the core had segregated.

We know that there are now sizable amounts of both iron oxide (FeO) and iron sulfide (FeS) within Earth's core. If these were present at an early stage, they would have

▽ Core formation in the inner planets involved the sinking of iron and other heavy metals, particularly nickel. It probably took less than 100 million years years. Mercury's core accounts for 40 percent of its volume; the Earth's nickel-iron core accounts for 15 percent. Mercury's large core is either due to its position close to the Sun, where the refractory elements were in greater abundance, or because a collision stripped off much of its mantle during its early history. The dimensions of Venus' core are not known, but may be 12 percent, similar to that of Earth. There is uncertainty about Mars. Geophysical data indicate that it may have a small dense core, about 7 percent of the planetary volume, or a less dense one which occupies around 20 percent.

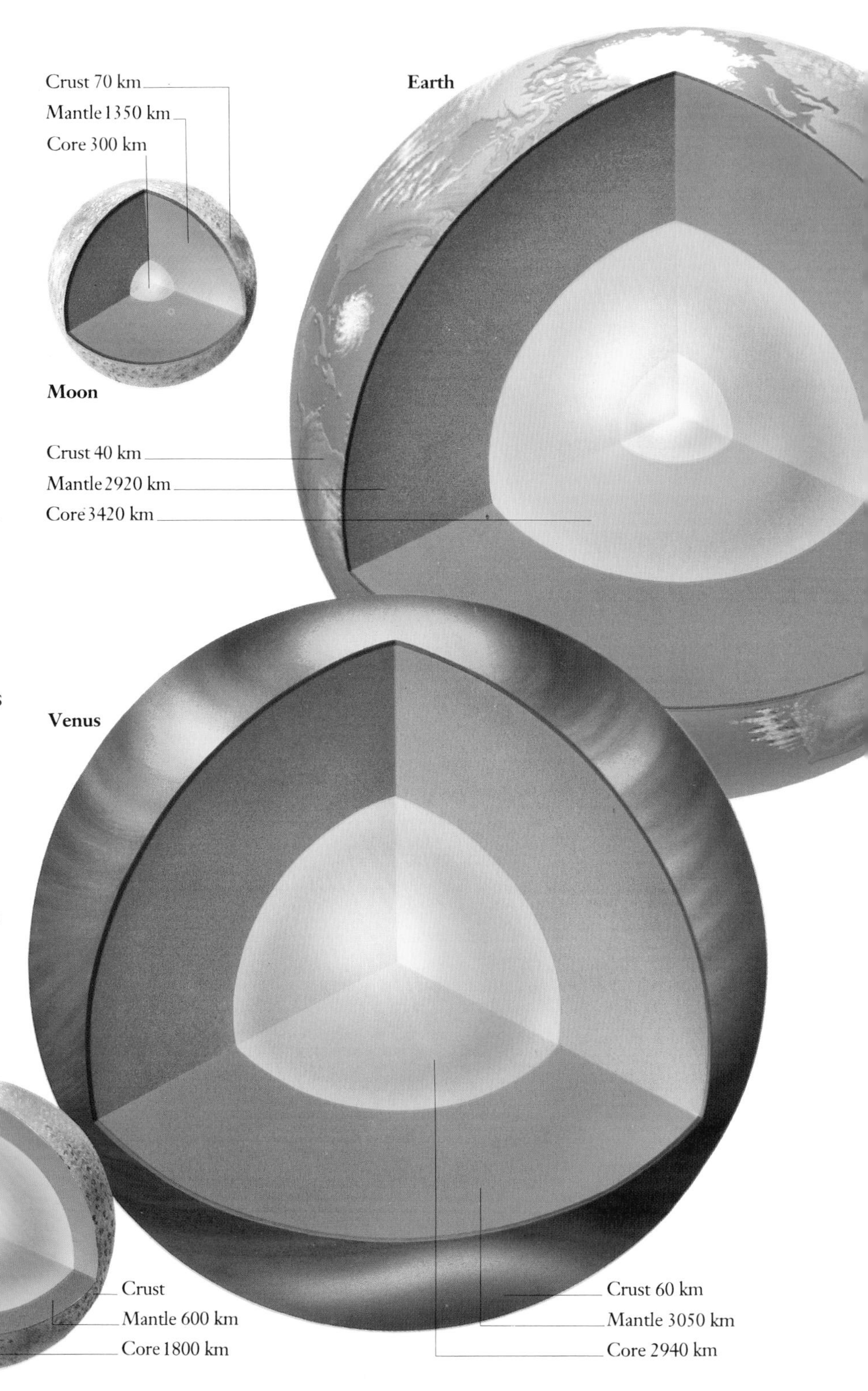

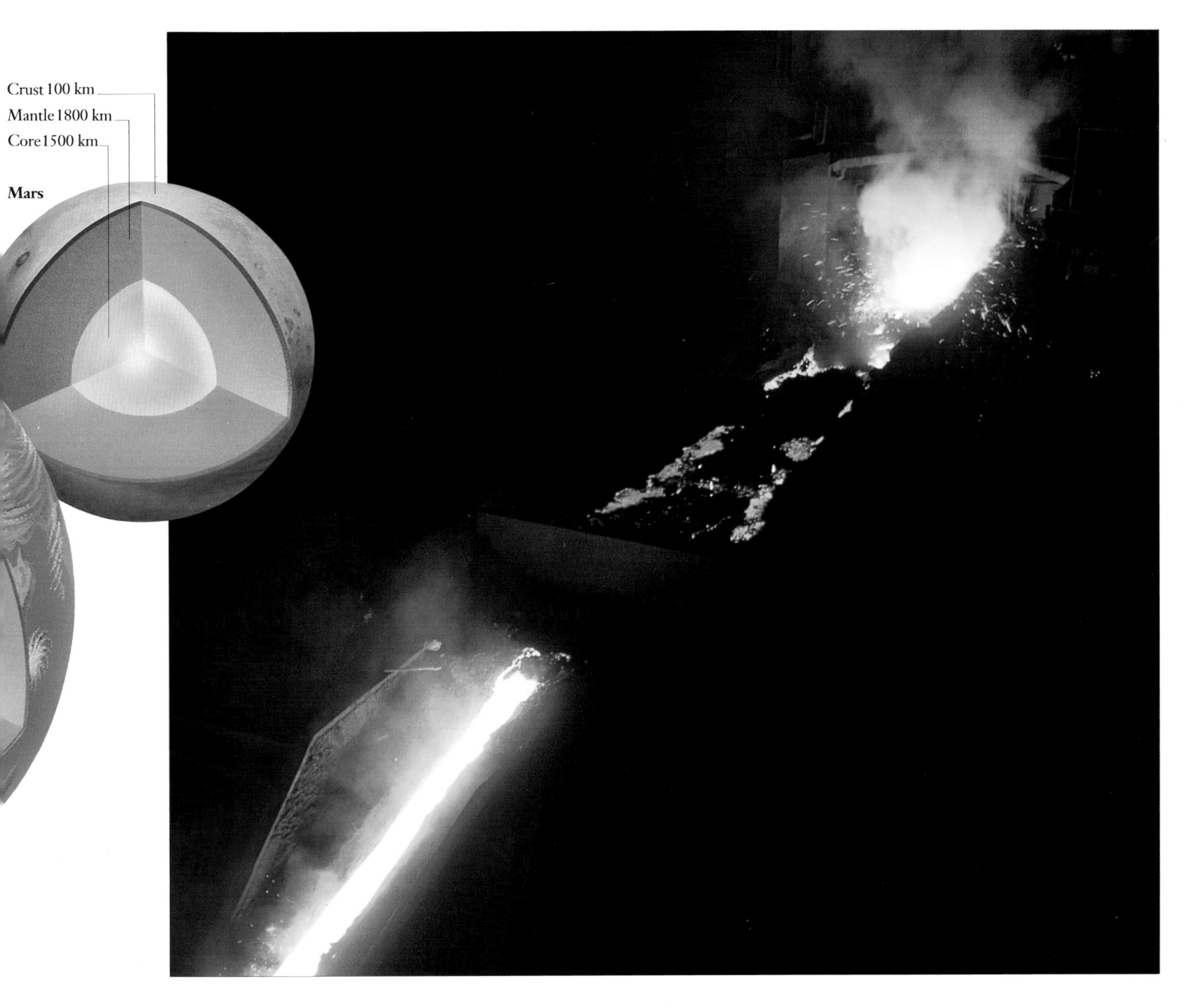

depressed the melting point of the metal phase as opposed to the silicate materials. As a result, the iron and nickel could sink toward the core as molten droplets, while the silicate materials (which eventually accumulated to form the mantle) remained solid.

The Moon and the Earth – which are the only planetary bodies accessible to thorough exploration so far – are the only two bodies for which there is comprehensive geochemical data. The Moon is a satellite, but it is large enough to be compared with the terrestrial planets. It has a small iron-silicate core which appears to have formed 4 billion years ago; its radius cannot be larger than 400 kilometers (the Earth's is 3485 kilometers). Core formation could have raised its temperature by a mere 10°C.

Among the other inner planets, Mars is of lower density than Earth and its core is either less dense or smaller than Earth's. There is some doubt, but most scientists believe that Mars' core accounts for between 7 and 21 percent of the planet's total volume. Venus, on the other hand, being similar to the Earth in terms of mass, probably has a very similar internal structure; its core is partly molten and slightly smaller than Earth's. This contrasts with Mercury, which is approximately 15 percent too dense to have been derived from the same parent material as the other inner planets. Its core is disproportionately large, having a radius of approximately 900 kilometers. Some scientists have conjectured that its original mantle was stripped away during its early history.

△ The formation of planetary cores and mantles is similar to the process in a blast furnace. Most iron ores are iron oxides. To produce pure iron, the oxygen must be removed by heating the ore to high temperatures with coke and limestone. The coke reacts with oxygen, forming carbon monoxide, which escapes. The iron trickles down and collects at the furnace base; earthy impurities float up as slag.

SEGREGATING THE ELEMENTS

WHEN the planets' cores first formed, little heat was lost into space, because it could be lost only by conduction through the solid outer layers; these provided effective insulation. Because the planets retained their heat, their temperatures eventually reached a point at which the Earth, at least, may have become entirely or almost entirely molten; the same was probably true of Venus. At this point, a more efficient mechanism for heat transfer took over: convection.

KEYWORDS

ATMOPHILE ELEMENT
CORE
CRUST
DIFFERENTIATION
LITHOPHILE ELEMENT
LITHOSPHERE
MANTLE
REFRACTORY ELEMENT
VOLATILE ELEMENT

Any body of liquid convects if it is hotter at the bottom than above: the hotter material expands and floats up through the cooler, denser material. It then cools, becomes denser and sinks. The cycle then repeats itself. By this process, heat was transferred toward the surface of the planets and dissipated into space.

Over time the Earth cooled sufficiently for the outer layers to solidify, forming the proto-crust, but the core remained molten, as it still partly is today – its temperature is about 4000 K, compared with 290 K at the surface and 700 K at the lower edge of the crust. Apparently 4.5 billion years is not long enough for the Earth's core to "freeze".

A similar process operated on the other planets, such as Venus; as its composition and structure are similar to Earth's, it is likely that Venus' core also remains at least partly molten. But for the smaller planets the rate of heat loss was much greater. There is little doubt that the Moon and Mars, which are relatively small bodies, now have frozen cores – and Mercury's, which is very large in proportion to the rest of the planet, is also mainly frozen.

Pockets of silicate material eventually formed inside the Earth. These rose, taking with them radioactive elements that had lithophile affinities. When these isotopes decayed, they released heat that sooner or later induced melting of the silicates in which they were trapped. In this way, small pockets of molten silicate were formed. These contributed to the segregation of the mantle layer, and the outer crust. Radiometric dating of ancient terrestrial crustal rocks reveals that significant volumes of relatively light crustal material was being produced by around 3 billion years ago.

△ A geyser in Iceland sends up a jet of superheated water and steam many tens of meters into the air. Geysers are produced when underground water comes in contact with hot volcanic materials, causing the water to boil. Steam is produced, and the increased vapor pressure forces a violent explosion. Geysers also exist on the moons of the outer planets, spouting liquid nitrogen.

Similar events took place on the Moon, within the parent bodies of the meteorites and on Mars and Venus. In other words, chemical fractionation – the redistribution of the chemical elements – was well underway by the geological era known as the Archean, at the beginning of the Precambrian time. The development of the Moon's highland crust probably occurred between 4.5 and 4 billion years ago.

Convection through the mantle played an important role in the subsequent evolution of the terrestrial planets. Their later history was dominated by slow cooling. Gradually the convective motions within the molten mantle gave way to convection in crystallized rocks (known as subsolidus convection); this continues on Earth (and probably Venus) to the present day. The realization that warm but solid substances can convect revolutionized the science of geophysics. This contribution came from the German-born US seismologist Beno Gutenberg (1889–1960) in the early 1920s.

Core formation and mantle segregation effectively brought about a separation of the elements that sank to the core, such as the dense metallic (siderophile) elements and those with an affinity for sulfur (chalcophile), from lighter silicate ones (lithophile) which reside in the mantle and crust of the Earth. Thus, iron, nickel and sulfur were concentrated in the core; silicates of magnesium and iron now reside in the mantle layer, while silicates of the alkali metals and aluminum are in the lithosphere, along with the volatile elements.

▽ Siderophile elements, particularly iron and nickel, and chalcophile elements (particularly sulfur) sank at an early date to form the core. The mantle layer was enriched in lithophile elements, and contains magnesium, calcium, aluminum, oxygen, silicon, sodium, potassium, barium, chromium, rubidium and lithium. Convection later brought about a further separation, so that silicates of the alkali metals and aluminum concentrated in the lithosphere.

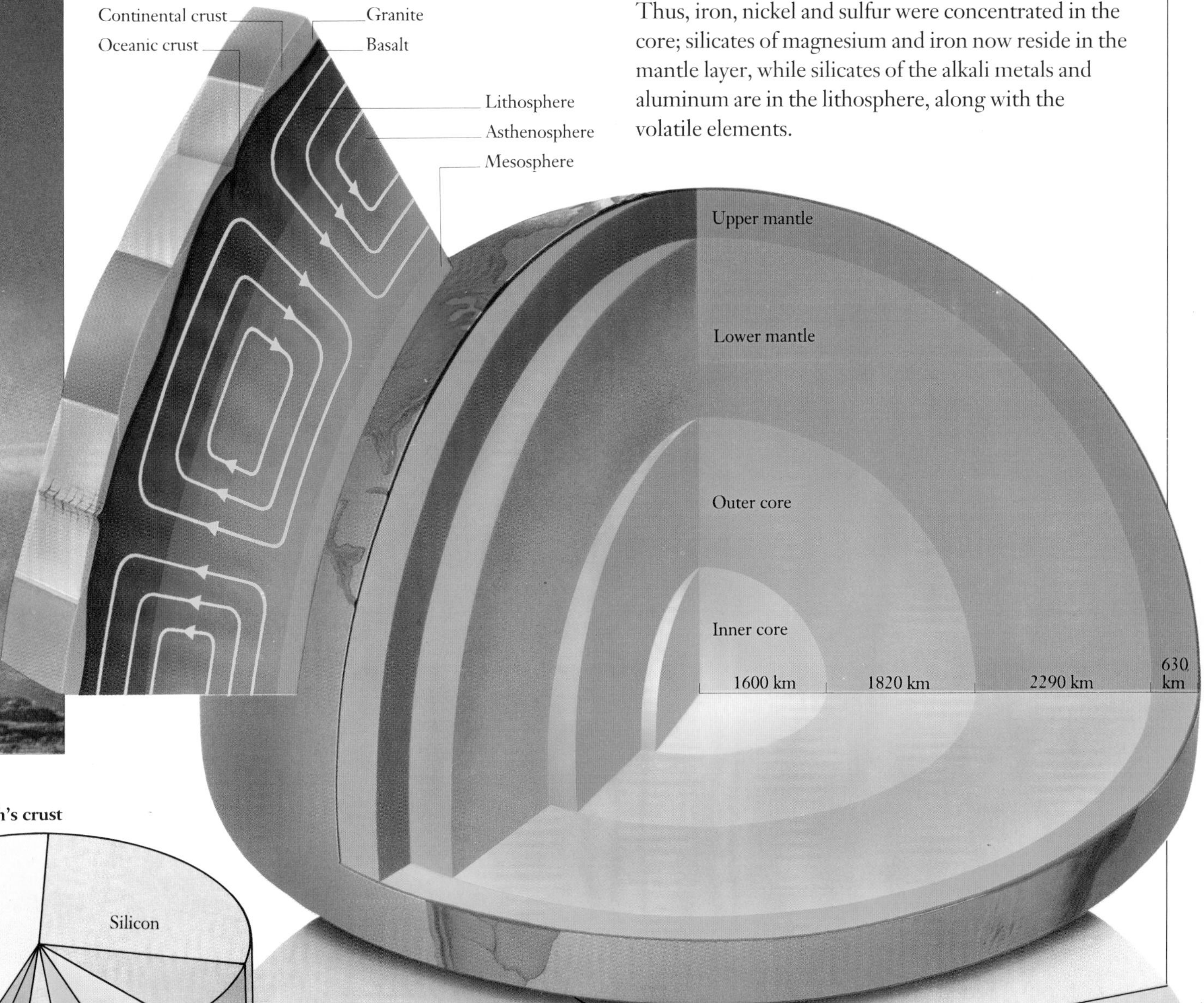

Composition of Earth's crust

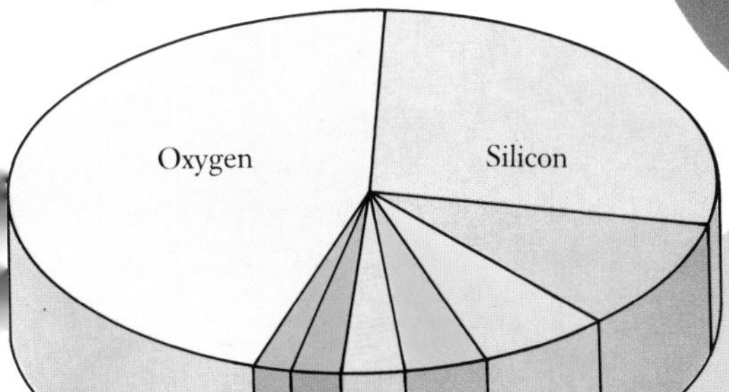

Composition of whole Earth

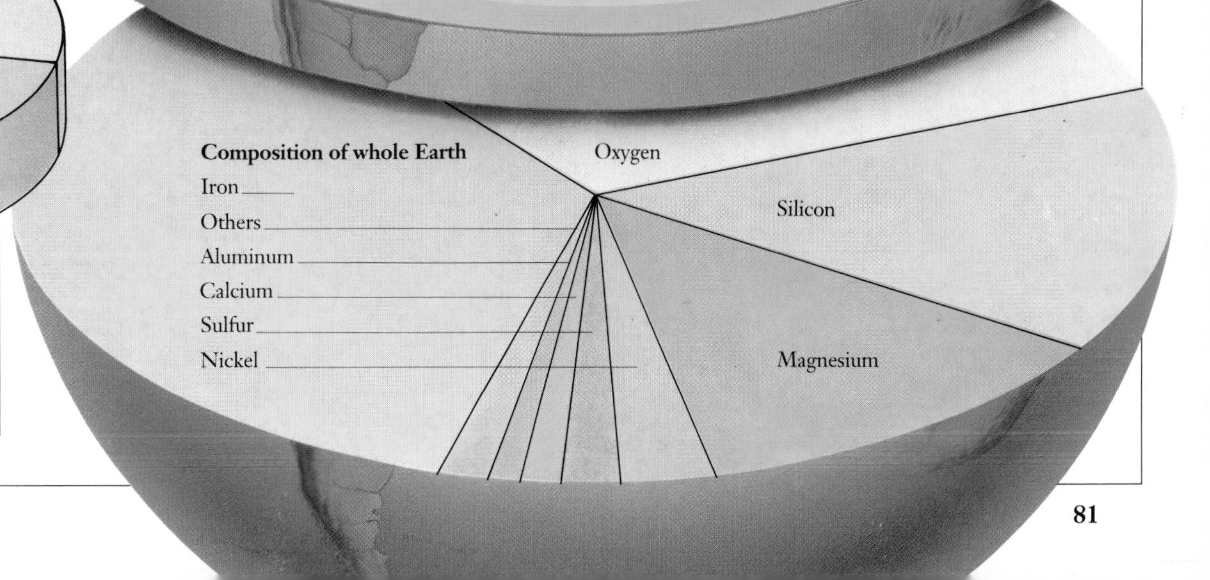

MAGNETIC FIELDS

DURING the early 17th century, the English physicist William Gilbert (1544–1603) first suggested that the Earth behaves like a huge magnet. Modern research shows that Earth and several other planets behave like vast electromagnets. The electrical currents that support them are maintained by self-exciting dynamos, in which the rapid spinning of the planet induces fluid motions in the core, causing it to behave like a dynamo. Once it begins to work it gathers momentum, regenerates the weak galactic field, and produces the field of today.

KEYWORDS

AURORA
COSMIC RAYS
DIPOLE
IONIZATION
MAGNETIC FIELD
MAGNETOPAUSE
MAGNETOSPHERE
SELF-EXCITING DYNAMO
VAN ALLEN BELTS

The Earth's magnetic field originates in motions within the fluid outer core which are partly maintained by the Earth's rotation. Earth behaves as if a giant bar magnet lies within, its length aligned nearly parallel to the spin axis. The associated magnetic field lines form a magnetic dipole, in which a magnetic needle, if suspended freely, aligns itself with the local field. At different latitudes its magnetic inclination differs; at the Equator it lies horizontal, but at the poles it dips vertically. The needle also aligns itself with the north magnetic pole, which is about 11° away from the spin axis. However, the magnetic axis slowly moves around the rotational axis, giving rise to slow variation. The field is vital to life. It shields the Earth's surface from harmful radiation, such as cosmic rays, and the steady stream of plasma or ionized particles that emanate from the Sun, by trapping them outside the Earth's atmosphere.

The United States' satellite Explorer 1 was launched in 1958 to check on cosmic ray counts in the atmosphere. Each time Explorer reached an altitude of 350 kilometers, its Geiger counter became so saturated with radiation that it cut out. These measurements proved the existence of zones of intense radiation surrounding the Earth, known as Van Allen belts. They are zones in which highly charged particles become trapped, and include cosmic rays and solar wind particles. The belts are made of charged particles, electrons and protons, contained within the magnetosphere – a teardrop-shaped region around the Earth whose tail extends far away from the Sun. Its shape is the result of the solar wind – rapidly-moving ionized particles that stream outward from the Sun. On hitting the magnetic force field of the Earth or any other planet, these are slowed down, producing a bow

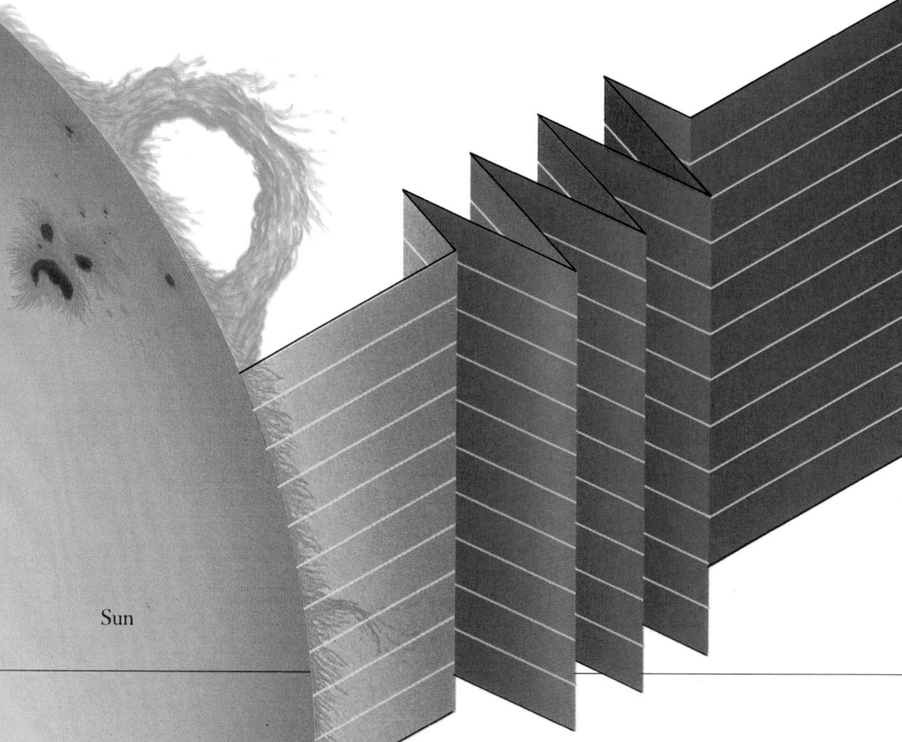

shock, inside which is an extremely turbulent region called the magnetosheath, where the force of the solar wind and magnetic field are roughly equal. Inside this again is an interface called the magnetopause, which encases the magnetosphere proper.

Mercury and Venus have weak fields, while Mars and the Moon effectively have none. Saturn, Uranus and Neptune have fields comparable with the Earth's, but Jupiter's field is more than 19,000 times more powerful than Earth's. Its magnetosphere is so large that it extends well beyond the orbit of Saturn. Extremely complex particle activity occurs inside it.

Many rocks contain iron-bearing minerals which are magnetic; most crystallize from magma or molten rock from the Earth's core. They behave like tiny magnets and therefore bear the imprint of the magnetization which operated at the time that they first crystallized. Such "fossilized" magnetization is called paleomagnetism. It shows that the magnetic poles have not always been in their present location, and that Earth's continents have moved in relation to the magnetic field and each other. Paleomagnetic studies further indicate that the positions of the north and south magnetic poles have interchanged in the past, giving rise to polarity reversals. During the past 4 million years alone, there have been at least nine reversals. These have been recorded in lava on the ocean floors, proving that new sea-floor is constantly being produced, and that, once formed, this new crust moves slowly away from its source. This process is a fundamental part of the theory of plate tectonics, which explains the movements of a planet's crust.

◁ Auroras develop most strongly around Earth's magnetic poles, caused by the precipitation of solar wind particles which descend into the upper atmosphere along magnetic lines of force.

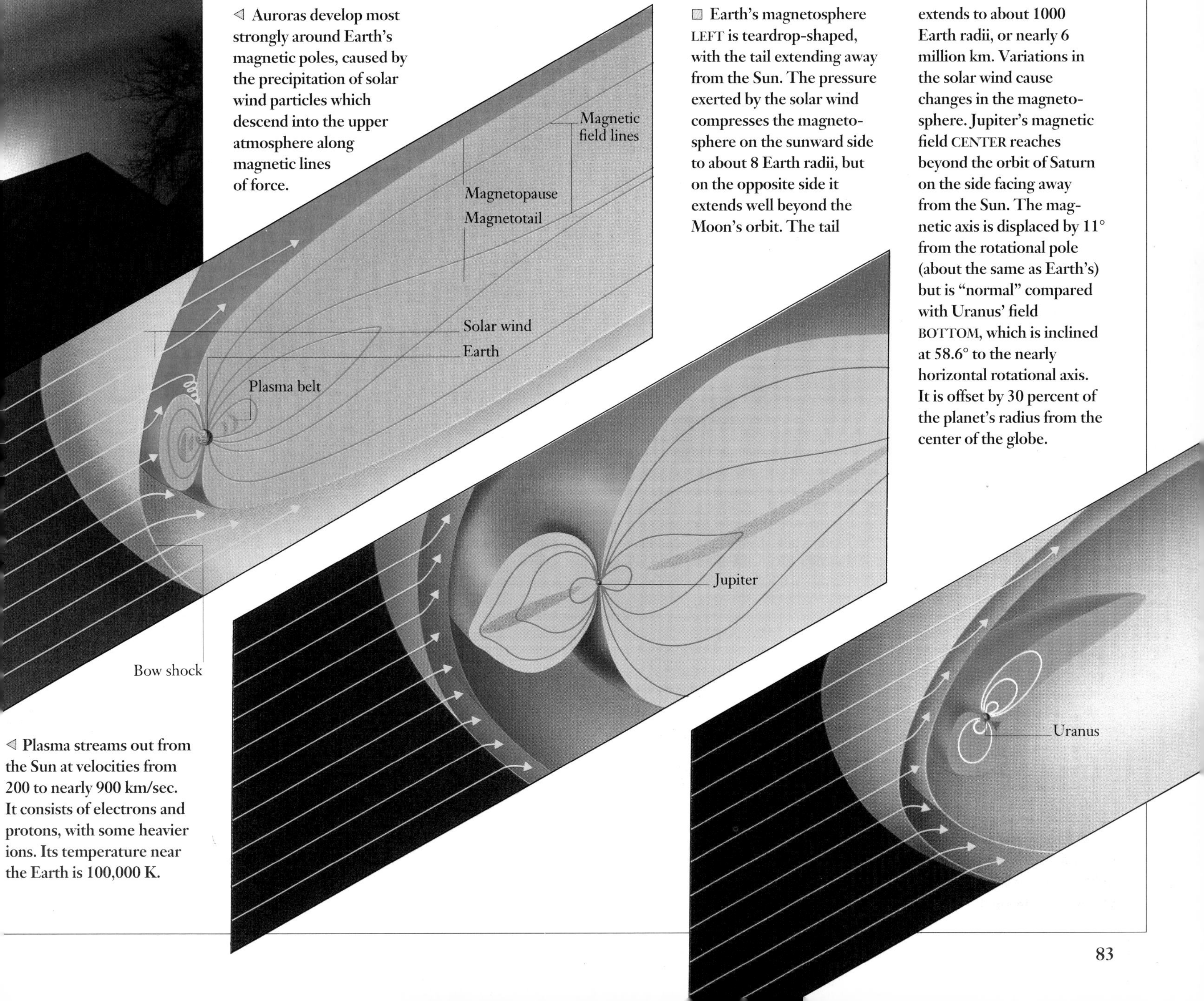

□ Earth's magnetosphere LEFT is teardrop-shaped, with the tail extending away from the Sun. The pressure exerted by the solar wind compresses the magnetosphere on the sunward side to about 8 Earth radii, but on the opposite side it extends well beyond the Moon's orbit. The tail extends to about 1000 Earth radii, or nearly 6 million km. Variations in the solar wind cause changes in the magnetosphere. Jupiter's magnetic field CENTER reaches beyond the orbit of Saturn on the side facing away from the Sun. The magnetic axis is displaced by 11° from the rotational pole (about the same as Earth's) but is "normal" compared with Uranus' field BOTTOM, which is inclined at 58.6° to the nearly horizontal rotational axis. It is offset by 30 percent of the planet's radius from the center of the globe.

◁ Plasma streams out from the Sun at velocities from 200 to nearly 900 km/sec. It consists of electrons and protons, with some heavier ions. Its temperature near the Earth is 100,000 K.

How Atmospheres Evolved

The earliest atmospheres resulted from the accumulation of gases that escaped from the interiors of rocky planets. All the planets except Mercury and possibly Pluto have – or once had – atmospheres, although they vary greatly from planet to planet, due to the different conditions found on each.

Silicate minerals in the Earth's crust, such as micas and amphiboles, contain hydrogen and oxygen within their crystalline structure, usually in the form of hydroxyl groups. These hydrated silicates occur widely in crustal rocks and have also been found in blocks believed to have originated in the upper mantle. As the planet heated up, silicate materials of the mantle partially melted to form magmas. These contained volatiles, including nitrogen, carbon dioxide, water and others. Together they rose toward the surface; as the gases expanded, they provided pressure for the eruption, which could often be explosive and caused the outpouring of lavas and eruption clouds containing hot vapor, especially water. Enough water would have been released in this way to fill the Earth's ocean basins, assuming that volcanic eruptions on the early Earth were as common as they are today. In addition to the volatiles from the interior, some light gases may have arrived from infalling comets. These icy bodies, which formed in the far reaches of the Solar System, are rich in hydrogen, nitrogen and carbon compounds. No doubt many of them collided with the Earth during the early years of its history.

Earth's early atmosphere was very different from today's. Research on volcanic gases suggests that water vapor, carbon dioxide, carbon monoxide, nitrogen, hydrogen chloride and hydrogen were most abundant. Hydrogen, which is very light, quickly escaped into space. Some of the water vapor in the upper atmosphere was broken down by sunlight into hydrogen and oxygen, the latter escaping and combining with gases like methane (CH_4) and carbon monoxide (CO) to form water (H_2O) and carbon dioxide (CO_2) . Free oxygen, so vital to life on Earth, was not found until the beginning of photosynthesis, by which carbon dioxide is taken in and free oxygen released as an end product of photosynthesis.

KEYWORDS

ARGON
ATMOSPHERE
DEUTERIUM
GREENHOUSE EFFECT
HELIUM
INERT GAS
PHOTODISSOCIATION
ULTRAVIOLET RADIATION
VOLATILE ELEMENT

▷ By about 2 billion years ago, photosynthesis had become sufficiently common for the accumulation of free oxygen in the air. By 300 million years ago, this gas accounted for about 20 percent of the total of Earth's atmosphere.

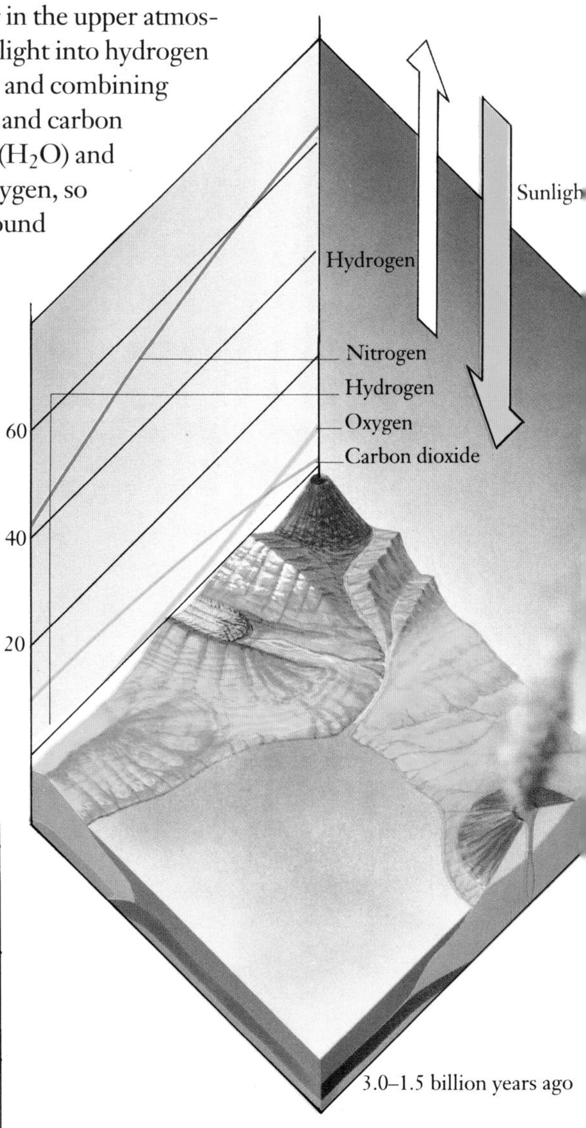

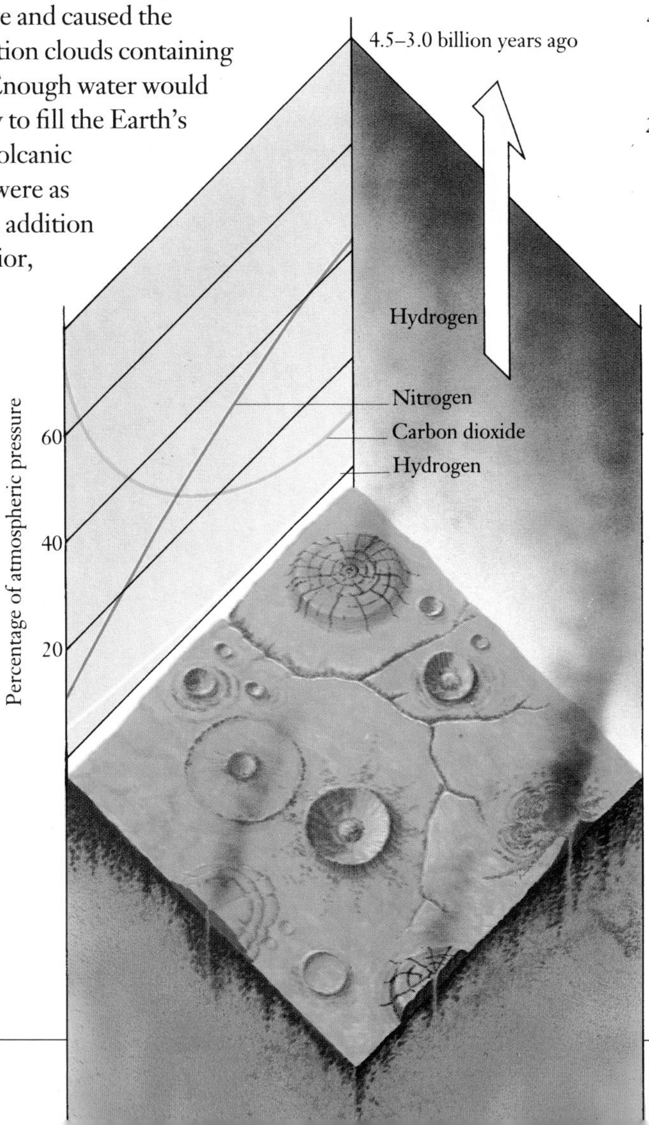

▷ The Earth's primeval atmosphere was very different from today's. Hydrogen was present from the planet's formation until about 3.5 billion years ago, but was gradually lost into space. Carbon dioxide made up about 80 percent of the total atmosphere at first, but as it became fixed in limestone rocks, it gradually diminished.

△ Nitrogen became more abundant as it escaped from the Earth's interior. Between 3 and 1.5 billion years ago, carbon dioxide concentrations fell; nitrogen reached its maximum level. The earliest bacteria came to life at this stage, followed by blue-green algae (cyanobacteria); these had the ability to carry out anaerobic photosynthesis.

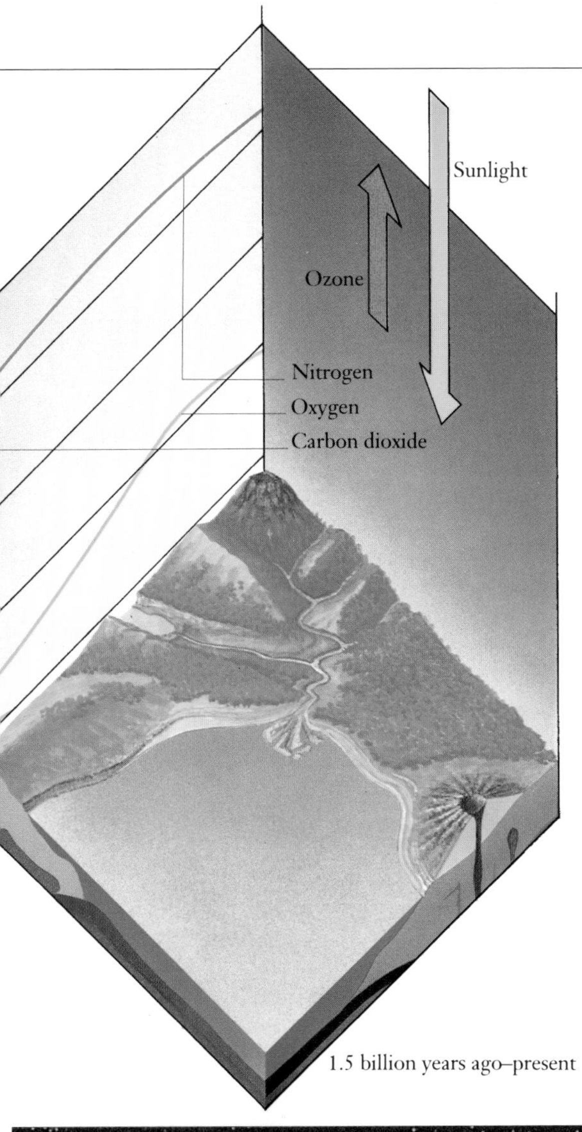

▷ The atmospheres of the rocky planets evolved as their interiors "degassed", and volcanic activity released volatiles. Subsequent interactions between atmosphere and crust may have fixed certain atmospheric components in crustal rocks such as limestones. The gas and ice giant planets, on the other hand, were able to retain much of the primordial hydrogen that was present in the solar nebula, forming hydrogen compounds that evolved into a complex "soup", whose structure in some ways resembles the atmosphere of Earth.

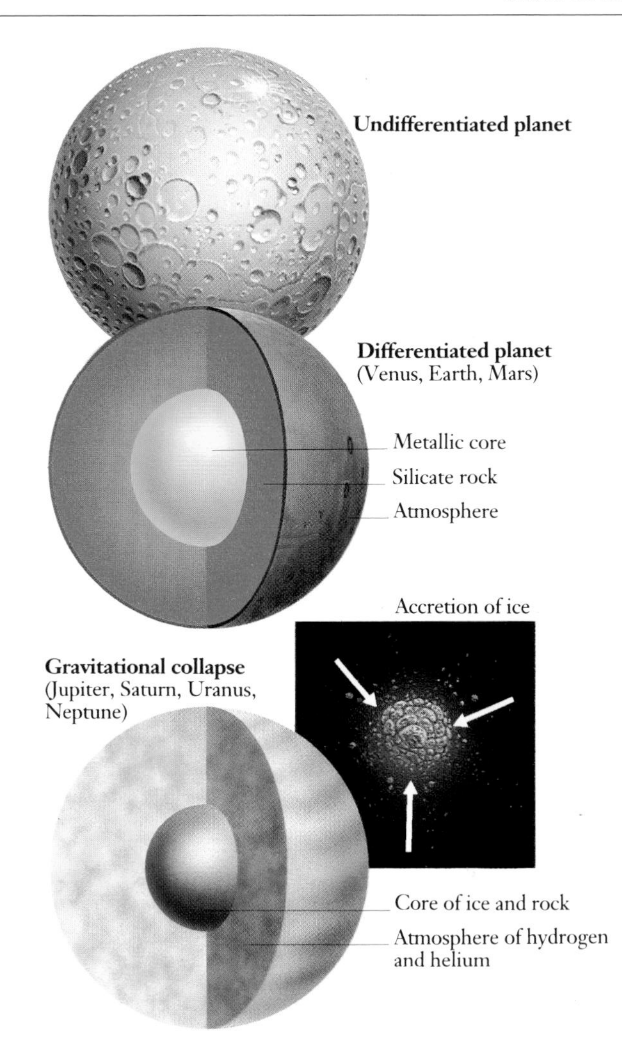

▽ Venus is surrounded by heavy yellowish-white clouds of carbon dioxide. Its atmosphere would not support the life forms that flourish on Earth.

Geological and other evidence suggests that in the distant past the inner planets were much less different in their volatiles composition than today, and that the present diversity is a result of differences in planetary mass and temperature. Mercury, which has weak gravity and high surface temperature, could never hold onto its volatiles. Venus may once have had a large amount of water, but the dense carbon dioxide atmosphere that built up caused surface temperatures to rise and the water to evaporate; the hydrogen and oxygen ions then dissociated under the strong sunlight, and the hydrogen escaped into space. On Mars, running and standing water once existed in sufficient amounts to have left many surface features; the planet exhibits large ice caps formed of water and carbon dioxide, and the atmosphere today is mainly carbon dioxide.

The gas giants, on the other hand, are themselves made up of light elements, their atmospheres forming a kind of chemical soup which precipitates clouds of varying colors. The dominant components are hydrogen and helium, with much smaller amounts of methane and ammonia. These are the gases that the warmer, less massive inner planets could not retain, and which were swept outward by the solar wind until they were captured by the strong gravitational fields of the giant planets.

INTERNAL CLOCKS

EACH chemical element is composed of identical atoms held together by electrical forces. An individual atom consists of a nucleus, which accounts for about 99.9 percent of the atom's mass, with one or more negatively charged electrons in orbit around it. The sum of the protons and neutrons in the nucleus is the element's mass number, while the number of protons gives the atomic number. Atoms with the same atomic number but different numbers of neutrons are called isotopes. Most elements have at least two isotopes. Some isotopes are radioactive, and they may decay spontaneously to lighter isotopes of other elements. The decay process takes place at a fixed rate. For example, when potassium-40 decays to argon-40, it takes 1.3 billion years for half of the amount of the parent isotope of potassium to convert to the daughter isotope of argon – a period called the half-life. Knowing this, a geophysicist can measure the proportions of the isotopes in a rock, and is able to derive the absolute age of the rock itself. This is called radiometric dating.

KEYWORDS

ABSOLUTE AGE
DAUGHTER ISOTOPE
GEOCHRONOLOGY
ISOCHRON
ISOTOPIC RATIO
METAMORPHISM
PARENT ISOTOPE
RADIOGENIC ISOTOPE
STABLE ISOTOPE

The decay sequence samarium-146 to neodymium-143 has a half-life of 130 billion years; uranium-238 to lead-206, a half-life of 4.5 billion years; and rubidium-87 to strontium-87, a half-life of 4.7 billion years. There is also one isotope of carbon (carbon-14) which behaves in this way. This is useful for dating archeological remains and pieces of organic matter incorporated in recent geological deposits. If the same rock is dated by at least two of the sequences, there is usually a high degree of consistency in the results. However, this is not always so, and it was not until considerable research was undertaken that such discrepancies were explained. Most of these reflect the fact that metamorphic rocks are deformed and refused through reheating, and this process resets the radiogenic "clock" for one mineral of a pair.

Absolute ages of rocks can only be obtained for planets from which rock samples have been returned to Earth; the measurement requires sophisticated and bulky instruments which cannot be carried on board a spacecraft. Such ages, therefore, are available only for the Earth, Moon and meteorites; rocks from the other planets cannot yet be collected for study. There is, however, another method by which the relative age of a surface can be assessed. This relies on the simple thesis that the longer an area of a planet is in existence, the greater will have been the number of meteorite impacts in its cratering record. For the Earth, which recycles its crust continually, this method cannot be used; for the other planets, which are less tectonically active (if at all), this problem is absent or reduced.

The technique of crater-age dating was first applied to the Moon, where the incidence of cratering varies widely from region to region: the dark maria basins have far fewer craters than the brighter highlands. By laboriously recording how the density of cratering varies from place to place, scientists have assigned relative ages to different regions. It is also possible to calibrate the different cratering densities by obtaining absolute ages for those same regions. Using the lunar data, it is possible to assess ages for other planets.

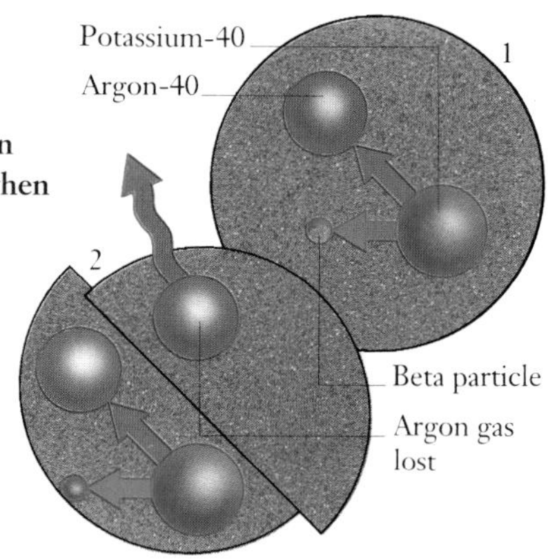

▷ **Radioactive decay in igneous rock begins when magma crystallizes 1. If the rock later is reheated and metamorphoses, it may lose some of its daughter isotopes 2, giving a different radiometric date.**

◁ **Impact craters on the Moon's surface have much lower density than the highlands. Smoother regions between many of the larger highland craters indicate the different relative ages of the craters. Samples of Moon rock range from 3.1 billion years old (basalt from the dark maria) to 4.5 billion years (anorthosite from the highlands).**

□ **Arizona's Grand Canyon ABOVE is up to 1500 meters deep, carved out by the Colorado River through a rising tableland. The geological strata it reveals tell the geological history of the region. At the bottom of the canyon, in the Inner Gorge, deformed metamorphic rocks and granites are 2.2 billion years old. These are separated by the "Great Unconformity" from the oldest Cambrian strata, the Tapeats Sandstones (570 million years old). The sequence through the Canyon RIGHT includes 1200 meters (4000 feet) of**

Fossil trilobite

horizontally-bedded Paleozoic and Mesozoic rocks, capped by the 240 million-year-old Kaibab Limestone. The Ordovician and Silurian are missing – presumably they have been worn away. Geologists may assign relative ages to strata on the basis of the fossils they contain. Each fossil species has a specific age range, which for some organisms is quite narrow. This trilobite (a long-lived species) is found in rocks that are of Cambrian origin, confirming that it derives from about 500 million years ago.

4

DYNAMIC *Planets*

THE EARTH is a truly dynamic world, to which active volcanism, tectonism and atmospheric phenomena bear witness. The outer gas and ice giant planets, with their short rotation periods and large masses, have also maintained vigorous atmospheric circulation and sufficiently active core motions to generate magnetic fields. Yet some planets, although active during the earlier stages of the Solar System's history, have become inert. Others experience only modest geological or atmospheric activity.

The differences in the degree of dynamism of the inner planets is a reflection of their mass and distance from the Sun. Planets with small mass accumulated modest amounts of accretional energy and lost this relatively quickly. Small bodies such as the Moon "froze" geologically during the first billion years or so of geological time. Mercury, being so close to the Sun, has been stripped of any atmosphere it did have.

More massive worlds, such as Venus, Earth and Mars, originally gained more energy from accretion and subsequently from decay of long-lived radioactive isotopes. They maintained relatively high levels of geological or atmospheric activity for much longer – for Venus and Earth, up to the present. On Mars, geological activity is now the result of winds, glaciation and atmospheric processes.

An aerial view of the landscape of Wyoming, in the western United States, shows where layers of rock have been forced up from below the Earth's surface to form anticlines. These have then become eroded by the action of the atmosphere to create lines of hills and outcrops. Many of the Earth's most dramatic landscapes are the result of dynamic movements in the crustal rocks, or in the lithosphere beneath.

The First Crusts

At a relatively early stage, all of the inner planets developed solid crusts made from silicate rocks. These were gradually built up as material from the planetary mantles beneath was extruded to form basaltic lavas. The Earth's primordial crust has long since been returned to the mantle. But clues as to its likely appearance are forthcoming from our nearest neighbor, the Moon, which (because its geological activity terminated over 2 billion years ago) retains most of its ancient features.

KEYWORDS

COESITE
CRUST
EJECTA
GREAT BOMBARDMENT
IGNEOUS ROCK
KOMATIITE
MAGMA
MANTLE
PAHOEHOE
RAREFACTION WAVE
SHATTER CONE
TECTONICS
TRANSIENT CAVITY

The Moon's ancient highland crust is made up of an igneous rock called anorthosite, which is composed in large part of the aluminum silicate mineral plagioclase feldspar. It is chemically enriched in lithophile elements, which have high melting points. This rather thick crust appears to have crystallized around 4.4 billion years ago from an "ocean" of magma rich in aluminum, calcium and silica. It bears the scars of an intense bombardment by meteoroids and asteroids which continued up to about 4.0 billion years ago. This left the imprint of huge impact basins and myriad impact craters, ranging in size from tiny microcraters measuring only a few microns across to great ring structures several hundreds of kilometers in diameter. The final size of the craters, however, is less than the cavity which was created on impact, as it was modified by decompression after the event.

Associated with the craters and basins are extensive blankets of material thrown out in response to the cratering process. These now form a complex and extensive succession of interleaving layers on top of the crustal base. The highly shattered surface layer is called the regolith. On Earth, this has become soil.

The composition of the Earth's early crust must have differed from that of the Moon. It was probably richer in iron and magnesium and had a lower percentage of high-temperature silicates. It was built

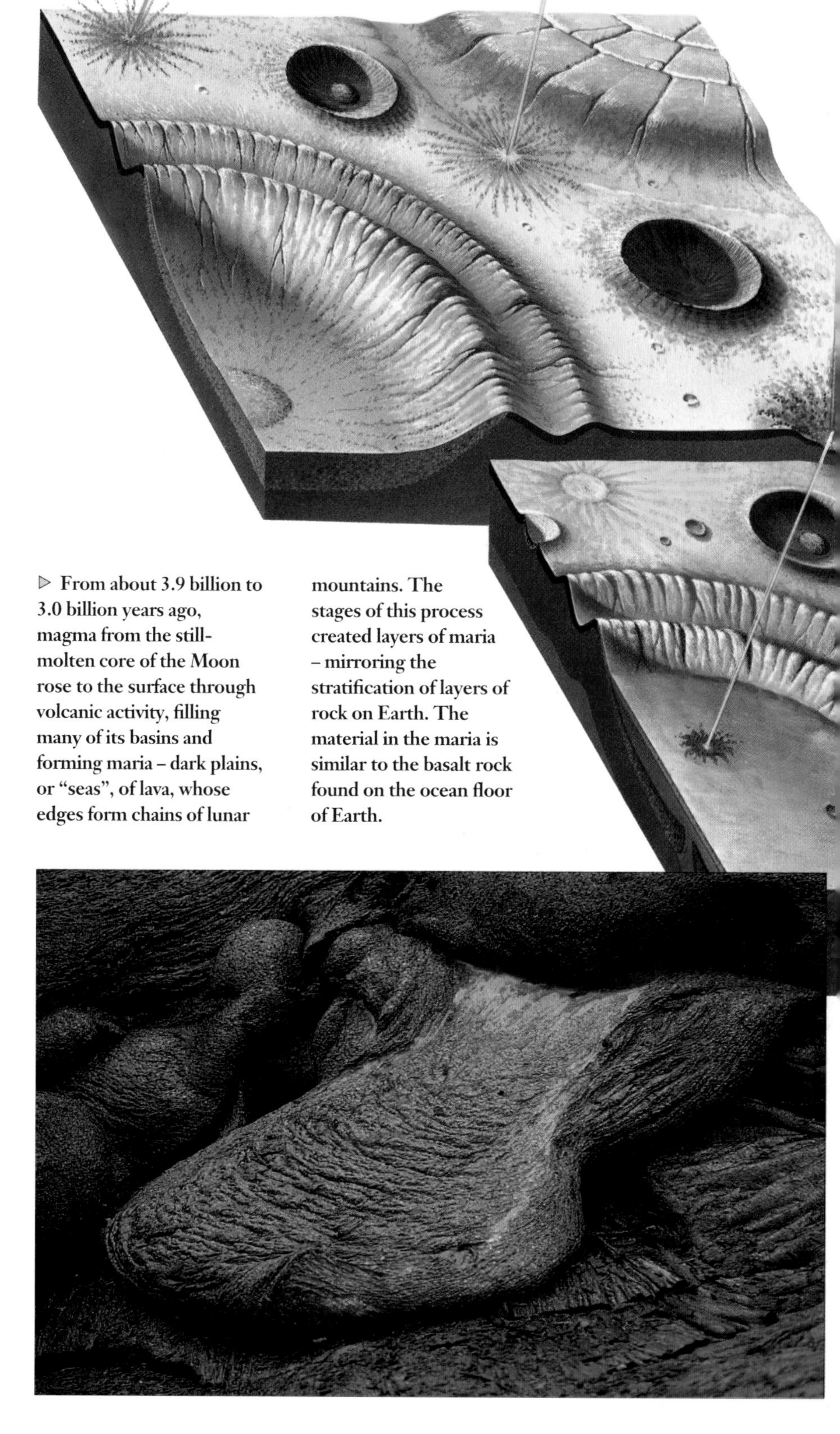

▽ In the early stages of its history – between 4.5 and 4.0 billion years ago – the Moon, like the other inner planets, underwent catastrophic bombardment by meteorites and asteroids, destroying most of the recently formed crust. Huge impact craters and basins were left as scars on the surface.

▷ From about 3.9 billion to 3.0 billion years ago, magma from the still-molten core of the Moon rose to the surface through volcanic activity, filling many of its basins and forming maria – dark plains, or "seas", of lava, whose edges form chains of lunar mountains. The stages of this process created layers of maria – mirroring the stratification of layers of rock on Earth. The material in the maria is similar to the basalt rock found on the ocean floor of Earth.

▷ Fast-moving lava from a Hawaiian volcano cools in the air and solidifies, forming ropy strands called pahoehoe. Planetary crusts were thickened by frequent extrusions of such fluid lava – a process that continues to occur on Earth, although it happens mostly at fissures in the crust underwater, rather than by the more obvious eruptions above ground.

▷ The early cratered crusts of the rocky inner planets were subsequently resurfaced by volcanic activity. This Soviet Venera 13 panorama of the surface of Venus shows a broken plain of slabs similar in appearance to slab pahoehoe (broken crust of lava). Analysis of the rocks also showed that they had undergone stratification (layering) and weathering, as do rocks on Earth. The orange color is due to Venus' light.

up by the repeated outflow of highly fluid magmas often found in connection with volcanoes. Soon after it began to crystallize, the crust was thin, and magmas easily pierced the fragile skin. In time, as layer upon layer accumulated, it became more difficult for this to happen and eruptions became focused where the crust was thinner or weaker.

This crust suffered the same intense bombardment as did the Moon, despite the shielding effect of the Earth's early atmosphere. Like those of the Moon, the rocks were intensely shattered and contained high-pressure forms of silicate minerals such as coesite. Subsequently this early crust was "recycled" due to plate tectonic activity, which moves the ocean floor and the continents with it, pushing down surface layers of crust and allowing new material to be pushed up. This process is still occurring on Earth. Because the material is recycled, the size of the lithosphere remains constant; it neither expands nor contracts.

The primeval crusts of the other inner planets must have been similar to those of Earth, but unlike Earth's, each retains an extensive record of impact. On Mercury, Venus and Mars, the most ancient cratered regions are partly obliterated by younger volcanic plains (intercrater plains), which indicate that extensive volcanism continued after the main phase of cratering had ended. The number of small impact craters on the Venusian crust is significantly less than on Mercury or the Moon. This is because the planet's dense atmosphere had a shielding effect, winnowing out smaller meteoroids as they entered it.

The crusts of Earth and Venus have never become sufficiently thick to prevent magma reaching the surface from inside. However, the smaller planets eventually reached a point where the thickening of their crusts prevented further volcanism. At this stage, modification of the surface by processes from within the planet effectively ceased. The Moon and Mercury are considered geologically dead.

◁ After an intervening period of minor volcanic activity and cratering, the Moon's surface has changed very little in the past 1 billion years. Plate tectonic activity is prevented by the thickness of the lithosphere and limited mantle convection.

THE RISE OF MAGMAS

THE temperature inside the Earth near the surface rises, on average, 3°C with each 100 meters in depth; about 50 kilometers below the surface, it is about 1000°C. The gradient decreases at greater depths, and the core probably does not exceed 4300° C. Despite these temperatures there is only a narrow zone within the upper mantle, known as the asthenosphere (70 to 250 kilometers below the surface), where molten rock or magma is found. Below this is a solid region of the mantle, the mesosphere, where enormous pressure prevents the rock from melting. The iron-rich outer core is also molten, at between 3700° and 4300°C, but the high pressures in the inner core make that solid too.

KEYWORDS

DIAPIR
FRACTIONAL CRYSTALLIZATION
GRAVITY SEGREGATION
HYDROSTATIC PRESSURE
OLIVINE
PARTIAL MELTING
PERIDOTITE
PYROXENE
VESICULATION

Molten rock may be forced up to the surface as lava. It may flow quietly or explode to form an eruption cloud. Whether the lava is explosive depends on how deep it formed, what sort of volcano is involved, how viscous the magma is, what volatile elements it contains and so on.

Most types of magma are silicates and give rise to igneous rocks built from minerals with the silica molecule (SiO_2) in them. As they rise toward the surface, they expand and cool, and eventually begin to crystallize. In addition to silica, terrestrial magmas contain elements such as aluminum, iron, magnesium, calcium, titanium, manganese, phosphorus, sodium and potassium. These combine with the silica to form silicates. Silica accounts for between 35 and 75 percent of terrestrial magmas. Those with low silica content are basic, while those with a high silica content are called acidic or silicic. Also present in smaller amounts are trace elements such as rubidium, strontium, zinc, sulfur and cobalt. Finally, there are many volatile substances. The most important of these is water, accompanied by elements such as boron, chlorine , fluorine and compounds containing sulfur, hydrogen, oxygen and nitrogen.

While the magma is pressurized deep inside the Earth, these elements and compounds are dissolved within it. As the magma rises and the pressure drops, the more volatile components expand and are released as gases. Thus, when a melt reaches the surface, it consists of a mixture of crystals, liquid and gases. Indeed, the expansion of the gas component is in large part responsible for forcing the magma to the surface, and possibly causing an explosive eruption.

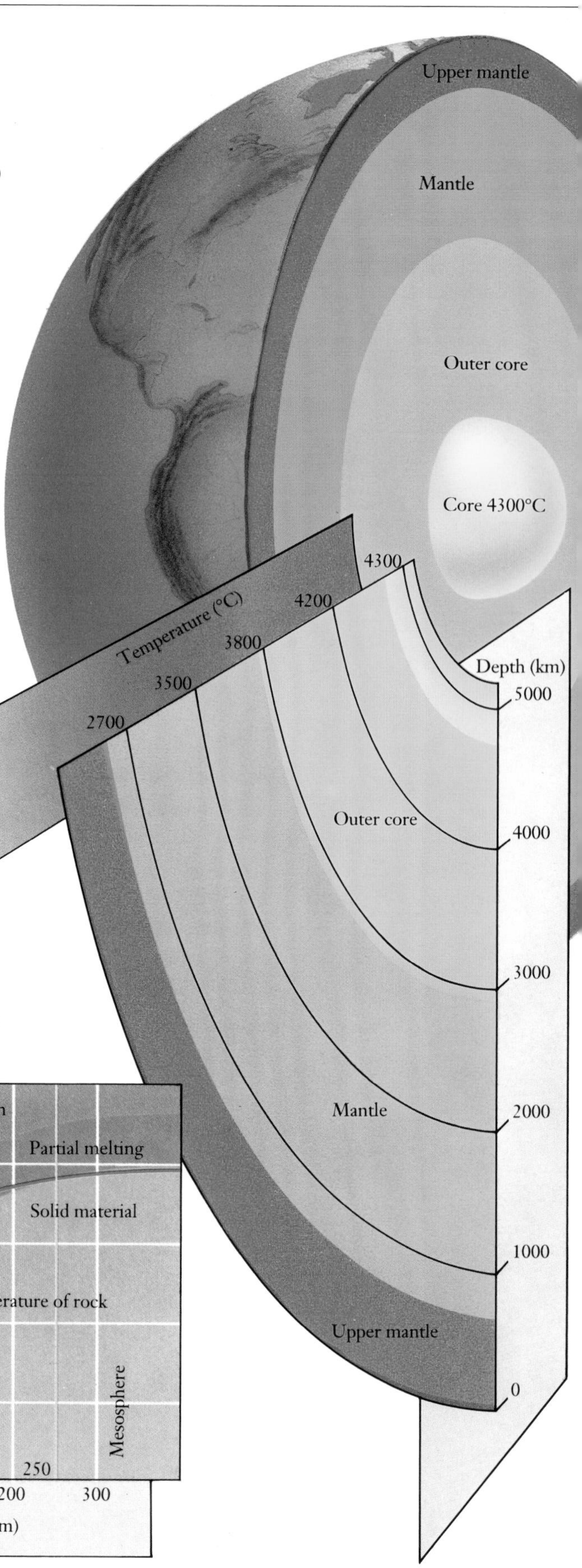

▷ The temperature rises very quickly in the first 100 kilometers below the Earth's surface, but the curve flattens up to the boundary between the outer and inner cores, about 5000 kilometers below the surface. Scientists have estimated these temperatures by studies of the melting points of mantle material and iron at pressures equivalent to those thought to prevail at various depths. Seismic surveys have indicated the zones where melting is found. These are associated with temperatures that give rise to phase changes (solid to liquid, or liquid to solid). At the base of the crust, the density is believed to be 3000kg per cubic meter, pressure is 200 kilobars and temperature 1000°C. At the base of the mantle the equivalent figures are density 5500 kg/ m^3, pressure 1400 kbar and temperature 3000°C.

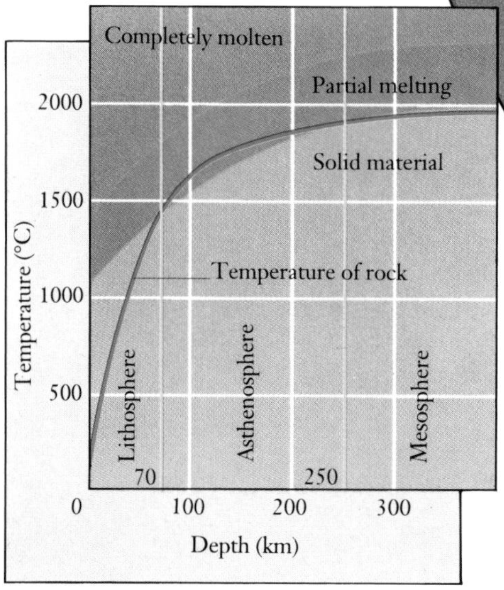

▲ As depth increases, pressure rises and higher temperatures are needed to cause rock to melt. If the geothermal gradient, or known temperature of the rocks in the oceanic crust and mantle, is plotted against this melting curve, it is only in the region between 70 and 250 kilometers – the asthenosphere, or low-density zone – that melting occurs. It is therefore in this region that magmas arise. These magmas may reach the surface through volcanoes.

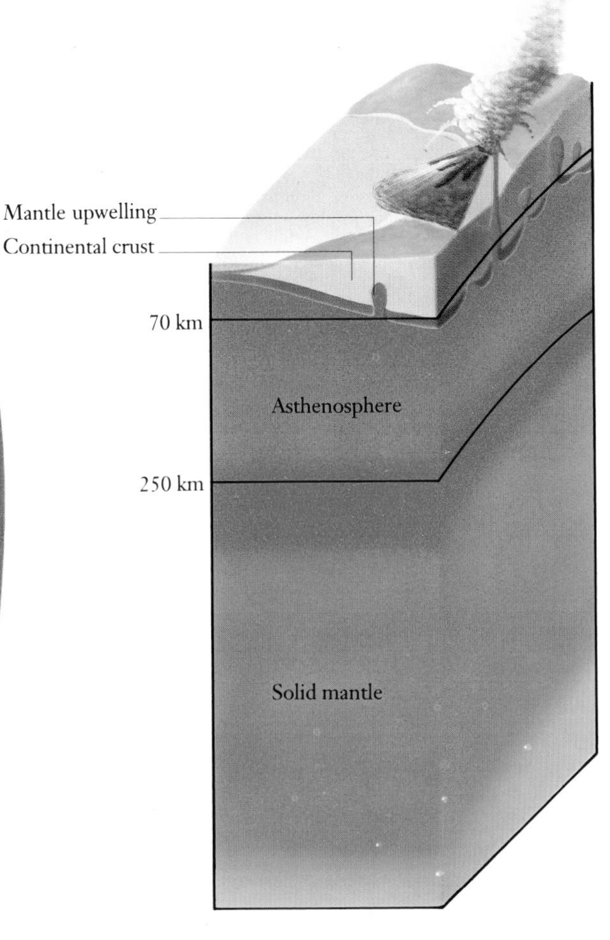

◁ **As the mostly solid mantle convects, hot material rises and cool material falls. Semi-molten material in the asthenosphere is less dense than the lithosphere on top of it. The hot, less dense material collects in localized bulges (Raleigh Taylor instabilities) and forms diapirs or "plumes" that become sources of magma.**

▽ **Vast volumes of fluid basalt lavas were extruded in the northwest British Isles between 65 and 50 million years ago – most extruded from fissures in the Earth's crust. Such lavas occur in the first stages of continental rifting at divergent plate margins. The beautiful hexagonal joints formed when the lava cooled and contracted.**

Even at the enormous temperatures and pressures found in the asthenosphere where the magmas form, the "dry" mantle material cannot actually melt. Volatile elements, however, can lower the melting point of silicate minerals by as much as 500°C, and water in this region of the mantle allows melting to occur. This process is called partial melting.

Studies of silicates such amphiboles and micas, which are sometimes brought up as blocks within magmas from the mantle, indicate that they contain significant amounts of water. More water reaches the mantle as the Earth's crust, including water-laden sediments, is recycled and returned to the mantle through tectonic processes.

Partial melts are less dense than the surrounding mantle and tend to rise as bulbous bodies called diapirs. When crystallization begins inside the diapir, the first crystals that form are more dense than the surrounding liquid, and tend to sink within it. If the deeper, crystal-laden part of a diapir is tapped, the melt that rises has a different composition from the largely liquid upper region. This process, called fractional crystallization, explains how different types of magma can arise from the same region of mantle.

VOLCANOES

VOLCANIC regions are concentrated along the boundaries of tectonic plates – vast blocks of moving crust. Where plates diverge, as in mid-ocean, magma and gas escape unnoticed from rifts in the sea floor rocks and generate new oceanic crust out of basaltic rocks. Where sub-oceanic ridges rise above sea level, volcanoes may produce new islands.

Oceanic basalt is low in silica, which means it tends to be fluid and can escape easily. A quite different magma is found in regions where two tectonic plates collide and one dives below the other, such as along the western border of the Pacific. Here, the typical magmas are more siliceous, gas-rich and viscous, and their escape is marked by violent explosive activity, as at Mount St Helens (in the United States) and Mount Pinatubo (in the Philippines). These volcanoes have the classic cone shape. They are built from lavas and pyroclastic rocks, which are composed of particles blasted out from vents. Dangerous ground-hugging mixtures of gas, lava and pyroclastic materials may sweep down the flanks of the volcanoes, burying or asphyxiating anything in their path. These *nuées ardentes* ("glowing avalanches") are deadly, as are lahars (mudflows), which are triggered by storms and torrential rains after eruptions, particularly in tropical regions.

KEYWORDS

AA
AIR-FALL DEPOSIT
ASH FLOW DEPOSIT
CALDERA
ERUPTION PLUME
LAHAR
LAVA
MAGMA
NUÉE ARDENTE
PAHOEHOE
PUMICE
PYROCLAST

Some of the largest but least dangerous volcanoes on Earth sit above what are called hot spots – uprising "plumes" of hotter-than-average mantle material. Such massive lava shields have gently sloping profiles which, in the Hawaiian Islands, rise as much as 10 kilometers above the ocean floor, and have diameters of more than 100 kilometers. Some shields on Mars and Venus are ten times larger than those on Earth.

Shield building is associated with fluid basaltic magmas which escape rapidly and in huge volumes. On Mars, where individual flows may be anything up to 300 kilometers long and 50 kilometers wide, magma may be erupted at rates as high as one million cubic meters per second. On Earth, because of plate movements, individual shields may grow for two million years before the plumes drift away from beneath them. On Mars and Venus, however, hot spots may feed the same volcanoes for hundreds of millions of years, allowing them to reach their huge dimensions. The narrow linear zones of intense volcanic and seismic activity typical of plate margins on Earth are not found on other planets.

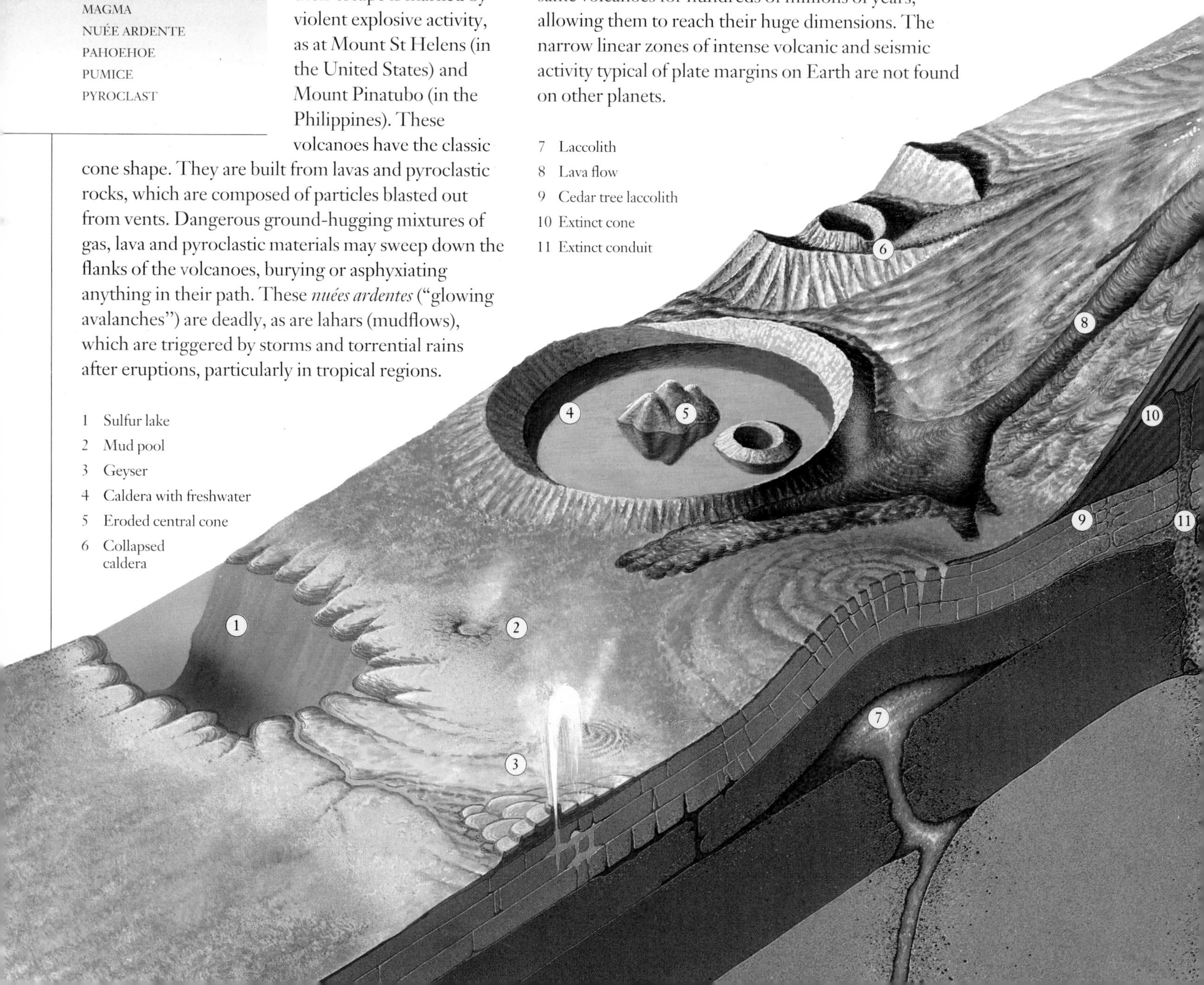

▽ **A volcano cone is built up of layers of lava (extruded magma) and pyroclastic material (fine dust, ash and blocks) cut by sills, dikes and plugs. Eruptions occur through a central vent and side vents. The blast leaves a depression which may become filled with a lake, or remain as a caldera more than 1 km in diameter.**

1 Sulfur lake
2 Mud pool
3 Geyser
4 Caldera with freshwater
5 Eroded central cone
6 Collapsed caldera
7 Laccolith
8 Lava flow
9 Cedar tree laccolith
10 Extinct cone
11 Extinct conduit
12 Central vent
13 Magma
14 Magma chamber
15 Side vent
16 Ash, cinders and volcanic bombs
17 Small lava cones

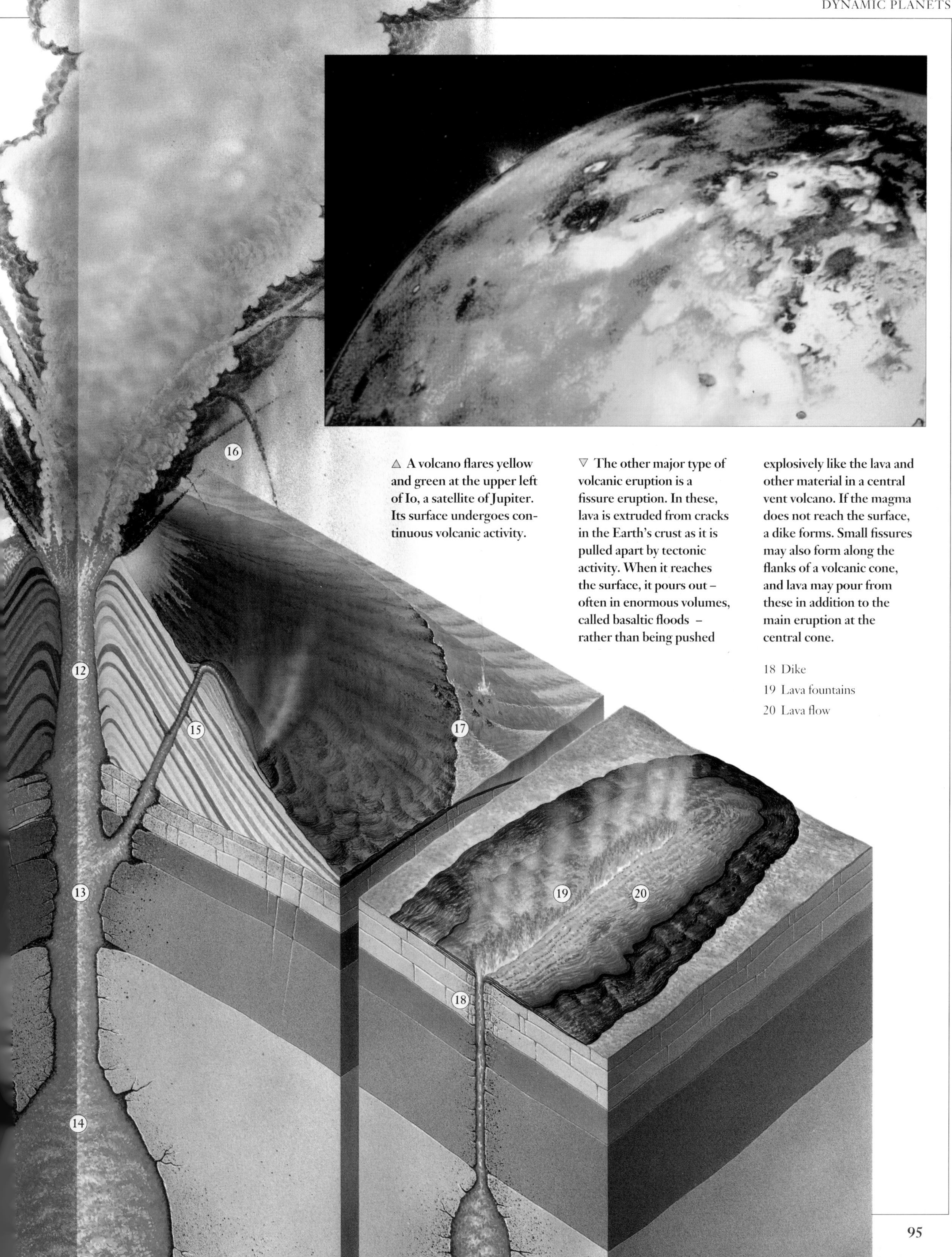

△ A volcano flares yellow and green at the upper left of Io, a satellite of Jupiter. Its surface undergoes continuous volcanic activity.

▽ The other major type of volcanic eruption is a fissure eruption. In these, lava is extruded from cracks in the Earth's crust as it is pulled apart by tectonic activity. When it reaches the surface, it pours out – often in enormous volumes, called basaltic floods – rather than being pushed explosively like the lava and other material in a central vent volcano. If the magma does not reach the surface, a dike forms. Small fissures may also form along the flanks of a volcanic cone, and lava may pour from these in addition to the main eruption at the central cone.

18 Dike
19 Lava fountains
20 Lava flow

Seismic Waves

Dynamic planets are marked by their internal activity, which is expressed at their surfaces in volcanoes and the tectonic movement of plates, and in earthquakes. This activity is significant in other ways: the passage of seismic waves through the interior enables geophysicists to probe parts of the Earth that are otherwise inaccessible. By doing so they have been able to demonstrate that our planet has a layered structure. This is true also of the other inner planets, and simple seismic studies have been made of both the Moon and Mars.

KEYWORDS

ASTHENOSPHERE
EARTHQUAKE
LITHOSPHERE
LOW-VELOCITY ZONE
L WAVES
MOHOROVICIC DISCONTINUITY
PHASE CHANGE
P WAVES
SEISMIC DISCONTINUITY
SEISMOMETER
SHADOW ZONE
S WAVES

Each time rising magma displaces the rocks of the crust, or each time brittle rocks fail, sound or "seismic" tremors are set off. They are propagated through the Earth with velocities proportional to the density of the medium through which they travel. They can be recorded on delicate instruments called seismographs and chains of these have been set up throughout the world, to record the quakes that occur daily.

Not all seismic waves are the same. P waves (primary or pressure waves) are compressional in type and move with a "push-pull" motion through solids, liquids and gases. Each molecular disturbance they effect displaces atoms over distances of around 10^{-13} meters. In contrast, S waves have a shearing motion and travel at about 60 percent of the velocity of P waves. They can be transmitted only through solids.

When an earthquake occurs, seismic waves are sent out from the focus; the point on the surface immediately above this is termed the epicenter. The intensity of the quakes is measured on the Richter scale; the largest have a magnitude of about 9. The position of the epicenter from a seismograph station is obtained by measuring the difference in arrival times of P and S waves. Records from at least three stations are needed to do this.

The internal structure of the Earth is revealed by studying the reflections and refractions of earthquake waves as they travel through it. When P- and S-wave velocities are plotted as a function of depth, a clearcut velocity decrease is seen at a depth of around 100 kilometers; this corresponds to a layer of low seismic velocities known as the asthenosphere. From this point to around 700

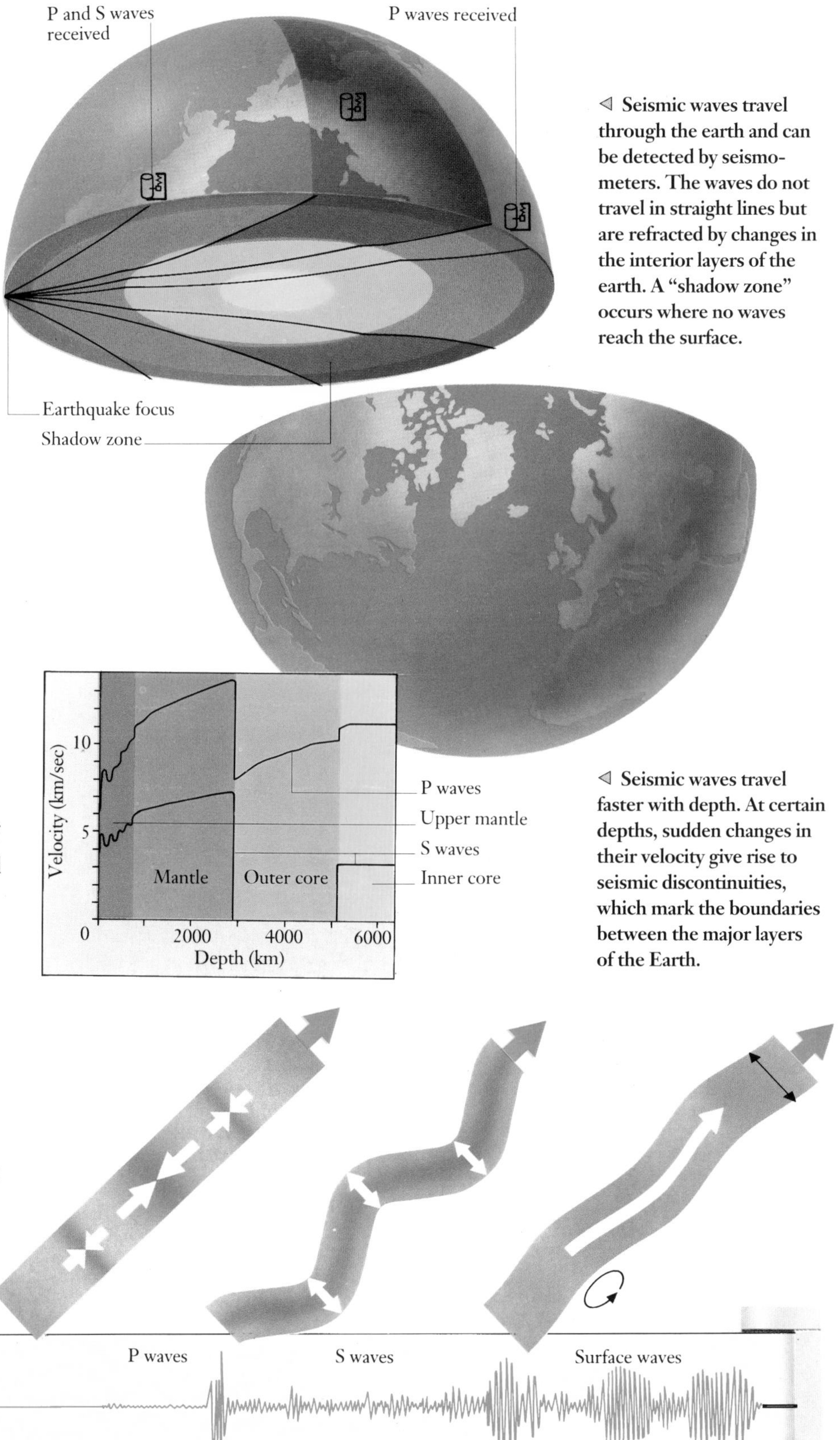

◁ **Seismic waves travel through the earth and can be detected by seismometers. The waves do not travel in straight lines but are refracted by changes in the interior layers of the earth. A "shadow zone" occurs where no waves reach the surface.**

◁ **Seismic waves travel faster with depth. At certain depths, sudden changes in their velocity give rise to seismic discontinuities, which mark the boundaries between the major layers of the Earth.**

△ **P (primary or pressure) waves have a push-pull motion and displace the atoms in solids and liquids. They travel faster than S-waves and arrive first at a seismic recording station.**

△ **S (secondary or shear) waves are produced by shear deformation which occurs perpendicular to the direction of travel. They travel at about 60% the velocity of P waves.**

△ **Both P and S waves (surface waves) may reach the surface of the Earth and travel along it. Unlike seismic shocks, nuclear explosions do not produce surface waves.**

▽ P and S waves radiate away in many different directions from an earthquake focus. The waves are refracted as they pass through materials of diffferent densities. Because they are deflected by the Earth's core, P waves do not reach the surface at certain points, giving rise to a "shadow zone".

kilometers, velocities increase but with several sharp "bumps" which represent seismic discontinuities brought about by differences in rock chemistry or structural state. Beyond this point and down to 2885 kilometers they increase again smoothly. Then S waves disappear while P waves show a large decrease in velocity. The latter creates a shadow zone between 103° and 143° from an earthquake focus. This marks the boundary between the mantle and the outer core, and is sometimes known as the Gutenberg discontinuity, after its discovery in 1909 by the German–American geologist Beno Gutenberg. The S-wave behavior indicates that the outer core itself is liquid.

The Mohorovicic discontinuity (Moho for short) lies where the velocity of P waves changes abruptly. It is found about 7 kilometers below the oceanic crust, and as much as 40–70 kilometers below the surface of the continents. It defines the boundary between the crust and the mantle, where there is a change in rock type. The dominant rock type below the Moho is peridotite; above it, the oceanic crust is mainly basaltic, whereas the continental crust is closer in composition to granite. Two further seismic discontinuities occur within the mantle, at depths of 400 and 670 kilometers. They correspond to phase changes brought about by increasing pressure which has a severe effect upon the structures of the minerals that occur there. The core itself is believed to have the same composition as iron meteorites.

The study of Moonquakes by Apollo seismographs landed during the 1970s shows that the lunar crust is about 60 kilometers thick on the nearside of the Moon, but may be as thick as 120 kilometers on the farside. The upper mantle is rigid but the lower mantle may be partly molten. The small solid iron-rich lunar core is believed to be found at a depth of approximately1500 kilometers.

▽ The Richter scale expresses the intensity of seismic events by measuring the frequency and amplitude of their surface waves. The scale, from 0 to 10, is logarithmic: each point represents an increase of a factor of 10, so that magnitude 5 is 100 times stronger than magnitude 3. A few hundred quakes of magnitude 6 or more occur every year.

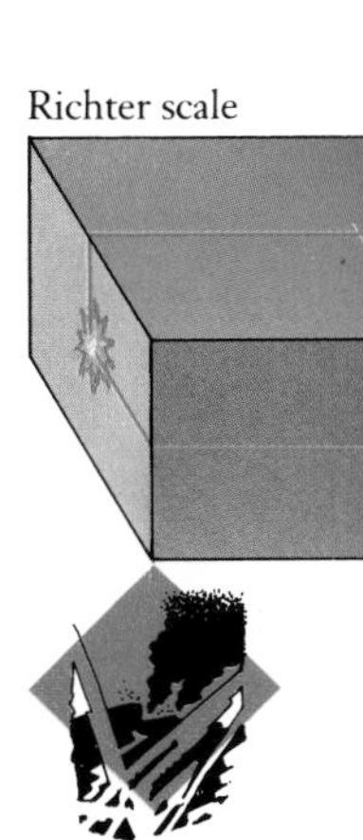

PRESENT-DAY ATMOSPHERES

THE gaseous envelope surrounding a planet forms its atmosphere. Earth's earliest atmosphere consisted of gases that escaped form the planet's interior. In decreasing order of abundance, these would have been methane, water, ammonia and hydrogen sulfide. The water molecules were split by the Sun's energy into hydrogen (which escaped to space), and oxygen, which oxidized methane to carbon dioxide. Carbon dioxide began to react with silicates, forming carbonates, and was removed from the air. If this atmospheric carbon had not thus been fixed – something which seems not to have occurred on Venus – the evolution of the Earth's atmosphere might have taken a very different course.

KEYWORDS

ATMOSPHERE
EXOSPHERE
GREENHOUSE EFFECT
IONOSPHERE
OZONE LAYER
STRATOSPHERE
THERMOSPHERE
TROPOPAUSE
TROPOSPHERE

When, about 3.5 billion years ago, bacteria evolved on Earth, they began to exploit the energy contained in carbon dioxide through photosynthesis, giving off free oxygen as a by-product. This began the process of releasing the Earth's oxygen that had been bound up in compounds.

Today, nitrogen accounts for about 80 percent by volume of the air, with oxygen making up most of the remainder. Although water vapor, carbon dioxide and ozone are present in much smaller amounts, they are vitally important, because they have the ability to absorb infrared radiation and thus affect atmospheric and surface temperature. Overall, the chemical composition of the atmosphere is remarkably uniform.

Clouds play a role in maintaining the delicate balance between solar (incoming) and thermal (outgoing) radiation levels. At any one time, approximately half of the Earth's surface is covered by cloud. In contrast, Venus has a total cloud cover; its clouds are opaque to outgoing longwave radiation, and the surface of the planet in consequence has heated up through a runaway "greenhouse effect". On Mars, on the other hand, the atmosphere is very thin and cloud cover minimal. Outgoing radiation escapes readily, and temperatures are very low.

The Earth's atmosphere occurs in four main layers, which are caused by variations in temperature and pressure resulting from the distribution of solar heating. These are three levels near ground level at which temperatures are highest: one near the surface in the troposphere, where visible and infrared radiation is absorbed; one about 50 kilometers above the surface, in the stratosphere, where ozone absorbs

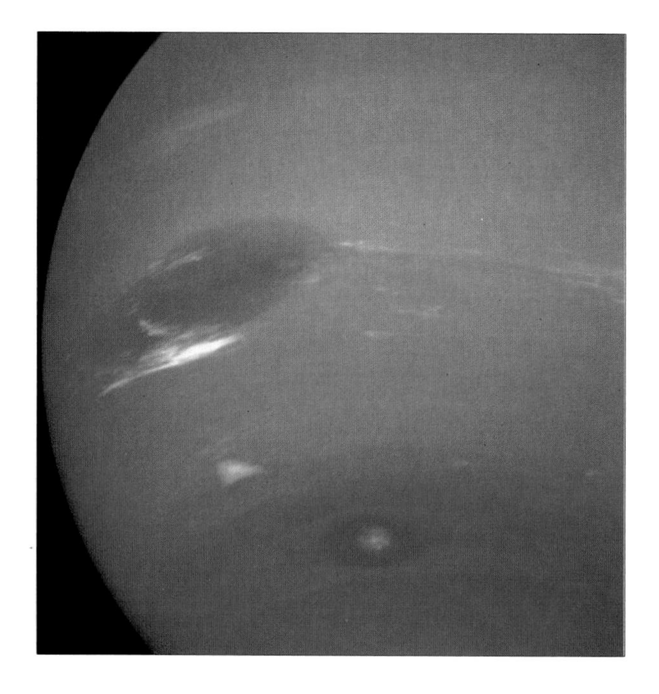

▲ Neptune and its clouds, from an image taken by Voyager 1989. Notable is the Great Dark Spot, with several white clouds which change rapidly. Cloud-top wind velocities can reach 650 km/h. Neptune's atmosphere contains hydrogen and methane, with a haze of ice crystals.

▷ The atmosphere of Venus consists almost entirely of carbon dioxide. The main cloud layer lies between 45 and 60 km above ground. In Earth, like Venus, the densest, lowest layer of atmosphere (troposphere) is heated by infrared radiation from the surface. However, nitrogen and oxygen make up 97 percent of the total air. Mars has a very tenuous carbon dioxide atmosphere, with clouds of water vapor at low levels. Jupiter's immense atmosphere is largely hydrogen and helium, with subsidiary methane. Saturn's atmosphere has a similar structure and make up to that of Jupiter but with a higher proportion of helium present.

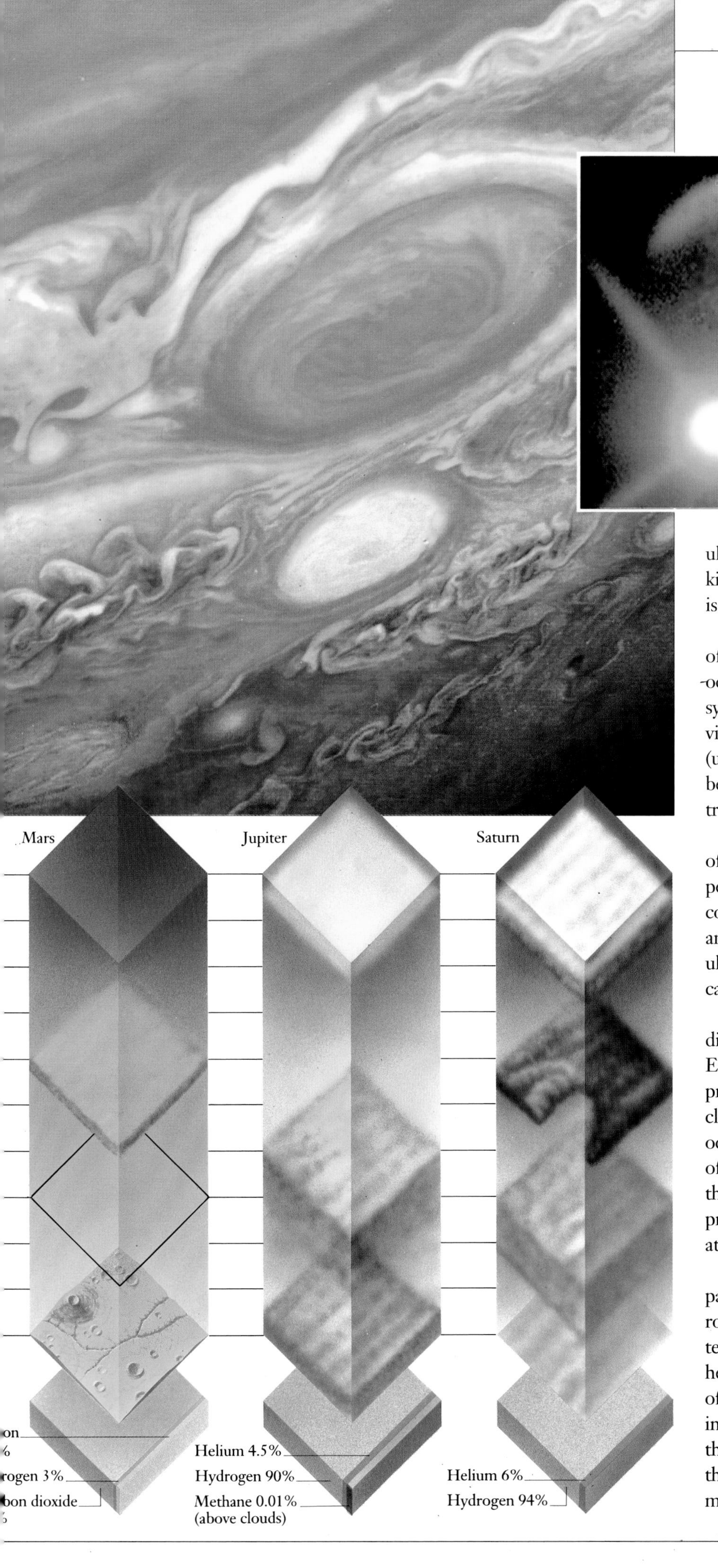

◁ **Jupiter's Great Red Spot** FAR LEFT **is the most prominent feature of the Jovian atmosphere. It measures 23,000 km across and rotates every seven days. It behaves as a free-floating body and is a large atmospheric phenomenon. The entry of the largest of the 20 fragments of the comet Shoemaker-Levy 9 into Jupiter's atmosphere in 1994** LEFT **produced a fireball almost as large as the Great Red Spot.**

ultraviolet radiation, and the third several hundred kilometers up, in the thermosphere, where ultraviolet is absorbed by photoionization processes.

Clouds and weather systems on Earth are the result of interactions between the atmosphere and the oceans. Winds are responsible for moving these systems over large distances, sometimes extremely vigorously. Most weather occurs in the troposphere (up to 11 kilometers above the surface). The boundary between this and the stratosphere is known as the tropopause.

Because there are no oceans on Mars, the dynamics of its thin atmosphere are simpler than Earth's. Ninety percent is composed of carbon dioxide, but there is a considerable amount of water locked in the polar caps and subsurface rocks. With no protective ozone layer, ultraviolet radiation breaks down water vapor and carbon dioxide.

The Venusian atmosphere is virtually all carbon dioxide and the surface pressure 90 times that on Earth. The lower layers of the atmosphere are probably fairly static, although the upper layers of cloud circulate freely. Venus may once have had an ocean, but the greenhouse effect led to the evaporation of a large amount of water, perhaps equivalent to one-third of that in the Earth's oceans. Mercury has probably never had either oceans or atmosphere. Its atmospheric pressure is so low as to be a vacuum.

The atmospheres of the outer planets are in large part composed of hydrogen compounds, and they rotate rapidly. The relatively small difference in temperature between equator and poles means that heat transfer processes are rather different from those of Earth. Unlike the other giants, Jupiter has an internal source of heat, and does not have to rely on the rather weak solar radiation to supply it with thermal energy. The thick atmosphere of Uranus is mainly hydrogen, helium and methane.

EARTH'S OCEANS

LOWLYING and filled with water, the oceans are geologically distinct from the continents on Earth, and their characteristics are the result of tectonic plate movements. Under the oceans are extensive linear submarine ridges and deep trenches, separated by flat abyssal plains. Oceanic crust forms at divergent plate margins and is destroyed along convergence zones. The plates are "recycled" quickly enough to ensure that no oceanic crust forming the floor of a modern ocean is more than 200 million years old.

KEYWORDS

ABYSSAL PLAIN
CONTINENTAL SHELF
CONTINENTAL SLOPE
GABBRO
OCEANIC RIDGE
PLATE TECTONICS
SEAMOUNT
TIDE
WAVE

The crust itself has a mean density of around 3.1 grams per cubic centimeter and is veneered with sediments. The upper 2.5 kilometers is composed of basaltic rocks. These are underlain by coarser gabbro rocks, which are 5 kilometers thick. Underneath this is a thin layer of even denser rocks, then the mantle itself.

Marine sedimentary rocks more than 3.5 billion years old prove that oceans are at least as ancient as the first continents. At that date there must already have been water-filled basins in the outer skin of the Earth. The water originated from the gases and vapors released by volcanoes. Today the oceans cover more than two-thirds of the Earth's surface area; in the past the percentage was greater because the early continents were smaller.

Seawater contains a wide range of chemical elements – primarily chloride, sulfate, sodium and magnesium, with calcium and potassium next in importance. Its salinity (33 to 38 parts per thousand) is remarkably constant over enormous areas and only differs near ice-sheets. It represents a standard solution with varying dilutions. The salts derive from the weathering of continental rocks and are transported to the oceans by rivers. The early oceans were probably less saline than today's, because the smaller continents that existed in the past (particularly during the first billion years of Solar System history) would have supplied the oceans with less salt.

Another source of salt for the oceans are hydrothermal springs, which were only recently discovered on submarine ridges by the submersible research vessel "Alvin". At such places, water passing through newly formed crust carries virtually all of the salts of iron, manganese, lithium and barium found in sea water. Furthermore, large quantities of silica and calcium originate here, as well as carbon dioxide.

▶ Underneath the oceans' waters lie the most imposing physical structures on Earth. Huge mountains (ridges) of basalt – up to 4 km high, 4000 km wide and 40,000 km long – divide the oceans into sections. Two of the most important ridges are the Mid-Atlantic Ridge and the East Pacific Rise. On the ocean floor, trenches hundreds of miles long run parallel to the edges of continents and extend to depths lower than 7 km. The trenches of the Pacific are particularly deep – one point is 11,033 m below sea level.

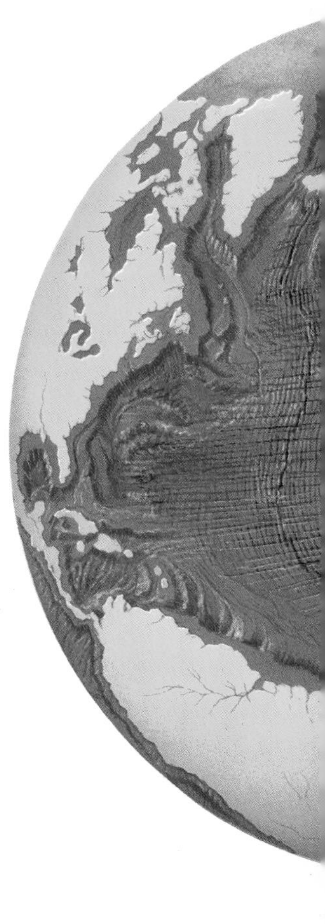

Others
Potassium
Calcium
Magnesium
Sulfate
Sodium
Chlorine

◀ The oceans have an average salt content of 3.472 percent (fresh water averages 0.3 percent). Tropical seas and enclosed seas, like the Caribbean and Mediterranean, have higher than average salinities; the heat causes seawater to evaporate faster. Salinity is lower in higher latitudes because it is diluted by higher rainfall and melting ice. Different salts arrive from rivers at different rates: sodium, for example, is supplied more slowly than dissolved aluminum.

◁ **Oceans are supplied with water by the Earth's rivers, which drain down from land to sea. Most of the planet's oceans exist beyond the edges of the continental shelves, where the water is deep and cold. Coastal waters are shallow and warm.**

▽ **The formation of ice effectively concentrates the liquid seawater beneath it, giving it a higher than normal salt content.**

The carbon dioxide content of oceans depends on an interchange of ocean water with the atmosphere. Thus, if carbon dioxide is added to the air, roughly one half of it enters the sea. Once in seawater, it exists in equilibrium with carbonic acid (H_2CO_3) and carbonate ions. In water of depths up to around 5 kilometers, carbonate tends to be precipitated; below this level it does not. This allows organisms to utilize it in the shallower waters to form their shells, without danger of depletion.

Oceanic circulation is driven by differences in the density of seawater, due to varying salinity and temperature. In general, the lower the temperature the greater the density; however, water reaches its maximum density at 4°C. Down to depths of between 100 to 200 meters the surface waters are heated by the Sun and stirred by winds and waves. Below this depth, or thermocline, there is a sharp fall in temperature; 2–4°C is typical.

The dominant factors in the present circulation pattern of Earth's oceans are the temperature and salinity differences created in the southern oceans by alternate freezing and thawing of Antarctic ice. In the past, the pattern of land and sea was different. For instance, when the supercontinent known as Pangea split in two during the Mesozoic era, an equatorial seaway opened around the world, allowing warm water to spread in both north and south latitudes.

It also appears likely that shallow seas may have existed on Mars, although it was never a truly watery planet like Earth, and that there were probably oceans on Venus; but these have long since evaporated.

The Early Continents

Earth's continents today cover about 30 percent of its surface. Made from rocks of lower density than the ocean floor, they "float" above the heavier rocks of the mantle. Continental crust varies between 20 and 90 kilometers thick, being greatest beneath major mountain ranges. The oldest reliably dated continental massifs are 3.9 billion years old. The structure of continental regions is much more complex than that of the younger oceans. Continents are oldest in their centers and younger toward their margins.

KEYWORDS

BASALT
CONTINENT
CRATON
GRANODIORITE
MOHOROVICIC DISCONTINUITY
OROGENESIS
PERIDOTITE
SIALIC CRUST
SIMATIC CRUST

Cratons, or shields, are found at the heart of most continents. They are composed of deformed metamorphic rocks intruded by granite rocks. Cratons are the remnants of ancient mountain chains. They are surrounded by stable platforms where a thick layer of horizontal sedimentary rocks has accumulated on top of cratonic rocks. Adjacent to stable platforms are younger orogenic or mountain-building belts: linear zones of compressed fold mountains produced by the collision between two continental plates, or between a continental and an ocean plate, as in the Andes mountains in South America.

The continents grew in stages rather than all at once. About 10 percent of the continental crust was produced in the Archean between 3.8 and 3.5 billion years ago; a further 60 percent also between 2.9 and 2.6 billion years ago; and the remaining 30 percent during major continent-building phases in the later Proterozoic era (1.9 to 1.7 and 1.19 to 0.9 billion years

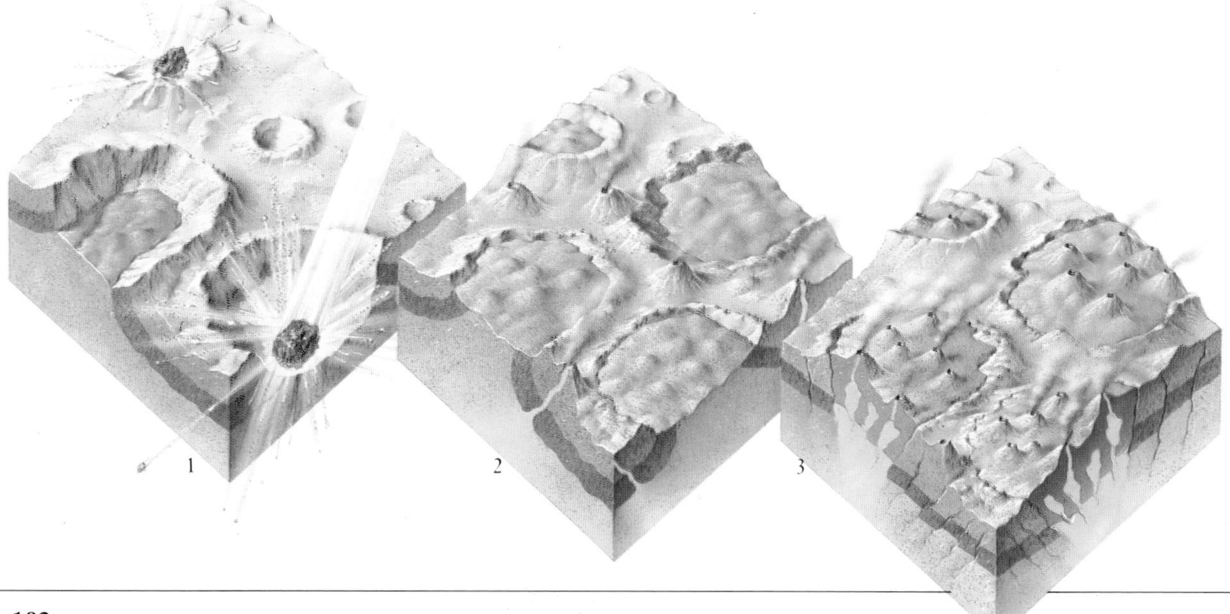

■ **The Pilbara block of Western Australia** ABOVE **shows how early continents may have formed. Cratons (red and orange) arch up the volcanic and sedimentary rocks of the greenstone belt (green) to make nearly vertical bedding.** LEFT **One theory of original continental growth begins with meteorites crashing into the newly-formed Earth 1, piercing the crust and causing magma to flow out 2. Igneous rock that formed in this region differed from that around it 3.**

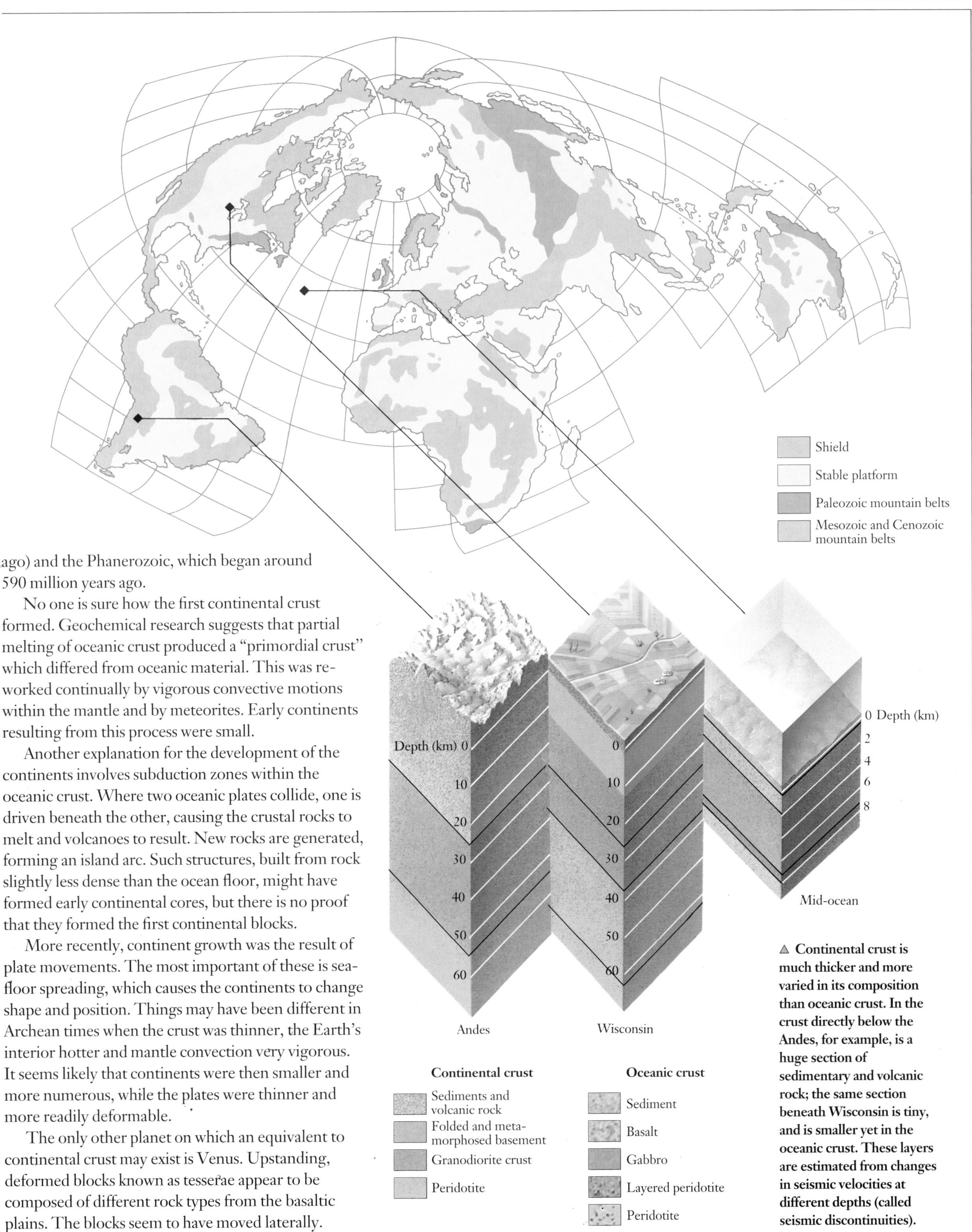

△ Continental crust is much thicker and more varied in its composition than oceanic crust. In the crust directly below the Andes, for example, is a huge section of sedimentary and volcanic rock; the same section beneath Wisconsin is tiny, and is smaller yet in the oceanic crust. These layers are estimated from changes in seismic velocities at different depths (called seismic discontinuities).

ago) and the Phanerozoic, which began around 590 million years ago.

No one is sure how the first continental crust formed. Geochemical research suggests that partial melting of oceanic crust produced a "primordial crust" which differed from oceanic material. This was reworked continually by vigorous convective motions within the mantle and by meteorites. Early continents resulting from this process were small.

Another explanation for the development of the continents involves subduction zones within the oceanic crust. Where two oceanic plates collide, one is driven beneath the other, causing the crustal rocks to melt and volcanoes to result. New rocks are generated, forming an island arc. Such structures, built from rock slightly less dense than the ocean floor, might have formed early continental cores, but there is no proof that they formed the first continental blocks.

More recently, continent growth was the result of plate movements. The most important of these is seafloor spreading, which causes the continents to change shape and position. Things may have been different in Archean times when the crust was thinner, the Earth's interior hotter and mantle convection very vigorous. It seems likely that continents were then smaller and more numerous, while the plates were thinner and more readily deformable.

The only other planet on which an equivalent to continental crust may exist is Venus. Upstanding, deformed blocks known as tesserae appear to be composed of different rock types from the basaltic plains. The blocks seem to have moved laterally.

The Ice Ages

At various times during its history, the Earth has been subject to glaciations, when vast layers of ice spread from the poles over the land and oceans. Such events have left their marks in the rocks of every continent and have given vital clues to geologists trying to understand the Earth's history.

During a glacial, it is not continuously cold, and the cold spells are punctuated by interglacials during which temperatures are far higher. Today we are living in an interglacial, toward the close of the Pleistocene ice age: this began some10 million years ago and its ice sheets retreated to their present position about 10,000 years ago.

KEYWORDS

ICE AGE
INTERGLACIAL
MORAINE
PANGEA
PRECESSION
RADIOMETRIC DATING

The earliest known glacial deposits are found near Lake Huron in Canada. Three layers of glacial deposits, dated between 2.7 and 1.8 billions years old, cover an area of 120,000 square kilometers. The glacial sediments are separated by interglacial deposits that formed during intervening milder spells. Typical rocks showing the features of glaciation – tillites and moraine deposits – of similar type and age are also found in northern Australia and in southeast Africa.

No further evidence for large-scale glaciation is then found until 940 million years ago; subsequent glacials occurred about 770 and 615 million years ago. After the Precambrian period, prominent glacial periods are known to have occurred during the late Ordovician and Permo-Carboniferous ages; then there was a long gap until the Pleistocene ice age.

Glacial rocks are preserved within rock sequences on continents such as Australia and in northern Africa – both currently enjoying dry, hot climates in low latitudes. Evidently the position of these continents has changed. For instance, the Permo-Carboniferous glaciation affected the whole of the huge "super-continent" known as Pangea, which existed at that time. Subsequently Pangea split into Gondwanaland (the southern continents) and Laurasia (the northern), and was later further fragmented into the present-day continents. The record of past glaciations allows geologists to establish how the continents have moved around with respect to one another and to the poles.

Exactly what causes a glaciation is not yet clear. Once it was thought that variations in the Sun's energy output were responsible; however, little is known

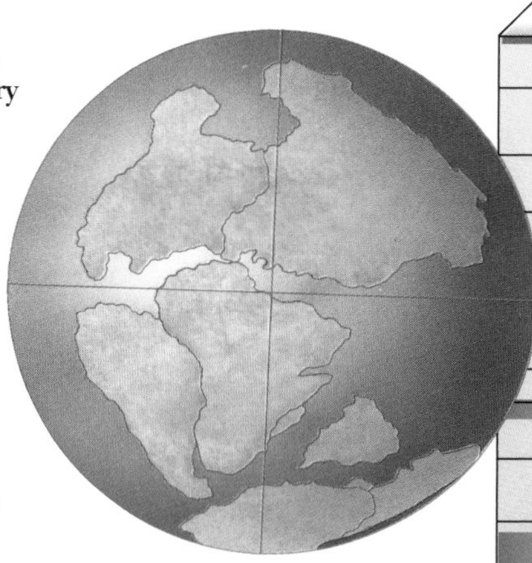

▷ **During the Permo-Carboniferous glaciation, the continents were in very different positions to today. This was shown by the discovery of glacial remains over Africa up to the Equator, southern India, the southern portion of Australia and the eastern regions of south America. All were then near the South Pole, and part of Pangea.**

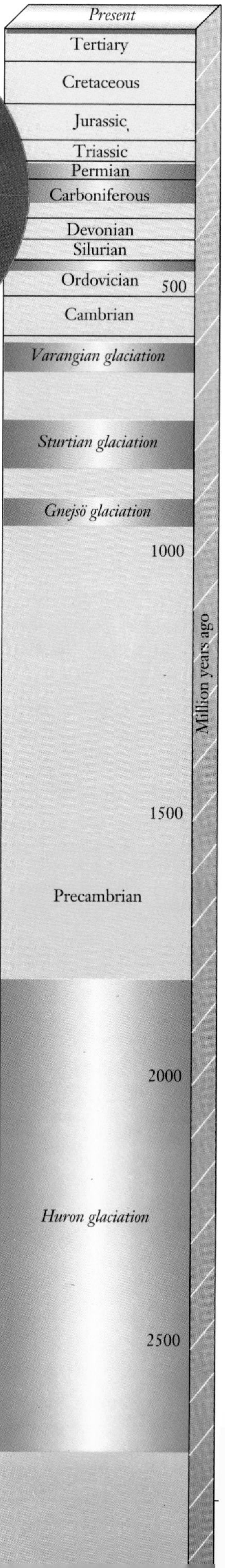

Present
Tertiary
Cretaceous
Jurassic
Triassic
Permian
Carboniferous
Devonian
Silurian
Ordovician 500
Cambrian
Varangian glaciation
Sturtian glaciation
Gnejsö glaciation
1000
1500
Precambrian
2000
Huron glaciation
2500
Million years ago

▷ **Ices ages are relatively rare in the Earth's history. They have been identified by the discovery of features of glaciation, including the fossilized beds of material transported by glaciers, known as tillites, and moraine, or deposits of boulders and other material deposited by the edges or ends of glaciers.**

◁ **Today glaciation is focused at the poles, where huge icebergs originate, and in mountainous areas such as the Himalayas and Alps. During much of the Earth's history, even the polar regions were free from ice, being heated by warm sea currents.**

about this. More likely are effects due to precession or changes in the Earth's axis. Over long periods, changes occur in both the Earth's orbit and axial inclination. Precessional cycles occur every 26,000 years, 40,000 years and 100,000 years. The English astronomer John Herschel first proposed precessional effects as an explanation in 1830. His ideas were refined in the 1930s by the Yugoslav Milutin Milankovitch.

Some support for such an idea comes from rock samples taken from deep ocean sediments and underlying crust, and also from the planet Mars. Precessional effects on Mars are much more significant than on Earth and might be expected to induce marked climatic variations. In the past, Mars enjoyed a milder climate during which there was running water and standing seas. Precession could explain the change.

Other possible causes for global changes of temperature include natural phenomena such as large volcanoes. These have the capacity to lower global temperatures by several degrees, as the clouds of ash and gas they eject absorb the Sun's heat.

△ Mars has icecaps which vary in size according to the seasons, and virtually disappear during the Martian summer. Here the southern polar icecap is clearly visible to the bottom of the photograph. The Martian icecaps comprise a mixture of water ice and carbon dioxide ice.

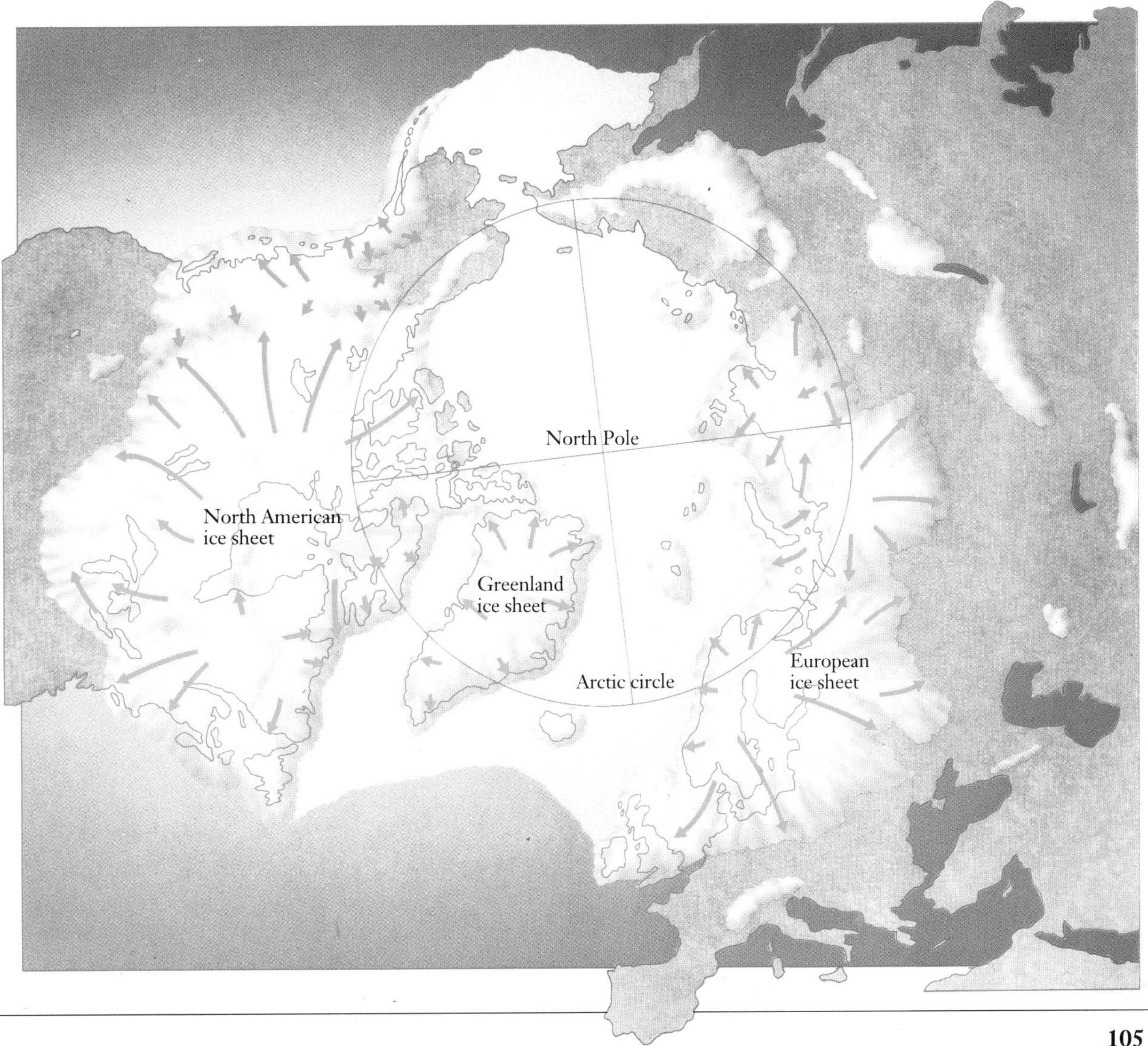

▷ At the height of the recent ice age, about 18,000 years ago, continental ice sheets covered much of modern North America, Europe and Asia. The glaciers originated in the mountainous regions, and advanced into lowland areas, their extent varying with the temperature. Land forms typical of glaciation include U-shaped valleys and fjords, low hills known as drumlins and gravel ridges or eskers.

5

GEOLOGICAL *Jigsaws*

THE EARTH HAS EXISTED for about 4.6 billion years. During this time, the patterns of land and sea have changed across the entire planet. New crustal rocks have been continually created; some have been on the surface for a very long time, others less so; still others have been destroyed and recycled.

A quick glance at a modern map of the world shows that if the continents were cut out like the pieces in a huge jigsaw, many could be roughly fitted together. This was first remarked upon as long ago as 1620. In 1912 the Austrian meteorologist Alfred Wegener published a book in which he noted similarities in ancient fossil remains between the rocks of western Africa and eastern South America. This, he argued, could not be mere concidence; might it not mean that the two continents were once joined?

This idea was not taken seriously at first. However, in the later 1950s and early 1960s, new techniques in geochemistry and geophysics – as well as the traditional methods of paleontology (the study of fossils) and stratigraphy (the study of rock layers) – produced a breakthrough. It became evident that Wegener had been right, and the continents had in fact drifted. A further new understanding was that Earth's lithosphere is segmented and that the individual pieces, or plates, are in constant motion.

The divides between the Earth's continents are created by the movement of those continents on rigid plates, which are moving apart in some regions and colliding in others. The Red Sea and the Gulf of Aqaba to the right, which separate the African plate from the Arabian, are northern extensions of the East African Rift Valley. The Arabian plate, an ancient continental shield, was originally attached to Africa but is moving northeastward.

Mobile and Stable Zones

Most of the Earth's crust is geologically stable most of the time. Intense geological activity is confined to narrow linear zones, called mobile zones, that correspond to plate margins; volcanoes, earthquakes and mountain-building occur here. Between them are extensive, relatively flat, stable zones.

Each of the stable continental regions is built from several components. Thus, extensive regions of interior Australia and North America are quite flat and have remained essentially undisturbed since Precambrian times – from the Earth's formation more than four billion years ago. The ancient cratonic core of Australia underlies the central and western part of the continent. Its components are separated by belts of past mountain-building activity. Sedimentary rocks covering the craton provide evidence for almost continuous undisturbed sedimentation for more than 1.5 billion years in this region. This is the hallmark of a stable zone.

KEYWORDS

CRATON
FOLD BELT
HEAT FLOW
IGNEOUS ROCK
METAMORPHISM
MOBILE BELT
OPHIOLITE
OROGENESIS
PRINCIPAL STRESS
TECTONICS
TRANSCURRENT FAULT
VOLCANISM

Volcanoes and earthquakes disrupt the crust on a local scale but form a part of a much more widespread phenomenon – orogenesis (mountain building). Terrestrial fold mountains are formed by complex processes that occur at the margins of plates in collision (destructive plate margins). Oceanic crust and its veneer of sediments are subducted or driven down into the mantle and buried. They then heat up, melt, deform and undergo metamorphism, leading to the eventual rise of new mountain chains from the sea.

Mobile zones are separated from one another by stable ones. An overview of the Earth reveals that orogenic belts form the boundaries of the continents and have been periodically accreted to continental cores. The history of mobile belts is cyclic. Periods of relative calm are offset by periods of mountain-building that have changed the appearance of the face of the Earth.

Mountain-building involves volcanism, magmatism and seismic activity – collectively called igneous activity. Because it involves all these processes, the history of an individual orogenic belt may be very complex. There have been peaks in igneous activity at certain times – 2.8–2.6 billion years ago, 1.9–1.6 billion years ago, 1.1 billion to 900 million years ago, and about 500 million years ago – suggesting that the Earth's heat engine needs to build up energy before another cycle can begin.

Divergent plate margins, where two plates move away from each other, also are active zones. Material from the mantle wells up to approach the surface beneath oceanic ridges; major rifting, hydration (swelling in minerals that have taken water into their structure) and eruption of lavas occurs. Magma is intruded into the subcrust as dikes and sheets. Recent evidence from the Magellan spacecraft indicates that similar activity has taken place on Venus, although along broader zones.

▼ The crust beneath the oceans is punctured by long ridges, where new crust is being created. Magnetic alignments "frozen" in rock across mid-oceanic ridges and radiometric dating both reveal that the age of the sea-floor rocks increases away from ridge axes. As the oceanic crust moves over mantle hotspots, chains of volcanic islands may form, and where it encounters the continents it is driven downward (subducted), to be recycled through the mantle. The subduction zones may involve regions of complex mountain-building, which eventually cause the area of the continental crust to extend. All these "crustal plate" movements are described by the theory of plate tectonics.

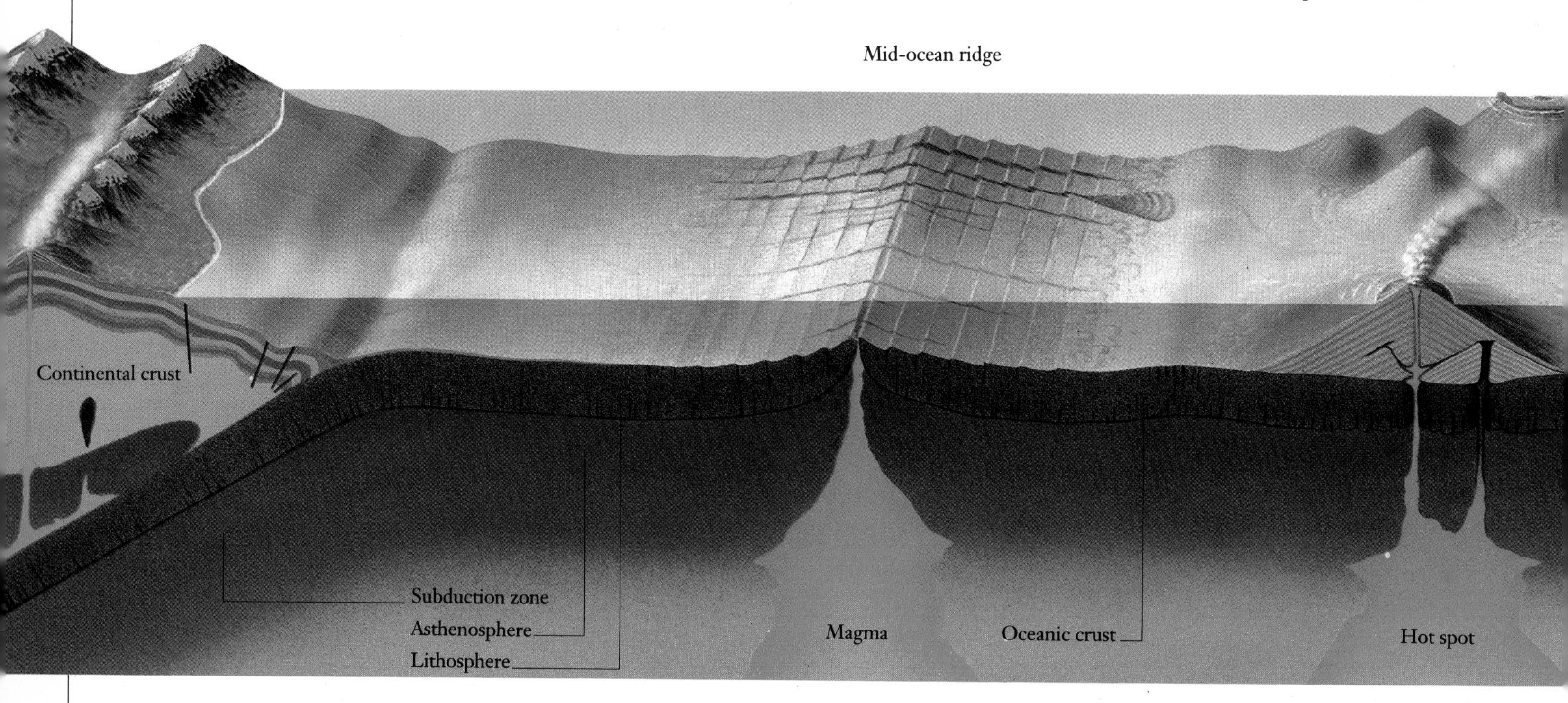

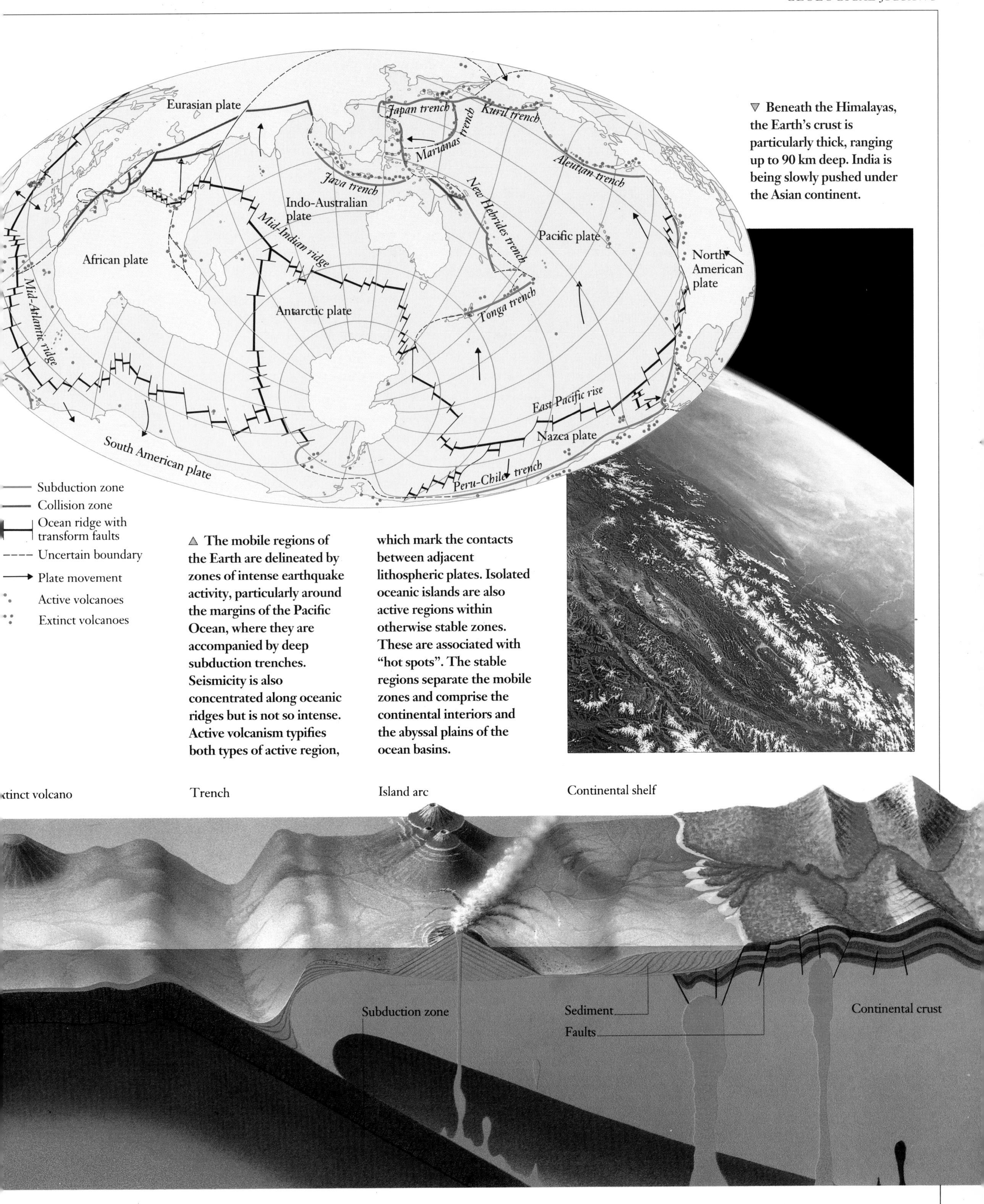

▽ **Beneath the Himalayas, the Earth's crust is particularly thick, ranging up to 90 km deep. India is being slowly pushed under the Asian continent.**

△ **The mobile regions of the Earth are delineated by zones of intense earthquake activity, particularly around the margins of the Pacific Ocean, where they are accompanied by deep subduction trenches. Seismicity is also concentrated along oceanic ridges but is not so intense. Active volcanism typifies both types of active region, which mark the contacts between adjacent lithospheric plates. Isolated oceanic islands are also active regions within otherwise stable zones. These are associated with "hot spots". The stable regions separate the mobile zones and comprise the continental interiors and the abyssal plains of the ocean basins.**

WANDERING CONTINENTS

THERE is good evidence for continental drift – the theory of the Earth's development that presents the modern continents as interlocking components of an ancient supercontinent that began to break up about 200 million years ago. One such piece of evidence is the structure of the Saharan shield in Africa. This 2-billion-year-old craton has a strong north–south grain in its interior, but this swings toward an east–west trend along the Atlantic margin. There is a well-defined junction between the ancient rocks and younger ones; this strikes into the ocean off the coast of Ghana. The geological features along the eastern coast of South America reveal almost identical relationships in Brazil. From these, it is clear that the two continents were once joined and have drifted apart. Similar evidence occurs on other continents.

KEYWORDS

FIELD REVERSAL
CONTINENTAL DRIFT
GLOSSOPTERIS FLORA
MAGNETIC STRIPING
PALEOPOLE
POLAR WANDERING

Then there is paleontological evidence. Fossil remains collected from Africa and Greenland indicate that, during Silurian times (430 million years ago), Africa was in the grip of glaciation (cold temperatures and extended ice cover), whereas Greenland had a tropical climate. Each must have changed latitude – quite dramatically. Again, evidence of similar changes of climatic conditions is found on other continents.

One of the most convincing arguments for continental drift is from paleomagnetism. It is known that the Earth's magnetic polarity varies and sometimes reverses. Magnetic minerals trapped inside rocks take on the magnetic polarity of their era. Geophysicists can use this phenomenon to determine their paleolatitude by using simple trigonometry. Once they have this information, they can establish the past magnetic orientation of any continent. Plots of paleopole positions for older and older rocks from any one continent define a smooth curve (called a polar wandering curve) which leads away from the present pole. One possible interpretation is that the position of the magnetic pole has changed. However, wandering curves for different continents over similar periods of time do not coincide. This suggests that it is not the magnetic pole but the continents themselves that have moved.

Attempting to reconstruct the positions of continents at the start of the Paleozoic era is not easy. Nevertheless, most geologists agree that North America and Greenland should be put back together and placed alongside western Europe. This part of the jigsaw is called Laurasia. At the same time, Africa has to be placed alongside South America. Australasia, India and Antarctica can be convincingly pieced together on the basis of fossil, structural and paleomagnetic data during the Early Paleozoic – somewhat sooner than the pieces of Laurasia. By the late Paleozoic (around 200 million years ago), however, Laurasia and the southern supercontinent, Gondwanaland, were united and formed a single vast continent, known as Pangea. At this time the eastern parts of Pangea were separated by a great ocean, known as Tethys, which was to remain a feature of the Earth for many millions of years.

During Triassic times (220 million years ago), the north magnetic pole lay in Alaska, while the southern pole was situated just off the coast of Antarctica. Then, between 160 and 120 million years ago, the supercontinent gradually split apart as new oceans began to form between the Americas, Africa and India, and Africa and South America. By about 80 million years ago, Australia and New Zealand – previously joined – became separated. It was not until 40 million years ago, however, that Australia finally split itself from Antarctica and drifted away from the pole.

▽ At the start of Paleozoic times there was a single supercontinent called Pangea, which stretched from pole to pole. A single ocean, Panthalassa, encircled it. By the Carboniferous (350 million years ago), a southern supercontinent, Gondwanaland (ancestral Antarctica, Australasia, South America, India and Africa), had moved over the South Pole, while ancestral China, Laurasia and Siberia formed separate northern continents. Laurasia included pieces of what eventually formed North America. By the Permian (250 million years ago), the continents had come together again, reforming Pangea. The Tethys Sea separated the northern and southern parts, which opened to the east. In the Cenozoic, drift of the continents had occurred again. Gondwanaland separated from Laurasia, and both split up. Today's configuration is as shown.

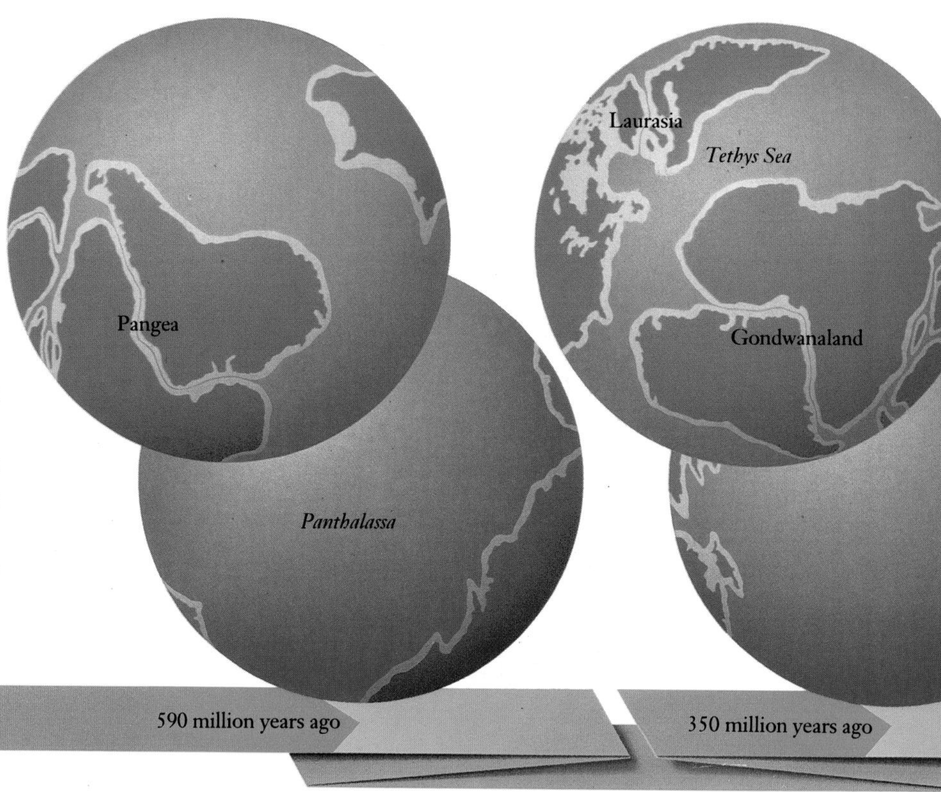

◁ Continental shields (central regions) have been stable for hundreds of millions of years. But they have moved large distances due to continental drift. Ayers Rock (Uluru), in the "red center" of Australia, is a remnant of Precambrian glacial deposits formed when Australia was close to the South Pole.

▽ Glossopteris seedfern were buried in rock on the supercontinent Gondwanaland. Their fossils are now found scattered throughout the southern continents.

PLATES AND PLUMES

NEW crust is continually generated at mid-ocean ridges and this causes the Earth's continents to move around. However, the Earth is neither expanding nor contracting; therefore, if a balance is to be kept, somewhere crust must be being destroyed. The theory of plate tectonics was developed to explain how this might come about and how the continents drift around the globe.

KEYWORDS

COLLISION ZONE
CONTINENTAL CRUST
CONTINENTAL DRIFT
CONVERGENT MARGIN
DIVERGENT PLATE MARGIN
HOT SPOT
PLATE TECTONICS
PLUME
SUBDUCTION ZONE

In the early 1900s, the idea of continental drift was viewed with skepticism, because no one could imagine how the mantle (which was proved by seismic waves to be solid) might carry slabs of lithosphere from place to place. Then the German-American geophysicist Beno Gutenberg (1889–1960) demonstrated that the mantle, although of extreme viscosity, could still have convective currents within it. This was a key step forward.

Scientists began to probe the ocean floors in the mid-20th century, using geophysical techniques and

△ Coronas such as Aine Corona on Venus are believed to develop above mantle plumes and are characterized by both concentric and radial fracturing. Volcanic activity is typically associated with them, either in the form of lava flows or fields of small shield volcanoes. Many coronas have a central rise which is surrounded by a depressed moat.

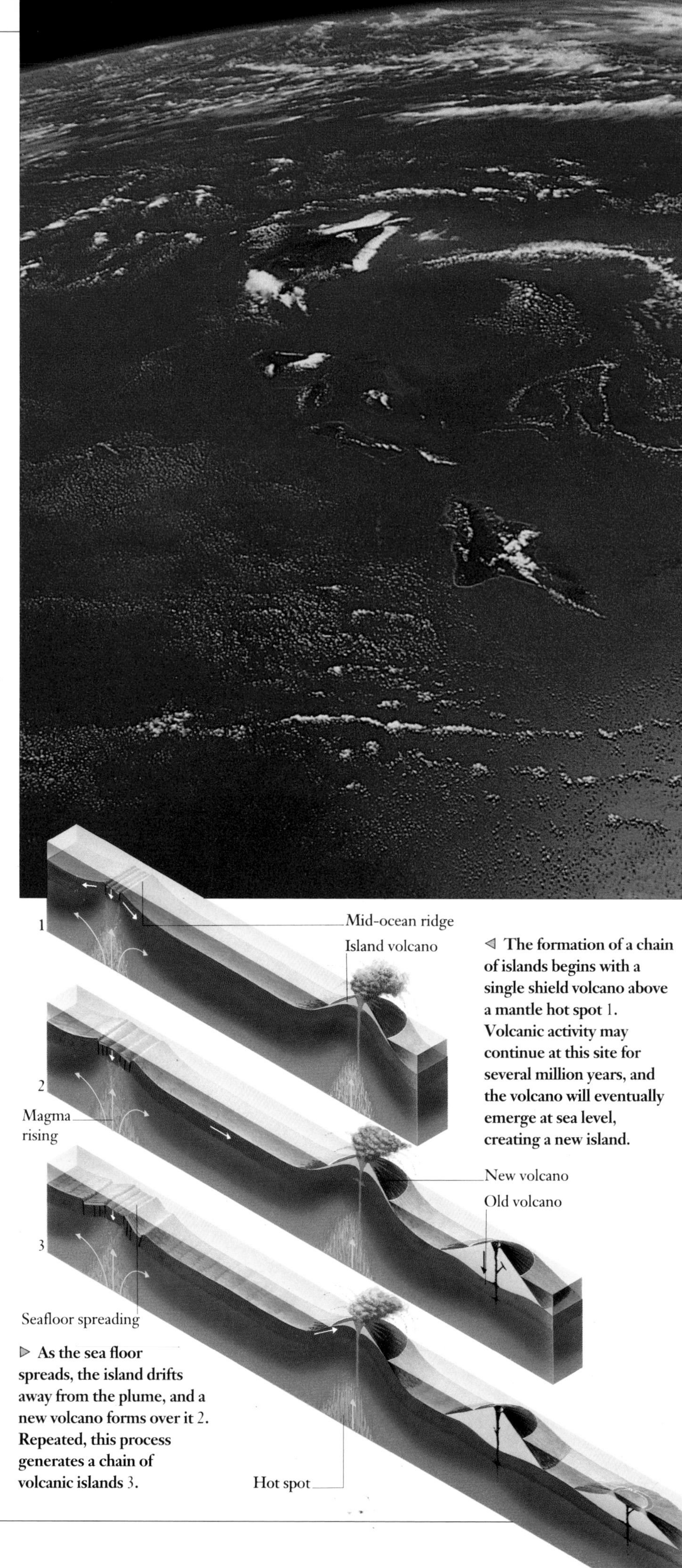

◁ The formation of a chain of islands begins with a single shield volcano above a mantle hot spot 1. Volcanic activity may continue at this site for several million years, and the volcano will eventually emerge at sea level, creating a new island.

▷ As the sea floor spreads, the island drifts away from the plume, and a new volcano forms over it 2. Repeated, this process generates a chain of volcanic islands 3.

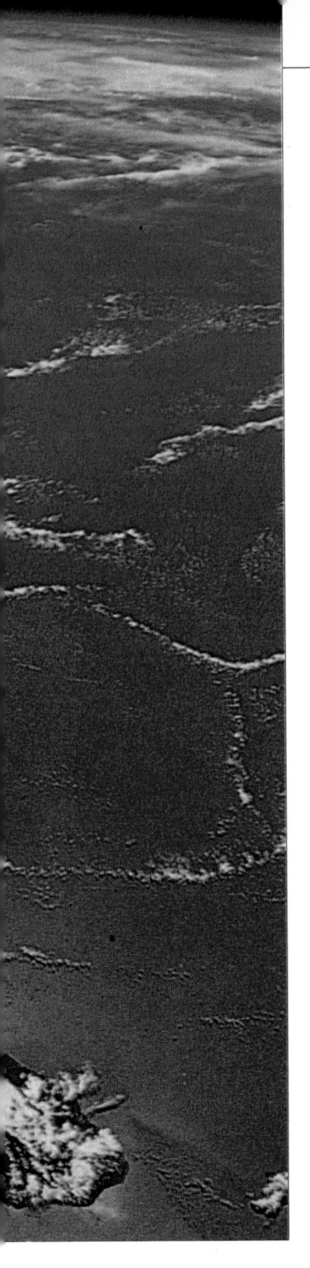

deep-sea coring and sampling. As a result, vital pieces of geological evidence came to light to show that the ocean floors, like the continents, were in motion. It could then be shown that the lithosphere of the Earth is composed of seven very large and a number of smaller plates which behave as if they were rigid and which are pushed around by movements within the mantle that are transmitted along the asthenosphere.

The mantle was considered to be in convective motion, with upwelling mantle found along the lines in mid-ocean, known as spreading axes. The hot mantle generates magma which is lighter than its surroundings and so rises toward the surface. As it does so it cools, crystallizes and moves sideways away from spreading axes. The cooling induces contraction, with the result that the spreading axes form ridges above the rest of the ocean floor, which subsides.

Destruction of the lithosphere takes place along long, narrow zones called subduction zones. Here, spreading lithosphere plunges down, at angles of about 45°, beneath an opposing plate, to be heated, melted and recycled in the Earth's interior. Sometimes, however, two continental plates converge. In this case the two may be thrust together to buckle and form mountain ranges, rather than subducted. India and Asia have done this along the line of the Himalayas, thrusting together slices of brittle crust.

Much more localized mantle upwellings can occur. These are called plumes or "hot spots". A long-lived plume lies beneath the Hawaiian Islands; others lie beneath the African continent. Plumes of varying size are the principal points of heat flow on Venus, which appears to lack plate boundaries like those on Earth. The larger Venusian plumes apparently have been active over lengthy periods, generating volcanic rises – regional-scale volcanic massifs often associated with crustal stretching. Smaller plumes produced circular structures known as coronas. Long-lived plume activity has also characterized Mars, where massive shield volcanoes have grown above major plume upwellings.

Earth appears to be the only planet that has more than one tectonic plate. Mars and Mercury lack any structures resulting from plate tectonics, as does the Moon; they are considered tectonically dead.

▽ As the Indian plate approached the Eurasian plate, the sedimentary rocks of the ocean crust were compressed and thrust up to form the rugged Himalaya mountain range.

Sedimentary rocks from old oceanic crust
Himalayan mountains formed
Indian plate
Eurasian plate

△ This photograph of the Hawaii island chain shows Hawaii in the distance, and Kauai in the bottom right. The Hawaiian chain is oldest at Midway (27 million years old) and is still active on Hawaii itself, and beneath the sea to the east.

Eurasian plate
Himalayas
Indian plate

▷ Collision between India and Asia began about 40 million years ago. Because the colliding plates were composed of buoyant continental rocks, subduction was inhibited. Huge slices of fractured strata, scraped off the advancing edge of the Indian plate, were stacked one above the other. Continued movement of the advancing plate then moved the individual slices many hundreds of kilometers. This tectonic activity formed the Himalayas. Elsewhere, faulting absorbed the compression, giving rise to features such as the Tibetan Plateau.

BENEATH THE OCEAN FLOOR

SINCE the late 1930s, new techniques have opened up the field of submarine geology. Gravity measurements and geotectonic imagery – in which very accurate measurements of the height of the sea surface allow the bottom structure of the oceans to be mapped – have greatly increased our understanding. The ocean floor, far from being smooth and flat, is crossed by enormous mountain ranges which rise 2–3 kilometers above the general level of the sea floor and form part of a global network which extends for more than 80,000 kilometers. These are the mid-oceanic ridges. In places such as Iceland, Ascension and the Galapagos Islands, ridges rise above sea level. The ocean floor is also cut by deep trenches which mark subduction zones and punctuated by isolated seamounts.

The discovery of what mid-oceanic ridge systems represented – the sites of crust formation, or constructive plate margins – was a major breakthrough in earth science. Basaltic volcanism – the upwelling of magma consisting mainly of basalt – characterizes oceanic ridges. Convective movements within the mantle force the overlying lithosphere move apart, allowing hot magma to reach the sea floor. At ridge crests, a zone of rifting separates regions of sea floor which are moving apart at 2–15 centimeters per year. Because the oceanic crust cannot withstand sufficient stress to allow for variations in spreading rate and changes in convection pattern, oceanic ridges consist of straight sections offset by transform faults, along which different sections of a plate slide past each other.

One of the key pieces of information came from paleomagnetic studies along the Mid-Atlantic Ridge. It was found that only half the rocks on each side of the

KEYWORDS

ABYSSAL PLAIN
BLACK SMOKER
HOT SPOT
LAYERED INTRUSION
MANGANESE NODULE
MANTLE NODULE
METASOMISM
OCEANIC RIDGE
PILLOW LAVA
PLUME
POLARITY REVERSAL
SEA-FLOOR SPREADING
SUBDUCTION ZONE
TRANSFORM FAULT

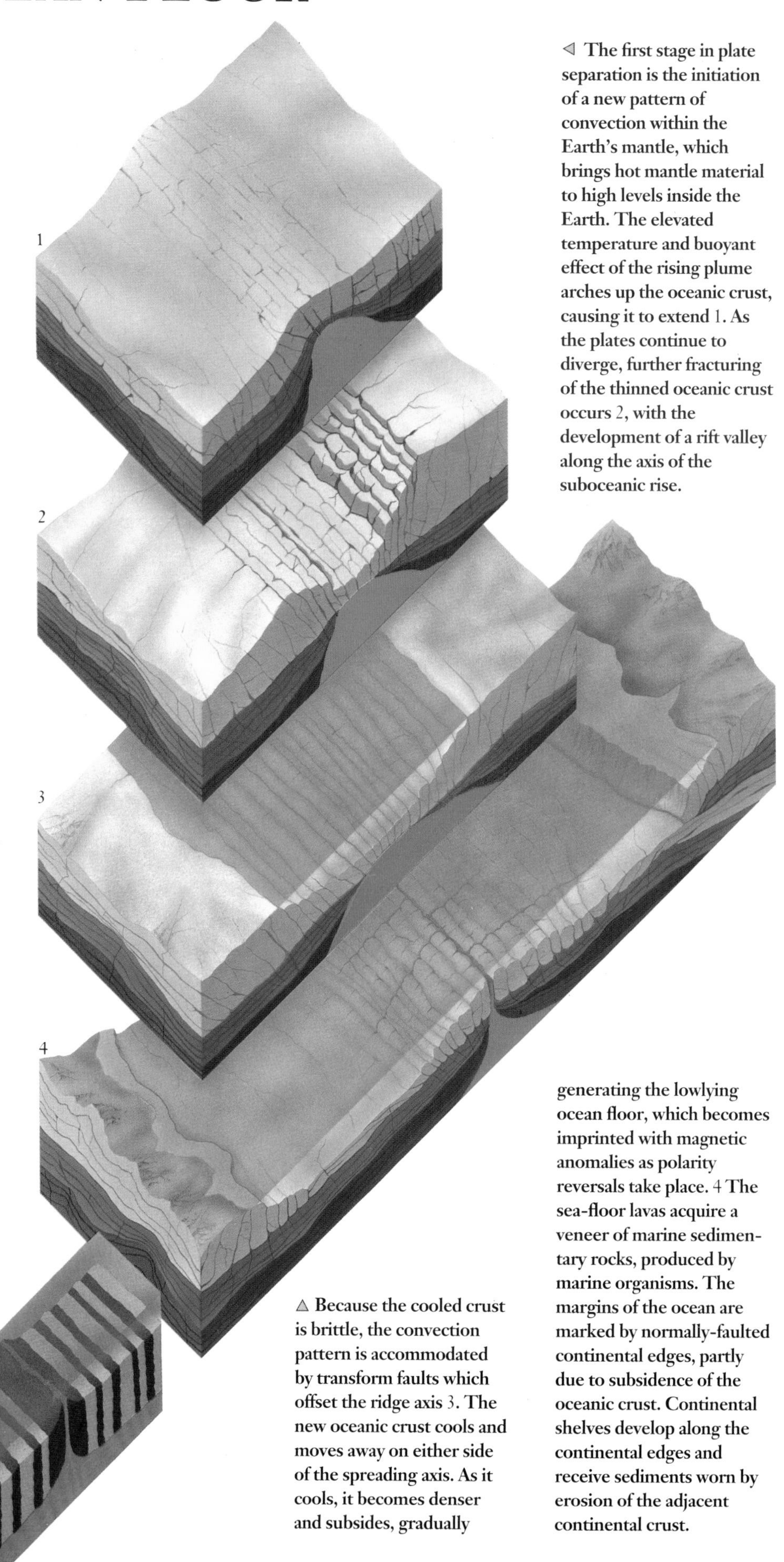

◁ **The first stage in plate separation is the initiation of a new pattern of convection within the Earth's mantle, which brings hot mantle material to high levels inside the Earth. The elevated temperature and buoyant effect of the rising plume arches up the oceanic crust, causing it to extend 1. As the plates continue to diverge, further fracturing of the thinned oceanic crust occurs 2, with the development of a rift valley along the axis of the suboceanic rise.**

△ **Because the cooled crust is brittle, the convection pattern is accommodated by transform faults which offset the ridge axis 3. The new oceanic crust cools and moves away on either side of the spreading axis. As it cools, it becomes denser and subsides, gradually generating the lowlying ocean floor, which becomes imprinted with magnetic anomalies as polarity reversals take place. 4 The sea-floor lavas acquire a veneer of marine sedimentary rocks, produced by marine organisms. The margins of the ocean are marked by normally-faulted continental edges, partly due to subsidence of the oceanic crust. Continental shelves develop along the continental edges and receive sediments worn by erosion of the adjacent continental crust.**

▷ **Magnetic anomalies in ocean-floor lavas reveal polarity reversals. The alternating pattern of normal and reversed polarity rocks is produced as successive belts of lava are extruded at the site of a a divergent plate margin. At the mid-oceanic ridges and associated rift zones, new sea floor is generated then carried away from the ridge axis by lateral mantle motions.**

Iceland LEFT **lies on the northern edge of the mid-Atlantic ridge, called the Reykjanes ridge. This is one of the only locations on Earth where a ridge rises to the surface, so that volcanic activity along the ridge results in eruptions visible above the ocean surface, rather than thousands of meters below. The center of Iceland remains volcanically active, and in 1963 an eruption off the southern coast formed the island of Surtsey** BELOW.

ridge axis near Iceland showed normal magnetic polarity; the remainder had a reversed polarity (a magnetic needle would point south). The pattern of normal and reversed polarity was manifested in a magnetic striping of the oceanic crust, mirrored on each side of the ridge crest. When individual stripes were dated, it was found that the rocks became older with increasing distance from the crest. In other words, the sea floor was spreading apart. Such spreading characterizes all oceanic ridges where lithospheric plate divergence occurs. During the past 80 million years, the Atlantic has spread at a rate of 2 centimeters per year. About 4 cubic kilometers of new crust is produced at mid-oceanic ridges every year.

Even more exciting discoveries have come from an international drilling project, the Deep Sea Drilling Project. Since 1968, a drill ship, the *Glomar Challenger*, has drilled nearly a thousand holes into the deep ocean basins, taking samples of deep-sea sediment and crust. One early discovery suggests that the Mediterranean dried up completely between 5 and 12 million years ago, leaving thick beds of sun-baked salts as evidence buried in today's ocean floor.

◁ **Only in recent years has it been possible to explore geological activity on the deep ocean floor. Now, however, by using submersible research vessels such as the United States' research vessel *Alvin*, scientists have been able to make personal visits to deep-sea sites rather than merely collect samples from drilling. One of the most interesting discoveries made was that of "black smokers" – mineral-enriched hot springs (sometimes colored white instead of black) which belch out from active regions of mid-oceanic ridges. They even have their own specially-adapted marine animals.**

Island Arcs

Chains of islands cover the Pacific Ocean, the largest on Earth. They stretch from New Zealand in the southeast, through Tonga, Indonesia, the Philippines and Japan, to the Aleutian islands off the coast of Alaska. This region is known as the "Ring of Fire" because of the intense seismic and volcanic activities that take place there; Indonesia alone has 150 active volcanoes. The arcs are formed by volcanic activity where two sections of crust collide. Most island arcs occur at the margins of the Pacific, though small ones exist in the Atlantic and in the Mediterranean Sea. On the sides of the islands that face the interior of the ocean are found deep trenches up to 1000 kilometers long. One Pacific trench, the Mariana trench, is more than 11,000 meters deep – the deepest point anywhere on Earth, and twice the depth of the average ocean basin. The trenches are usually steeper on the side that faces a continent, and those in the Pacific are usually deeper than those in the Atlantic. Trenches are formed when colliding plates bring oceanic crust into contact with less dense continental crust; the oceanic crust is pushed down toward the mantle.

KEYWORDS

ACCRETION
BENIOFF ZONE
COLLISION ZONE
DEEP-FOCUS EARTHQUAKE
EARTHQUAKE
IMBRICATE STRUCTURE
ISLAND ARC
MÉLANGE WEDGE
SEISMIC REFLECTION PROFILING
SUBDUCTION ZONE
TRENCH
VOLCANO

Discoveries about the processes that formed modern arcs have made it possible to recognize the rocks of island arcs that were active tens or hundreds of millions of years ago. This enables geologists to trace the pattern of plate movements and gain a clearer perception of how continents grow. They have shown that island arcs develop through volcanic activity, and tend to be composed of granodiorite (coarse igneous rock similar to granite), which resembles continental rock and is very different from the oceanic crust. This is true even in the Aleutian islands, off the coast of Alaska, where an island arc has been formed by the collision of two plates of oceanic crust. This implies that the processes that create island arcs are very different from other geological activity that occurs in mid-ocean, such as the volcanic islands that form over "hot spots" in the Earth's mantle.

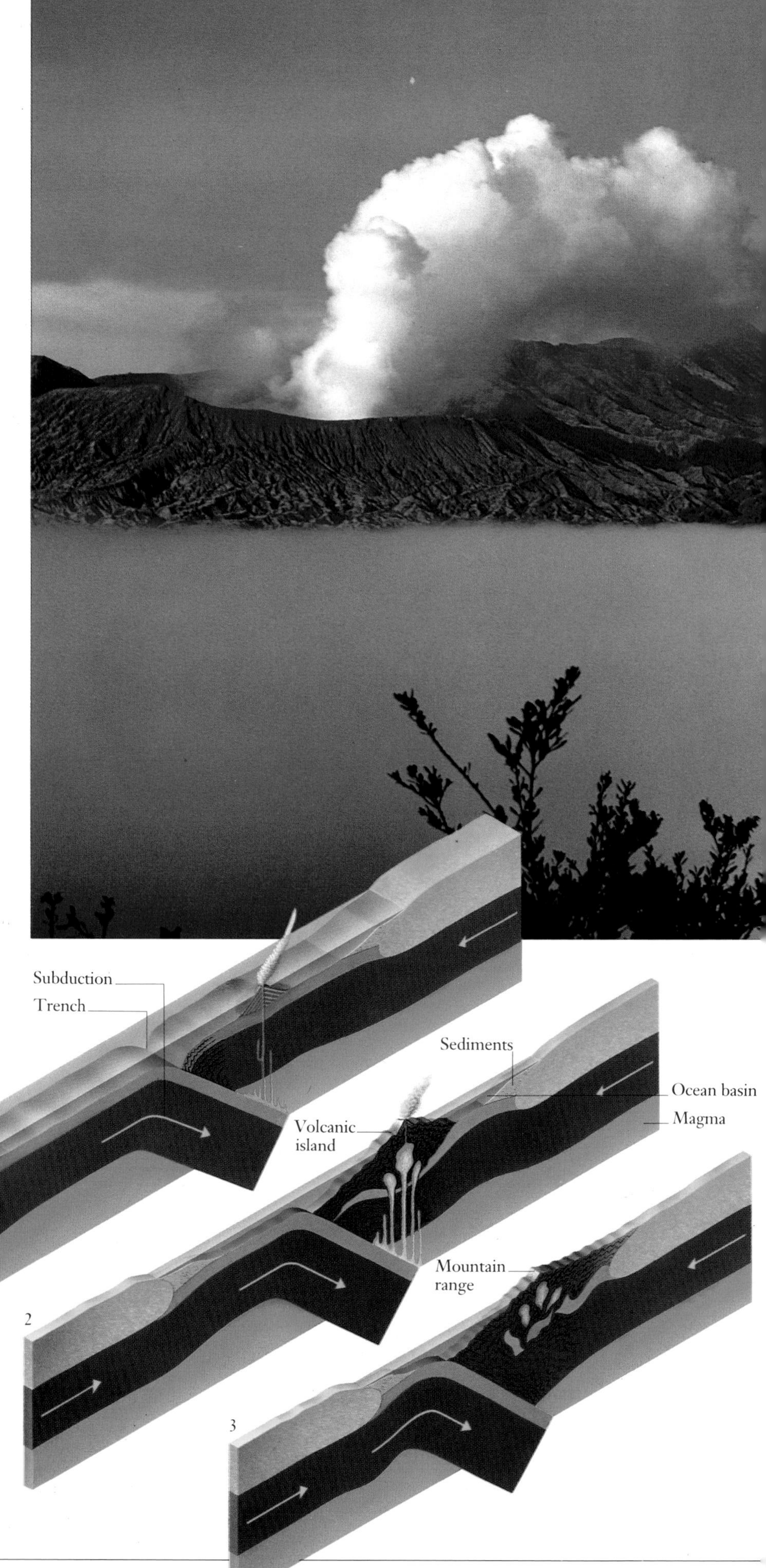

An island arc can eventually accrete onto the nearby continent. The Indonesian arc has grown where the plate carrying Australia northward is being subducted beneath the plate that is bringing Southeast Asia southeastward. Eventually, when the two plates meet, all of the intervening oceanic crust will have been consumed between them, forming a continental rather than an island arc. The rocks of the new arc will then collide with mainland Asia, leading to a complex sequence of events, involving deformation, magmatism and metamorphism, by which they will be accreted on to the edge of the continent of Asia.

The study of island arcs allows geologists to predict the inevitable eruptions and earthquakes in these regions. Earthquakes near a trench tend to be shallow, and the seismic foci become progressively deeper on the continental side of it. This observation has allowed Benioff zones – seismically active planes inclined at about 45° – to be defined. These planes represent the area in which oceanic crust is actively being subducted. The discovery of Benioff zones provided important evidence for subduction caused by plate collision.

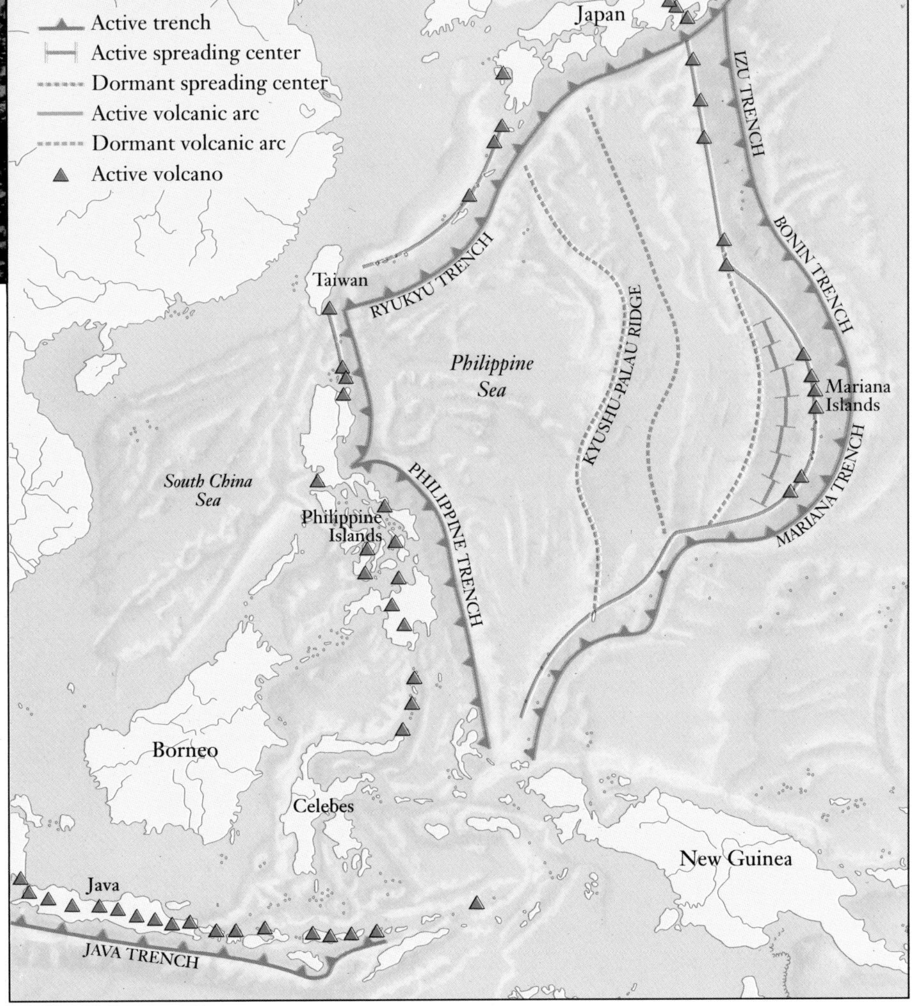

◁ Island arcs form when moving plates of the lithosphere collide. As one plate is pushed beneath the other (subducted) 1, some of the crust melts, forming magma (molten rock), which pushes up to the surface as a volcano, which eventually forms an island. Sediments collect, and as plate movements continue, the ocean basin diminishes as the continents on either side come closer together 2, with the island arc accreting to one side as more sediment builds up. Eventually the ocean basin may close 3 and a mountain range comprising deformed sedimentary and metamorphic rock is formed.

△ Mt. Bromo is only one of some 50 volcanoes on the island of Java. Off Java's southern coast lies the deepest trench in the Indian Ocean, where the Indo-Australian plate of the Earth's crust is slipping under the Eurasian plate.

▷ The Mariana, Philippine and Ryukyu trenches are moving west as the Pacific ocean crust is pushed under Asia. Between the Philippine and the Mariana trenches lies an inactive region where, about 25 million years ago, a new mountain ridge and ocean basin appeared and then became extinct.

MOUNTAINS FROM THE SEA

WHERE lithospheric plates converge, compressional stresses result. Sedimentary rocks that have accumulated near continental margins or alongside island arcs may be dismembered by massive submarine landslides triggered during subduction and collision. As they are carried deeper into the Earth, they become crumpled as if held in a huge vise. In this way, a complex sequence of events begins, culminating in the formation of fold mountain chains: the process of orogenesis.

KEYWORDS

ANDESITE
GNEISS
IGNEOUS INTRUSION
METAMORPHISM
OROGENESIS
OROGENIC BELT
SCHIST
SUBDUCTION ZONE
THRUST FAULT

The Andes mountain chain runs along the length of the western seaboard of South America. These comparatively young mountains have resulted from the Nazca plate – which carries a part of the Pacific ocean floor – being overridden by and subducted beneath the westward-moving South American plate which has continental South America on its leading edge. Deep trenches have formed offshore and the oceanic plate dips down at 25° beneath the continent. Some 300 kilometers east of the trench line, the foothills of the Andes rise to their summits, the sites of numerous active volcanoes.

The Andean ranges are characterized by linear chains of intensely deformed rocks which were laid down before Mesozoic times and folded in the Late Mesozoic. Deeply-buried sedimentary rocks becme deformed and recrystallized to form metamorphic rocks, while large volumes of magma pushed upward to create batholiths made of granite. These were less dense than their surroundings, and so added buoyancy to the continental margin. The latest phase of mountain-building took place during Late Neogene times (10 million years ago). Earthquakes and volcanic activity have continued until today.

Off the north Pacific, similar movements formed the Cascades and other northern cordilleras. Today, however, the northwest-moving Pacific plate has lost ground to

△ **The snow-covered peaks of the Andes' Cordillera Oriental (eastern range) give way to the altiplano, or high plateau, that separates them from the Cordillera Occidental (western range).**

▽ Adjacent to the Andean coast of western South America, the South American plate is advancing westwards and overriding the Nazca plate, located to its west. The latter is virtually stationary and carries part of the Pacific ocean floor. Deep trenches parallel the plate boundary and are located almost immediately above the contact. A Benioff zone slopes at around 25° eastward under the continent. Frequent earthquakes characterize this zone. Approximately 300 km east of this line, the foothills of the Andean cordilleras rise from the coastal plain. Active volcanism is a feature of this impressive chain, magmas being generated where the subducted oceanic plate has reached sufficient depths for melting to take place. The Nazca plate is being subducted at a rate of 10 centimeters per year.

the westward-moving North American plate. North of the Gulf of California, the boundary is offset by transform faults, such as the San Andreas Fault. There is no deep trench here, and earthquake foci are near the surface. These characteristics have arisen since North America overrode the eastern Pacific floor, so that the "mid-oceanic" ridge now lies beneath continental America. A triple junction formed where the East Pacific plate, the current Pacific plate and the North American plate meet. This unstable plate contact is slowly migrating up the continental margin.

The highest range of mountains on Earth, the Himalayas, was formed where the continents of India and Asia collided about 40 million years ago. Before this, fold mountains of similar age to the Alps (formed during Tertiary times) existed at the same site. The continental crust beneath the Himalayas is about 70 kilometers thick, about twice the average thickness. This is because continental crust does not subduct easily, and collision pressures were absorbed not by folding, but by the stacking up of huge rock slices along low-angled faults, which are called thrusts.

RIFT VALLEYS

FIFTY million years ago, the island of Madagascar began to split off from the African continent. Today Africa is still being torn apart from the Red Sea and Jordan valley, through Ethiopia and across Kenya as far south as the borders of South Africa. This area, known as the mighty East African Rift Valley, is a 5000-kilometer-long zone of fracturing which began to form during the mid-Tertiary (about 30 million years ago) by stretching associated with continental drift. The same process caused the supercontinent Gondwanaland to break up 100 million years earlier.

KEYWORDS

BASALT
CARBONATITE
CONTINENTAL DRIFT
FAULT
GRABEN
HORST
LITHOSPHERE
MANTLE
OCEANIC RIDGE
PHONOLITE
PLUME
RIFT VALLEY

Continents break up along the lines of faults – weak points in their structure caused by geological movement that fractures rocks. The subsiding of the land along the faults forms a rift valley. A block of planetary crust drops down between the faults. This is called graben. It is often accompanied by a block of upfaulted crust, called horst. These contrasting blocks produce the characteristic valleys and peaks of rift valleys.

A major feature of the eastern side of the African continent was a broad dome of planetary crust (which still occupies most of Kenya). This rose in response to hot mantle material rising up beneath the continental crust, which began to stretch during Tertiary times. The rift faulting peaked during Miocene times, 5 to 25 million years ago. It was accompanied by volcanic activity, which has continued to the present. Not surprisingly, the rift is an area of above-average heat flow from the Earth's interior, and volcanic activity and faulting continue to tear apart this side of Africa at a rate of a few centimeters per year.

The main rift faults are as much as 3000 meters deep in Tanzania, and the rift itself – which splits in two around Lake Victoria – is up to 200 kilometers wide. The movement of the crust has resulted in volcanic activity, giving rise to large volumes of flat-lying basalt and phonolite flows and a number of large stratovolcanoes (volcanoes with a distinctive conical shape), including Africa's highest mountain, Kilimanjaro.

Rifts and volcanic activity are typical of regions where mantle material is rising under the lithosphere. The most extensive of all terrestrial rift faults are found along oceanic ridge systems. If Africa continues to split apart at its present rate, eventually a new ocean will flow into the rifted region and a new continent will form.

Faulting has also affected both Mars and Venus. Mariner-9, which visited Mars in 1971, was the first probe to photograph Valles Marineris, a system of deep fault-bounded canyons along the Martian equator. It begins at a huge upwarp in the lithosphere known as the Tharsis Bulge and plunges more than 6000 meters below its rim. On Venus, one of the most spectacular systems is associated with Beta Regio, a region of intense volcanic activity where several major tectonic zones intersect. These may have developed above a long-lived rising plume of hot material from the planet's mantle.

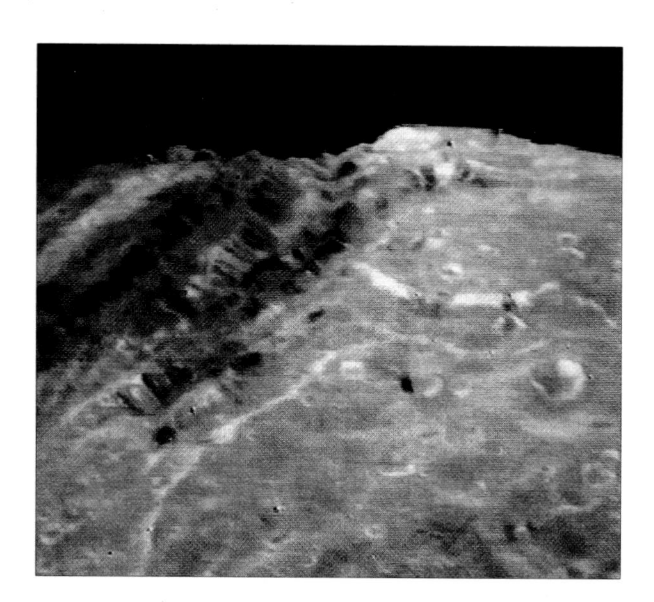

◁ **Miranda, the closest and smallest of Uranus' major moons (measuring 500 km), is one of the most-photographed worlds in the outer Solar System. Pictures taken by the Voyager 2 mission revealed it to be startlingly complicated, its cratered surface being transected by immense fault scarps. With slopes of at least 60°, some of the scarps are over 10 km high. The grooves and troughs may be several kilometers deep. The complexity of Miranda's surface reveals its long geological history.**

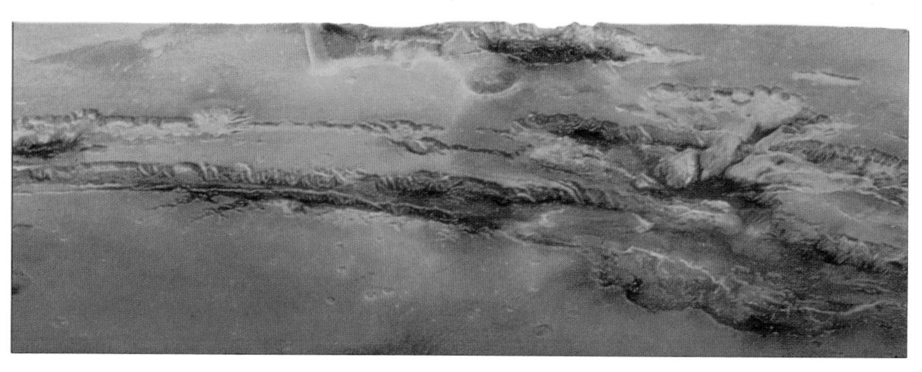

▷ **Valles Marineris on Mars is probably the largest rift valley in the Solar System. 4500 km in length, its complex of deep canyons is 600 km wide and 7 km deep. It formed by subsidence along major faults on the flank of the Tharsis Bulge, followed by scarp retreat due to collapse.**

◁ **The Great Rift Valley of East Africa is the result of continental drift that is still pulling Africa apart at a rate of a few centimeters a year. The surface appearance of the rift is relatively undramatic compared with the huge geological upheaval that lies below.**

▽ **The initial stage in rift valley development 1 is the upward arching of the crust caused by rising plumes of mantle material. The zone of potential splitting is one of high heat flow and seismic activity. 2 Melting of rocks below the crust generates magmas. Because the overlying lithosphere has been stretched and thinned, these gain easy access to the surface. Sheets of lava are extruded on the rift floor and flanks. Vertical faulting takes place. 3 As extension continues, the rift widens and series of normal faults develop. The throw on these may be 1 km or more. Volcanic activity becomes more centralized and, because magmas have to penetrate continental-type crust, any volcanoes tend to be highly explosive.**

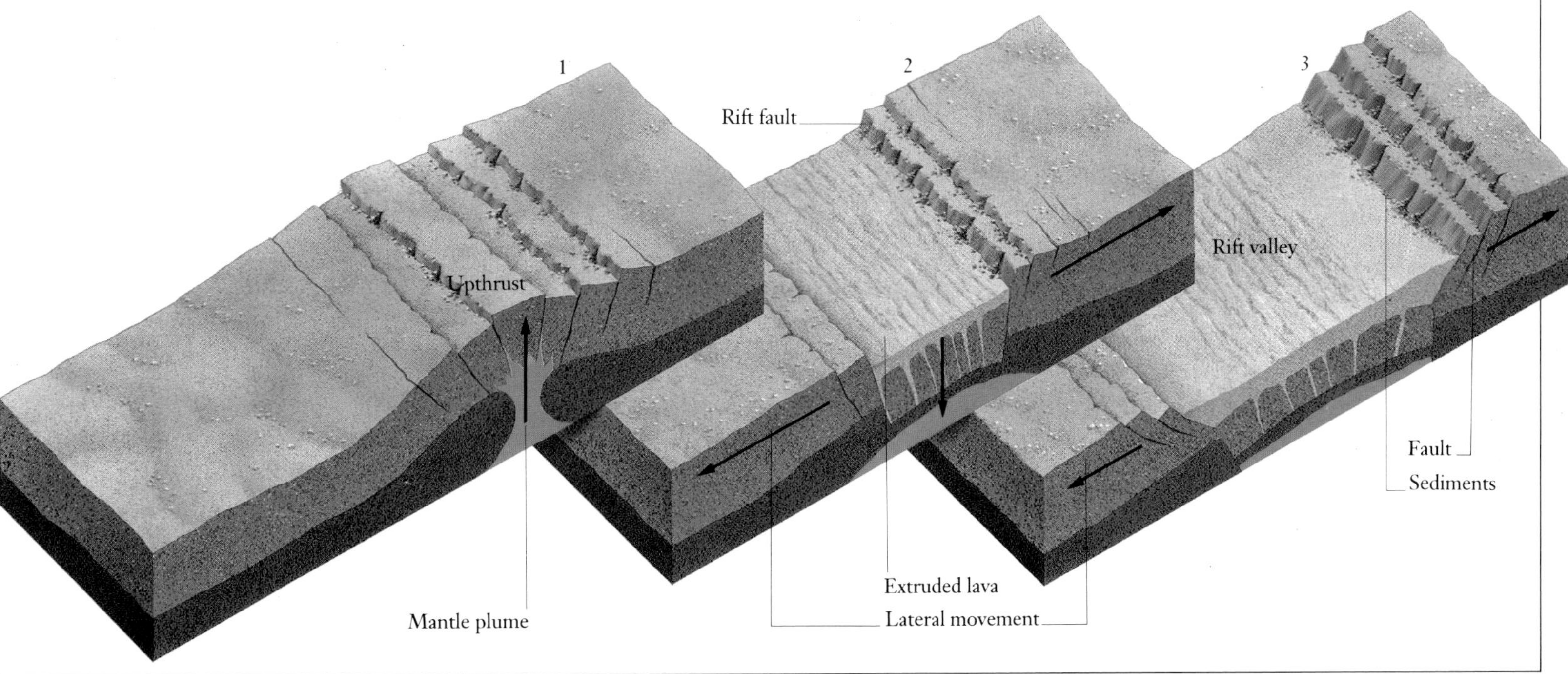

6

CHANGING *Worlds*

MODERN UNDERSTANDING of the history of the Earth has come largely from stratigraphy – the study of layers of rock. By studying the rocks being formed today, in known environments, geologists are able to infer what was happening in the distant past, interpreting each rock layer in the light of this knowledge.

Because life began in the sea and has continued to flourish there, the record of fossil remains within sedimentary strata also represents an evolutionary series. Thus stratigraphy is aided by paleontology, the study of fossils. Stratigraphic and paleontological studies of the rock record indicate the Earth's history moves in cycles, with periods of relative stability punctuated by upheaval. During the process of orogenesis, once-horizontal strata may be rucked up into fold mountains, while others may be completely destroyed by erosion, giving rise to gaps in the rock record. These gaps are called unconformities. These features can be seen in Grand Canyon of Arizona.

The geological processes of today are similar to those of the past. However, the rate at which the processes take place may have changed. Time is very important in geology. Most changes occur very slowly, but sudden storms or volcanic eruptions may change the face of a region very quickly. These events are recorded by rocks.

The distinctive sharp peaks of limestone karst mountains in Guilin province, south-central China, are a national landmark. They formed from a huge block of limestone which was gradually eroded by weathering. In the warm, humid climate of this region, even deep-lying granite and volcanic rocks undergo chemical weathering, extending as far as 60 meters (200 feet) below ground. The thickness and size of the limestone, together with the depth of the water table, produce the formation of the isolated towering shapes on an otherwise flat plain.

Nothing is Forever

A ROCK is a collection of minerals – solid materials with ordered atomic structures. Most rocks are silicates, consisting of various metallic cations (positive ions) in combination with silicon and oxygen, the latter usually in the form of the negative SiO_4^{4-} anion. Because of the different cations they contain, the various silicate minerals have diverse internal structures. Some are strongly bonded and therefore very hard; others have looser bonds and may be more susceptible to mechanical or chemical attack.

KEYWORDS

EXFOLIATION
FELDSPAR
HYDROLYSIS
MAGMA
MASS WASTING
MICA
MINERAL
QUARTZ
ROCK
ROCK-FORMING SILICATES
SOIL
WEATHERING

All minerals ultimately come from inside the planets, having crystallized during the cooling of molten magma. The crystals that form do so under considerable pressure and at temperatures ranging between 500° (in the case of granite) and 1100°C (in the case of basalt). As a result, if they reach the surface, they are not necessarily stable at the low temperatures and pressures found there. Furthermore, most have weaknesses in their structures, inherited from the way their atoms are bonded.

At the Earth's surface, outcrops of rock are attacked by water, wind and ice. Rocks are vulnerable to attack along inherent weaknesses such as bedding planes, joints and fractures, and eventually pieces are broken off. In time they are transported elsewhere by a combination of gravity, running water, ice and wind. As they do so, the fragments collide with each other and against other outcrops and are eventually broken into individual mineral grains. These destructive processes of weathering and erosion are followed by the deposition of the rock fragments as sedimentary layers, or strata.

The mineral quartz (SiO_2) is a major constituent of the granitic rocks which are typical of continental interiors. Quartz is also extremely abundant in many of the sedimentary rocks derived from these and later deposited as sediments. It has a strong atomic structure, making it resistant to chemical attack. In this respect, quartz contrasts sharply with feldspar, another component of granite, whose crystals are weakened by the presence of cleavage planes (planes of weak bonding in the mineral's atomic lattice) and of molecules that are readily broken down by the weakly acidic rainwater. Feldspar grains, therefore, are steadily broken down and form clays, while some of

▷ These steep limestone cliffs are among the jagged peaks of the Dolomites, at the northeastern edge of the Alps in Italy. The Alps formed when, about 100 million years ago, the continental blocks of Africa and Europe were pushed together. This caused the lime-rich sediment of the Tethys Sea (between the two) to be compressed into rock, which melted to igneous magma from the force of the impact of the continental plates. The area is still tectonically very active.

▽ Sandstone monoliths are found in the state of Utah, a region of North America that has been a stable zone in recent times. Recent stripping of the plateau-like landscape of Mesozoic sandy rocks has left residual mesas, buttes and pinnacles, such as those of Arches National Park. The final form of the landscape is sculpted by wind. The original joints have been exaggerated by the daily expansion and contraction of the rocks, freezing and thawing, and flash floods.

their constituents dissolve and are carried away by streams and rivers, eventually to be recrystallized elsewhere. All other rock-forming silicates except quartz undergo similar degradation.

The process of sedimentation, which occurs when the load of sediment carried by rivers (or glaciers) enters the sea or a lake, involves not only the deposition of layers of grains but also the eventual precipitation of mineral matter in solution. Sometimes this takes the form of a cement which crystallizes, after the sediment has been buried, from the water trapped between sediment grains. Processes occuring after deposition are important in converting separate grains into solid rocks, for instance, by the immense burial pressures that squeeze out pore water trapped between grains and push them closer together.

Many limestones are directly precipitated from seawater. They are made of calcium carbonate, which originated in reactions between atmospheric constituents such as carbon dioxide and carbonic acid in seawater. Carbonate is also fixed by marine organisms, which build their shells from it. The fixing of carbon dioxide into solid carbonate form has had the advantage of preventing buildup of the gas in the Earth's atmosphere; without it, a greenhouse effect like that found on Venus could have occurred on Earth and prevented the development of life.

On Mars, most erosion and transport of the widespread dust found there is carried out by wind; however, in the past running water and ice both were involved in sedimentation. On Venus, where there is no water, some impact debris is apparently moved by wind and may be laid down by turbid currents which are a mixture of solid particles and trapped atmospheric gases.

▷ In a granitic rock, tabular prisms of feldspar contrast with brownish sheets of dark mica (biotite). Colorless crystals of quartz fill the spaces between them. The opaque grains are iron oxide (magnetite). Weathering of granite produces feldspar, which converts to clays; quartz, which is highly resistant and forms grains in secondary rocks; mica, which also breaks down to clays; magnetite, the most common iron oxide, which is relatively insoluble and remains partly as opaque grains.

THE WORK OF RIVERS

RAIN may simply run off the Earth's surface or seep down into porous rocks to emerge at a lower level at a spring line; but eventually it feeds the tributary networks of rivers. These carry sediment and also deposit it along their courses, either as point bars within their channels or as debris spread out over the surrounding countryside in the form of flood deposits.

Water is collected into a river system by tributaries – a veinlike network of streams. In its upper reaches, slopes tend to be steep and the stream flows quickly; the river carries its load in suspension or by dragging larger debris along its bed. As it descends, it may pick up more load, but gradients lessen and the rate of flow diminishes; the rate of deposition increases. Where it finally disgorges into the sea or a lake, the river loses energy rapidly and deposits its load, resulting in a delta of sedimentary material through which the river must thread a new path.

KEYWORDS

CROSS-BEDDING
DELTA
REJUVENATION
SORTING
SUSPENDED LOAD
TRACTION LOAD

The world's largest rivers, such as the Mississippi and Amazon, deliver huge quantities of water to the oceans: the Mississippi discharges 17,715 cubic meters per second, but the Amazon disgorges ten times this volume. Each carries around a billion tonnes of sediments per year to the oceans. Beneath the modern delta of the Mississippi is a 6 kilometer-thick layer of sediments accumulated during the past 40 million years. At present, the river adds a further 1.5 millimeters of new sediment to its delta each year.

Evidence of erosion caused by rivers can also be found on Mars, though there is no running water on the planet now due to its extreme cold. Valley networks between the cratered regions of the planet must have been cut by rivers which may have flowed 1.5 billion years ago. Heavily scoured "channels", up to 200 kilometers wide and 1000 kilometers long, appear to have formed in response to the almost instantaneous release of large volumes of water – presumably by the melting of ground ice. They may be similar to the channeled scablands of the American state of Washington, where the breaching of a natural dam during the last Ice Age disgorged the contents of a vast lake, Lake Missoula.

The course of a major river system does not always remain the same. The Mississippi, for instance, during the Pleistocene Ice Age (up to 2 million years ago) when the sea levels were lower than today, cut itself a deep bed to compensate for the lowering in level. When the ice melted and sea level rose again, the river dumped a broad sheet of coarse debris, then built up its banks and developed a more sinuous course. As each stage passed, it entered the sea at different points through channels of varying profile; the position and form of its deltas changed with time, producing a complex interleaving pattern of sedimentation.

In arid regions, rivers tend to flow only at certain seasons and may never reach the sea from interior deserts, instead depositing their sediment load in temporary lakes. Relatively small volumes of material are transported, but during violent storms substantial amounts of erosion and rapid movement of sediment may occur. Seasonal river flow is also typical of areas next to the polar ice caps. The main season of sediment transport and deposition is the summer, when debris is freed from the ice.

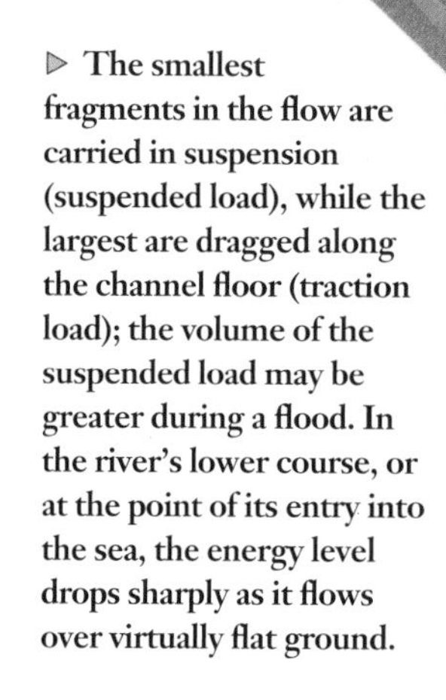

▷ **Rivers begin in high places and flow down to lowlands and the sea. They may begin with a lake or a waterfall high in a mountain, which feeds a stream. Tributaries (thin streams) may add their flow at any point. In the upper reaches, slopes are relatively steep, so that the runoff flows at high velocity. Under such conditions, large boulders can be carried in the stream's load. Lower in the course, the channel is at a lower gradient, although wider, and the available energy is somewhat less; consequently, only smaller fragments are carried.**

▷ **The smallest fragments in the flow are carried in suspension (suspended load), while the largest are dragged along the channel floor (traction load); the volume of the suspended load may be greater during a flood. In the river's lower course, or at the point of its entry into the sea, the energy level drops sharply as it flows over virtually flat ground.**

◁ **The Zambezi River, the longest river in southern Africa at 2650 kilometers long, has an average flow as high as 16,000 cubic meters per second. In Zimbabwe it spills over the magnificent Victoria Falls, shown here.**

▽ River channels may be braided, when there are several main channels; meandering, when the channel twists; and straight (rare, and usually not perfectly so). Meanders are associated with floodplains and oxbow lakes, which form when meanders loop and cut into one another, abandoning a bend and shortening the course.

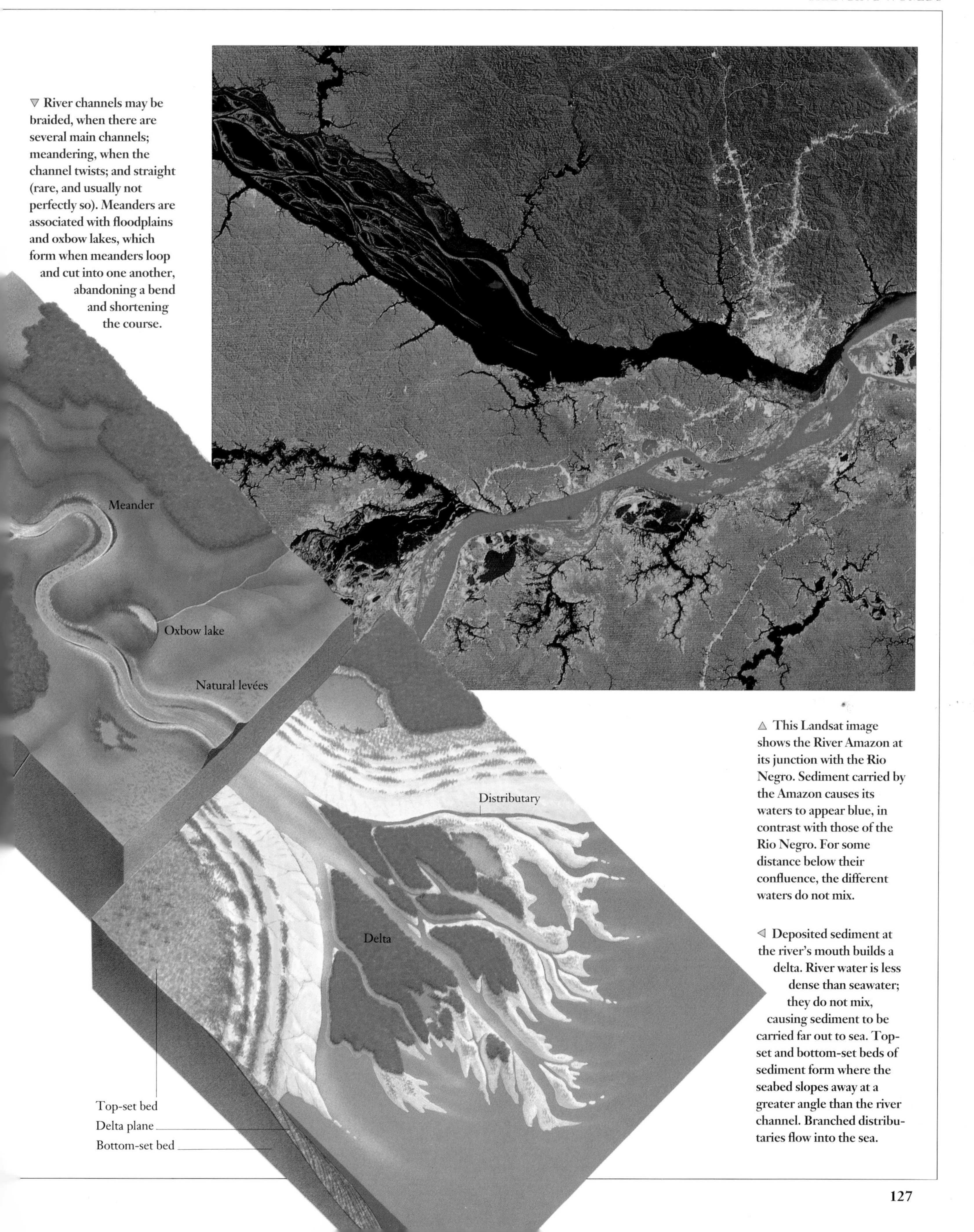

△ This Landsat image shows the River Amazon at its junction with the Rio Negro. Sediment carried by the Amazon causes its waters to appear blue, in contrast with those of the Rio Negro. For some distance below their confluence, the different waters do not mix.

◁ Deposited sediment at the river's mouth builds a delta. River water is less dense than seawater; they do not mix, causing sediment to be carried far out to sea. Top-set and bottom-set beds of sediment form where the seabed slopes away at a greater angle than the river channel. Branched distributaries flow into the sea.

COASTS AND OCEANS

SURROUNDING the world's oceans are coastlines that stretch for hundreds of thousands of kilometers. They have been molded by the forces associated with the sea and hewn out of rocks of varying age and resilience. The sweep of shorelines is broken by the mouths of rivers, which bring sediment from the continental interiors to the oceans, often building extensive deltas at their mouths. The action of waves and currents may redistribute a proportion of this sediment along the neighboring coasts by longshore drift.

KEYWORDS

ABYSSAL SEDIMENT
COASTLINE OF EMERGENCE
COASTLINE OF SUBMERGENCE
ESTUARY
ISOSTACY
LONGSHORE DRIFT
MARGIN
TURBIDITY CURRENT

Some shorelines are developing along rising continental margins. Isostatic adjustment or rebalancing that occurred after the retreat of Ice Age glaciers may have caused one part of the continent to rise; alternatively the margins themselves may be emerging from the oceans as a result of plate movements. Coastlines of emergence display active erosion as waves attack cliffs and raised beaches – the remains of old shorelines now lifted above sea level and backed by fossil cliffs. Conversely, coastlines of submergence occur where the land is sinking relative to sea level. They are found, for instance, along the southeast coast of Britain. Coastal plains may flood, leaving ridges and hills as islands near the shoreline.

The power of sea waves is immense. The hydraulic action of waves pounding into joints and weaknesses within cliffs has the capacity literally to blow them apart; waves have been known to blast the roof out from a cave. This pounding action is rendered even more powerful by the debris which accumulates at the junction between land and sea. During storms and exceptionally high tides, sand and gravel enhance the scouring effect of waves and increase coastal erosion.

Beach sediment is derived in part from rivers and their deltas, but also comes from cliff erosion. Where cliffs are built from soft rocks such as clays, slumping is common, erosion rapid and material is removed relatively quickly. Tides tend to carry small mud-sized grains offshore before depositing them. Harder rocks are more resistant to attack and tend to form headlands. The pebbles and sand produced from such rocks stay close to the cliff base, forming beaches. Shelving beaches protect cliffs from erosion, because they absorb much of the power of breaking waves.

◀ **Wave-cut rock platforms cut into the Jurassic rocks of Loch Slapin, Skye. As the sea cuts into the land, the cliff collapses and the debris is broken down and carried away. A rock platform develops at a cliff base; in sandstones shown here, the platform is nearly flat. The break of slope between the platform and the cliff is generally located a meter above the level reached by the highest waves.**

The margins of the continents slope gently under the waters of the ocean, forming continental shelves. In tropical latitudes, if sheltered conditions prevail, this shelf provides the ideal environment for the growth of coral which may build fringing reefs. Sometimes the coral growth on a sunken island produces an atoll or circular coral reef.

In unstable regions, land-derived sediments which have built up at the edge of the continental shelf may be set in motion by seismic disturbances, which generate flows of sand, mud and water called turbidity currents. These sweep down the

△ Japan's Inland Sea is a narrow arm of the Pacific that extends between Shikoku and Kyushu, two of Japan's four main islands. Most of this area is subsiding, both the sea itself and the coastline of Kyushu. As the land dropped and was flooded by the ocean, hills were transformed into many small islands. Faults lie on the borders of the rocky inlets and bays. Marshes are forming where sediment from the land is deposited.

△ The power of waves smashing into cliffs on the Atlantic coast of Ireland erodes the rock faces of the cliffs, carving out spaces and leaving sea stacks – free-standing rocks in the water. The average Atlantic wave in winter exerts hydraulic pressure of 10 tonnes per square meter. In a storm, this pressure may be as much as three times greater.

▽ In 1929 an undersea earthquake off the Grand Banks of Newfoundland caused offshore sediment to slump, forming strong turbidity currents. The relatively dense suspensions prevented mixing with the water surrounding them, and the sediment spread across the sea floor at speeds up to 70 kilometers per hour, cutting trans-Atlantic telegraph cables in many places. The time at which each cut was made marked the current's progress.

Submarine telegraph cables
Time and location of earthquake
Time at which cable broke
200 km
400 km
600 km

continental slope onto the abyssal plains that extend oceanward, laying down sedimentary rocks known as turbidites.

Turbidity currents off the coast of Newfoundland in the North Atlantic have broken transatlantic telephone cables several times; their activity may also help to cut deep canyons along the continental margin.

On the deep sea floor, far from land, the only sediments that accumulate are organic oozes produced by marine algae and diatoms. This sediment is supplemented by volcanic dust which falls out of the atmosphere and slowly settles onto the abyssal plains.

DESERTS AND WINDS

WHERE global atmospheric circulation brings dry air down from the upper atmosphere (troposphere), deserts may develop. These are found both in hot regions at low altitudes (20–30° north or south of the Equator) and in cold ones next to polar ice caps. Near the ice, the sediments are derived from shrinking glaciers, and surface features such as stone polygons may develop in response to freeze-thaw activity. In hot deserts such as the Sahara, extensive "sand seas" or ergs grow in response to accumulations of wind-blown sand, forming dune fields. These great seas of sand may cover areas as large as 500,000 square kilometers. Deserts also are found in continental interiors far from the sea.

KEYWORDS

BARCHAN
CROSS-BEDDING
DRIEKANTER
DUNE
DUNE-BEDDING
MILLET-SEED GRAIN
PEDIMENT
PLAYA
TALUS
THRESHOLD WIND SPEED
TUNDRA
WADI

Wind is the most powerful force that shapes the desert. It picks up sand grains and transports them across the desert floor. The largest features of the desert shaped by the wind are called bedforms. Dunes and ripples develop on the bedforms as the wind moves the sand around. Where sand is deposited, dune crests tend to be arranged at right angles to the prevailing wind direction. Individual dunes have a steep front face and shallow back slope. Due to the way in which dunes grow, akin to the tipping of unwanted rock over the edge of a slag heap, bedforms are characterized by surfaces inclined to the horizontal, a characteristic called cross-bedding.

In regions of fairly constant wind direction, widespread crescent-shaped dunes known as barchans form. Fields of such dunes often coalesce to produce a broad ripplelike swathe across the desert floor. On very bare rocky surfaces, the sand may be drawn out into elongated seif dunes, whereas in regions of changeable wind direction, dunes with star-shaped structures develop.

▷ The sandy desert of southern Namibia meets the rocky desert of the north in this natural-color Landsat satellite image. Dunes stretch 400 km along the coast of south-western Africa, accounting for a third of the desert's north-south extent.

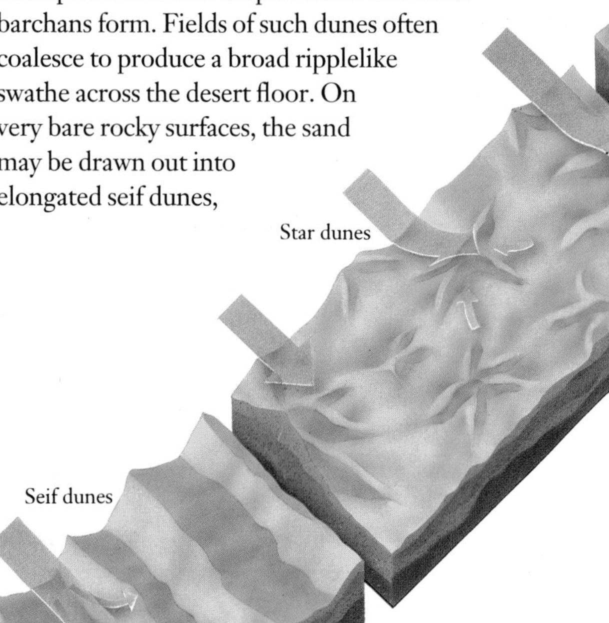

▽ Dunes occur in four variations. Tail dunes are created by small obstacles such as bushes or hills blocking the wind. Crescent-shaped barchan dunes form in regions of regular wind direction and limited sand. They may migrate downwind at rates of up to 25 m per year. Star dunes form where variable winds prevent regular deposition on any one face. Where winds are constantly shifting and there is a large supply of sand, seif dunes form. In the Sahara they reach 300 m high and 300 km long.

▷ Dune-bedding in Utah's Jurassic sandstones. Internal discontinuities represent successive slip faces as ancient dunes advanced across the desert.

Erosion is another effect of the wind. The constant abrasion of particles in the air tends to result in well-rounded grains, more spherical than those that develop in river and marine environments. Larger fragments which the wind is unable to pick up may simply be dragged along the ground, generating faceted pebbles called dreikanter. Sand removal by the wind may create areas such as the 300-km Qattara basin in Libya, 134 meters below sea level. Erosion on the face of individual rocks – both small and large – produces strikingly different features, such as honey-combed cliff faces, rock arches or pedestals.

Rain occasionally falls in even the dryest deserts, usually during short but violent storms. When these occur, there may be rapid erosion and movement of material over very short periods. In this way, fluvial

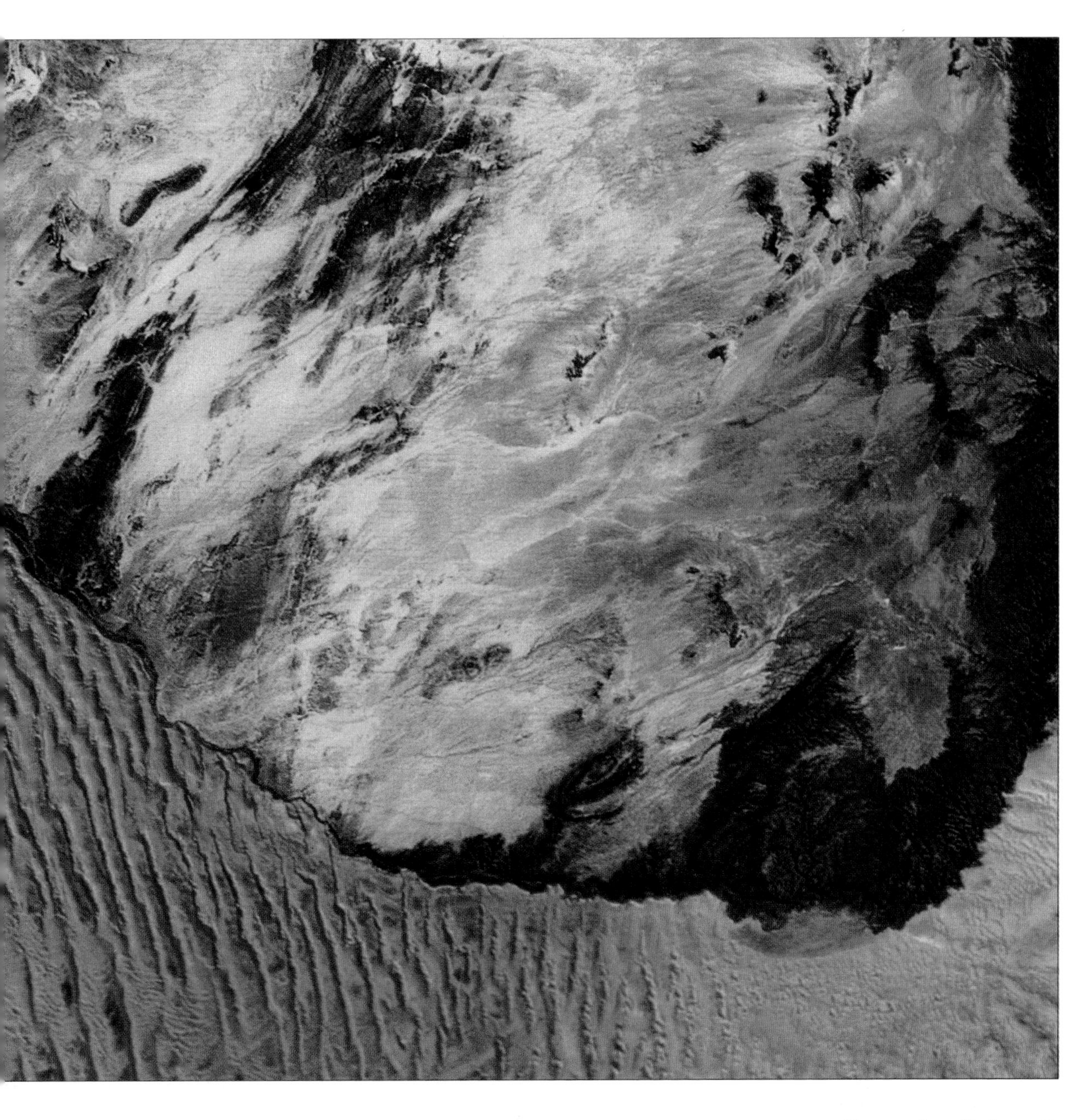

▽ Largescale dunes form on the interior of a large Martian impact crater. Wind is the major cause of sediment movement on Mars. Dune formation is concentrated in a zone around the polar regions. Wind-blown material on Earth is predominantly of sand grade, and composed of quartz. Such material is unlikely to be found on Mars and the predominant sediment is of dust grade.

valleys and wadis form. Seasonal lakes, called playa, often occupy desert interiors. These are collecting grounds for finer-grained deposits.

Desert interiors may be rather flat, but they often have peripheral plateaus and isolated hills. The rocky plateaus within the desert are attacked by wind, and generate large deposits of fragmented bedrock of talus formed from blocks, and pebble-sized fragments which accumulate at the feet of escarpments. These are too coarse to be moved by the wind, although the wind may blow finer sand into the spaces between the talus material.

Mars may be considered a large cold desert. Extensive dune fields have developed around the polar caps, while global dust storms – which typically develop near perihelion – may obscure all surface features, as they did in 1971 when Mariner-9 approached Mars. Dune structures and wind-generated streaks have also been identified on Venus.

GLACIERS AND ICE

POLAR ice caps have formed on both the Earth and Mars. On Earth the ice takes the form of frozen water, but the Martian caps contain both this and solid (frozen) carbon dioxide. Moving glaciers may have existed on Mars in the distant past, a conclusion based both on theory and observation of landform types which could be interpreted as glacial in origin. The Earth has active valley glaciers as well as polar ice sheets which retreat and advance with the seasons.

KEYWORDS

BOULDER CLAY
DRUMLIN
ESKER
GLACIAL
MORAINE
ROCHE MOUTONNÉE

Ice covers about 15 million square kilometers of the Earth's surface today – that is, about 0.003 percent. During the last Ice Age, ice sheets extended over large areas of North America and Europe. In these once glaciated regions, ice is now restricted to glaciers in the higher mountains. Evidence of their former extent is found in the glacial striations produced on rock surfaces by the scouring action of stones trapped at the base of moving ice sheets, and by the plucked form of rock outcrops which now are known as roches moutonées.

Then as now, glaciers advanced and retreated along valleys in response to changes in climate. As they advanced, the glaciers carried debris scoured from the land; when they retreated, this debris was left as terminal moraines. The location of these deposits has allowed geologists to trace the various phases of the glaciation. The valleys themselves deepened and became steep at their sides, generating flat-bottomed, steep-sided, U-shaped profiles – a typical feature of glacial erosion.

The glaciers are still in retreat, and it is possible to see the depositional landforms that are their legacy. These include low, hummocky hills called drumlins and long, winding gravel ridges called eskers. Both are aligned parallel to the direction of ice movement, and the pattern of their ancient counterparts may be used to trace earlier glacial geography. In addition, at the sides of a typical valley glacier, coarse debris accumulates and forms lateral moraines.

Although such features give vital clues to geologists, they do not form the bulk of glacial deposits. These are boulder clay, also known as "till", which is the morainic debris that accumulates beneath the ice. Within such debris may be blocks that have been transported from distant locations as well as a mixture of sand and gravel. Erratics left behind in lowland Britain after the retreat of the Pleistocene ice sheet include rocks as diverse as igneous rocks from Norway and chalk from East Anglia. Boulder clays that become compacted by burial are known as tillites. Identification of such rocks within the rock record has enabled geologists to show that glaciations occurred in the distant past in locations as unlikely as Australia, South America and Africa.

Beyond the margins of any ice sheet the ground is still very cold – indeed, frozen. Regions of permafrost, as this ground is known, may thaw in summer. Such alternate freeze-thaw cycles heave up the ground surface and sort the different-sized fragments into mounds, forming patterned ground and giving rise on flat surfaces to stone polygons. Collapse depressions may form where ice mounds up beneath the surface during winter and then melts during the summer months, causing collapse. Elsewhere, ice filled blisters pushed up by pressure from below may form pingos.

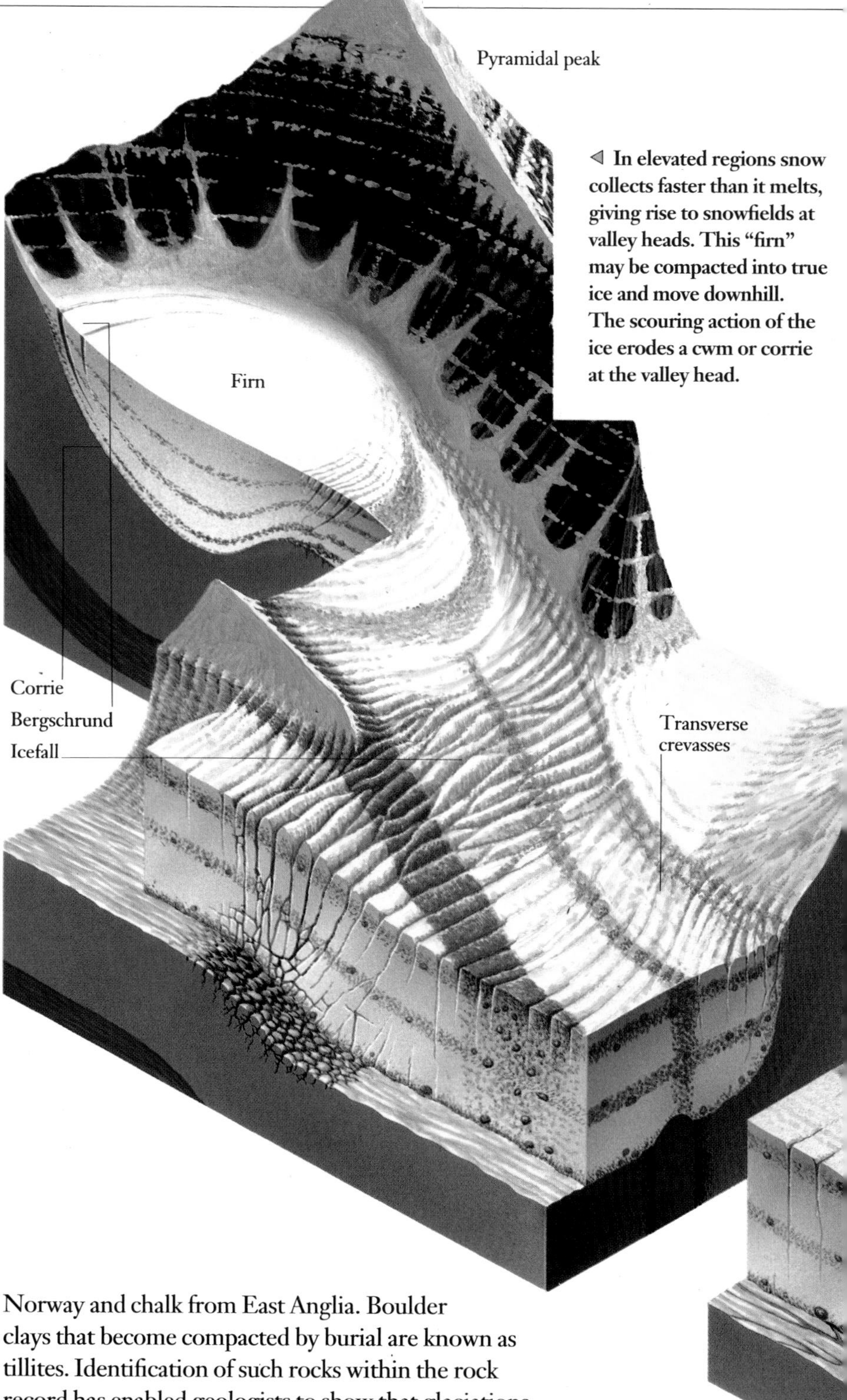

Pyramidal peak
Firn
Corrie
Bergschrund
Icefall
Transverse crevasses

◁ In elevated regions snow collects faster than it melts, giving rise to snowfields at valley heads. This "firn" may be compacted into true ice and move downhill. The scouring action of the ice erodes a cwm or corrie at the valley head.

▷ The debris carried by glaciers forms moraine. Some is carried along the margins, some at the moving glacier front, and some beneath the glacier. When the ice encounters irregularities, crevasses form, tending to lie at right angles to the flow. The toe of a glacier is characterized by meltwater ponds, seasonal streams and terminal moraine debris.

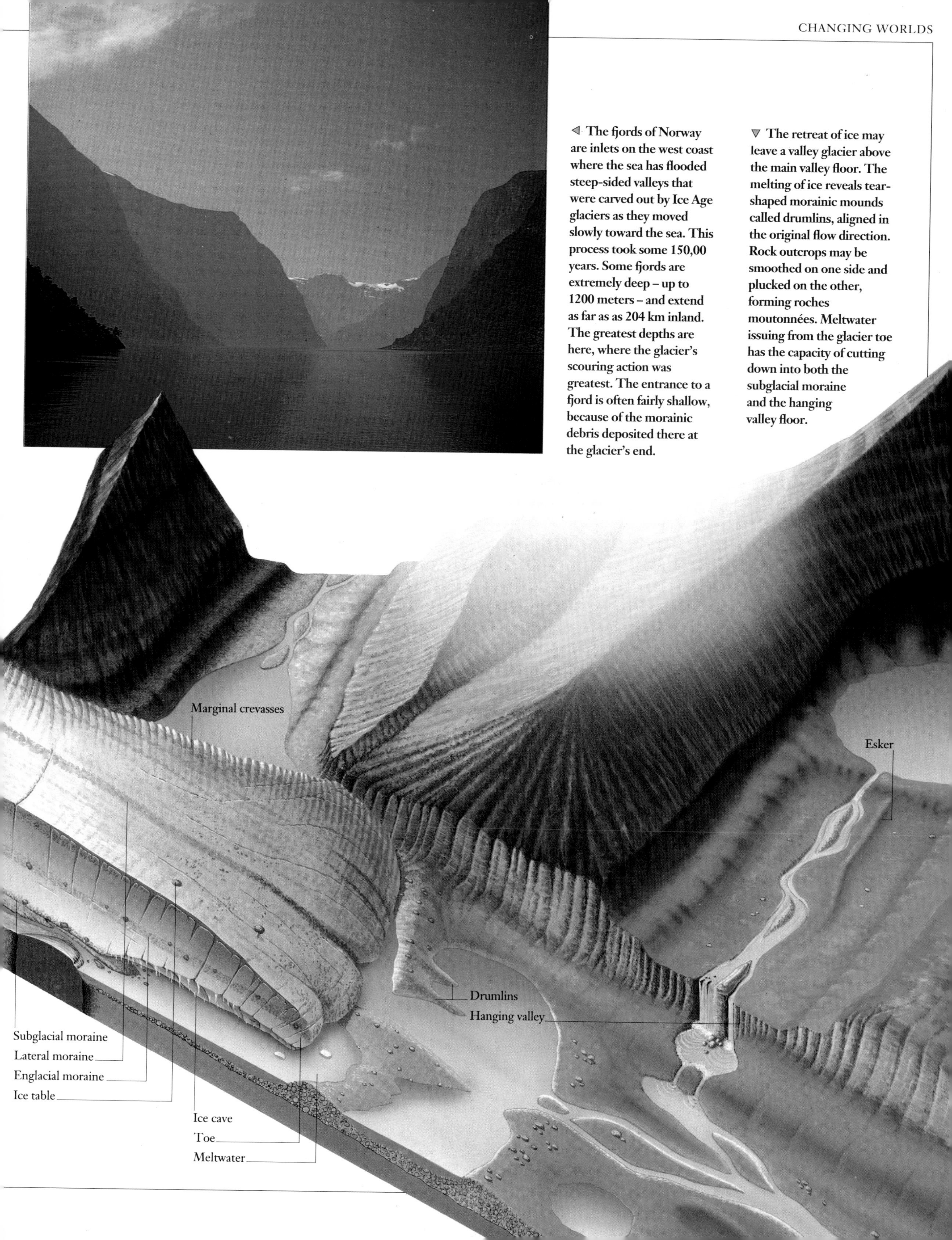

◁ The fjords of Norway are inlets on the west coast where the sea has flooded steep-sided valleys that were carved out by Ice Age glaciers as they moved slowly toward the sea. This process took some 150,00 years. Some fjords are extremely deep – up to 1200 meters – and extend as far as as 204 km inland. The greatest depths are here, where the glacier's scouring action was greatest. The entrance to a fjord is often fairly shallow, because of the morainic debris deposited there at the glacier's end.

▽ The retreat of ice may leave a valley glacier above the main valley floor. The melting of ice reveals tear-shaped morainic mounds called drumlins, aligned in the original flow direction. Rock outcrops may be smoothed on one side and plucked on the other, forming roches moutonnées. Meltwater issuing from the glacier toe has the capacity of cutting down into both the subglacial moraine and the hanging valley floor.

7

BEGINNINGS *and Endings*

SINCE THE SUN and its system of planets were formed, about 4.5 billion years ago, younger stars – and possibly other planetary systems – have been born. Equally, since the Earth first came into existence, other stars and any attendant planets will long since have reached the end of their active lives. Will this happen to the Solar System?

Our Sun is currently in "middle age". Although its supply of hydrogen is abundant, eventually it will run low; this will probably occur about 5 billion years from now. Energy production will then cease, and the helium-rich core will begin to shrink under the influence of gravity. In doing so it will heat up again; and, because there will still be some hydrogen left around the core region, it will start reacting once more. The inert helium core will become surrounded by an envelope of burning hydrogen.

At this stage the Sun will begin to expand, while its helium core is still contracting. The helium will be converted to heavier elements such as oxygen and carbon, and energy will be dissipated into space, lowering the Sun's temperature. By the time it has dropped to about 3000° K (about half its present temperature), the Sun will be too cool to warm the Solar System. The Earth and other planets will not survive this stage. This will be the end of our world.

In this small section of the Milky Way, the galaxy to which the Earth and the Solar System belong, the red supergiant star Antares (at the bottom center of the image) is the brightest star. A red supergiant is a cool, very luminous star in the late stages of its life. Its core has contracted and begun to burn helium, as its outer gaseous layers have expanded, increasing its radius; Antares is more than 1000 times the size of the Sun. Its temperature is about 3000 degrees – about half that of the Sun today, and too low to heat any nearby planets, if they existed. The Sun is likely to end its life as a star similar to Antares.

LIFE ON EARTH

THREE conditions appear to be necessary for life as it exists on Earth: no free hydrogen in the atmosphere; an ample supply of water; and the presence of hydrocarbons. These conditions do not exist on the other planets; nor have they have always existed on Earth.

Early in the history of the Earth, the planet's surface was not sufficiently shielded from harmful solar radiation to allow life to begin. The atmosphere was made primarily of nitrogen and carbon dioxide, and any water vapor was quickly split into hydrogen and oxygen by the effects of solar radiation. The hydrogen escaped into space, but oxygen reacted to form ozone (O_3) or remained as atomic oxygen (O). The ozone collected in the upper atmosphere to form a protective layer which shielded the surface from strong ultraviolet radiation, though this process must have must have taken a very long time.

KEYWORDS

AMINO ACID
AMPHIBIAN
ATMOSPHERE
BLUE-GREEN ALGA
EUKARYOTE
EXTINCTION
HYDROCARBON
ORGANIC MOLECULE
OZONE LAYER
PHOTOSYNTHESIS
STROMATOLITE
TRACE FOSSIL

Small amounts of metal–carbon compounds (carbides) probably also existed, although not abundantly; their presence is suggested by the fact that such compounds are found in meteorites. When carbides react with water they form hydrocarbons. Though the early Earth did not contain a great deal of methane (CH_4), the little that existed may have provided hydrocarbon molecules too.

Life probably began either beneath the sea or in damp sheltered places near to it. Blue-green algae were among the first types of life to evolve. Evidence for this comes from fossilized remains of colonial structures called stromatolites, the oldest of which are about 2 billion years old. These have a similar structure to modern algae and are found in Pre-cambrian rocks in Australia and elsewhere. Ancient bacteria 3 billion years old also have been found.

By about 2.5 billion years ago, prokaryotes (simple marine organisms whose cells have no differentiated nucleus) evolved the capacity to use sunlight to make food: photosynthesis had begun. The prokaryotic cells gradually became more complex until, about 1.2 billion years ago, eukaryotes – organisms with a nucleus in each cell – appeared. Eukaryotes gave rise to the diverse life that now exists on our planet.

In photosynthesis, a living cell (usually in a plant) takes in carbon dioxide and water, and gives off oxygen

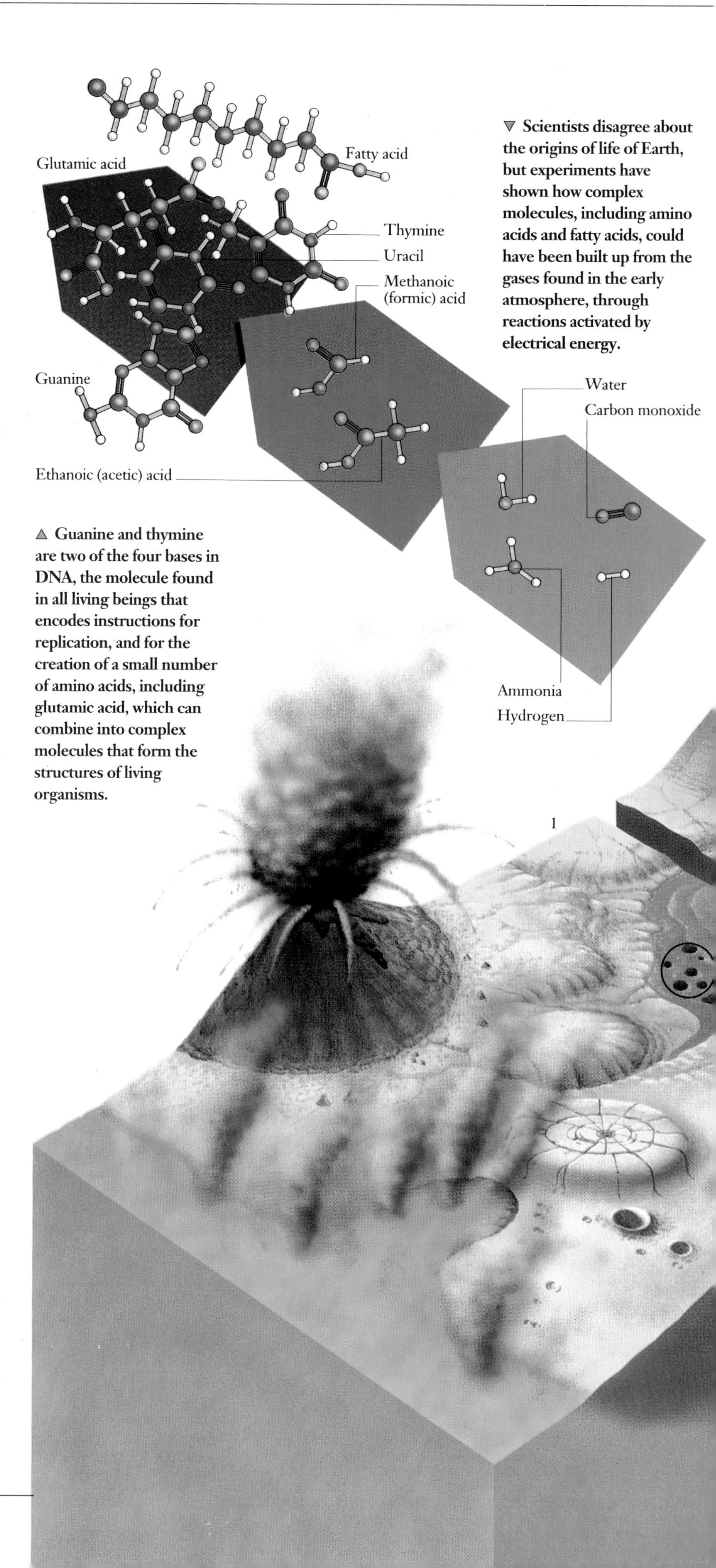

▼ **Scientists disagree about the origins of life of Earth, but experiments have shown how complex molecules, including amino acids and fatty acids, could have been built up from the gases found in the early atmosphere, through reactions activated by electrical energy.**

▲ **Guanine and thymine are two of the four bases in DNA, the molecule found in all living beings that encodes instructions for replication, and for the creation of a small number of amino acids, including glutamic acid, which can combine into complex molecules that form the structures of living organisms.**

▽ **By the Ordovician period (between 505 and 438 million years ago), algae were building reefs near to the shores, and the first corals were appearing, as well as the first land plants. Jawless fish, the ancestors of modern fish, were beginning to be found, as well as trilobites and other arthropods, ancestors of modern crustaceans. These may have resembled the life forms found today in the deepest parts of the ocean INSET, near geothermal vents.**

▷ **During the Carboniferous period (360–286 million years ago), substantial forests emerged both in the swamps and in drier regions, comprising tree ferns INSET, club mosses and giant horsetails. Amphibians and a range of invertebrates, including dragonflies, were also found on land.**

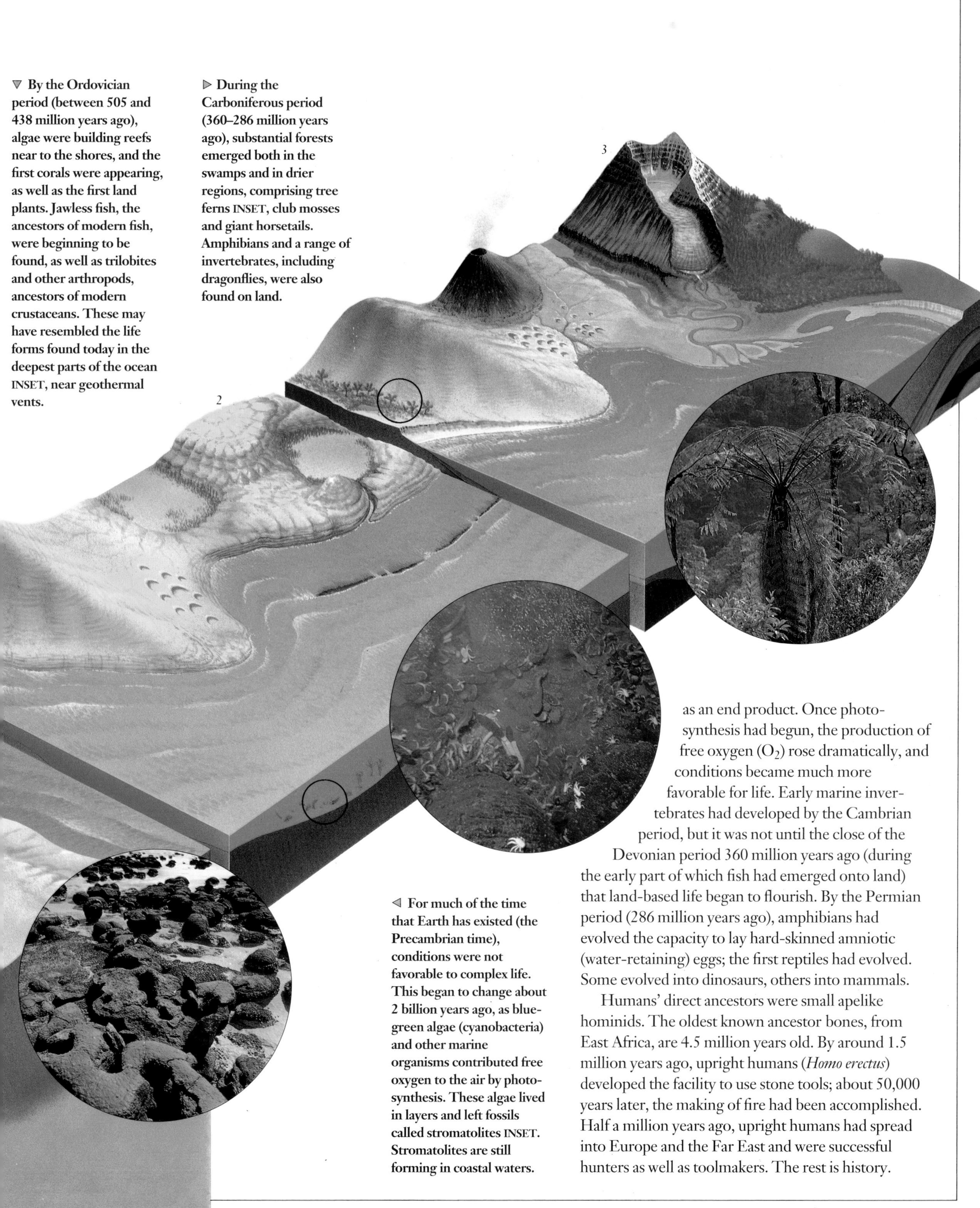

◁ **For much of the time that Earth has existed (the Precambrian time), conditions were not favorable to complex life. This began to change about 2 billion years ago, as blue-green algae (cyanobacteria) and other marine organisms contributed free oxygen to the air by photosynthesis. These algae lived in layers and left fossils called stromatolites INSET. Stromatolites are still forming in coastal waters.**

as an end product. Once photosynthesis had begun, the production of free oxygen (O_2) rose dramatically, and conditions became much more favorable for life. Early marine invertebrates had developed by the Cambrian period, but it was not until the close of the Devonian period 360 million years ago (during the early part of which fish had emerged onto land) that land-based life began to flourish. By the Permian period (286 million years ago), amphibians had evolved the capacity to lay hard-skinned amniotic (water-retaining) eggs; the first reptiles had evolved. Some evolved into dinosaurs, others into mammals.

Humans' direct ancestors were small apelike hominids. The oldest known ancestor bones, from East Africa, are 4.5 million years old. By around 1.5 million years ago, upright humans (*Homo erectus*) developed the facility to use stone tools; about 50,000 years later, the making of fire had been accomplished. Half a million years ago, upright humans had spread into Europe and the Far East and were successful hunters as well as toolmakers. The rest is history.

NATURAL CATASTROPHES

IN ARMERO, Colombia, South America, there was a sequence of earthquakes during November 1984. Steam emissions increased one month later; then, on September 11, 1985, a small explosion threw ash and rocks – parts of the old crater of Nevada del Ruiz volcano – into the air. Nobody appeared to be particularly worried, despite the fact that local officials had been warned that the town of Armero was built on top of a mudflow that had covered the region in 1845, killing 1000 people. The risk was clear, but nothing was done.

By the end of the day of November 13, 1985, 22,000 people in the area were dead. A relatively small eruption had ejected a blanket of hot pumice and ash which melted snowfields high on the volcano above. The meltwater rushed downhill at speeds of more than 35 kilometers per hour, gathering soil, rocks and trees as it did so, and generated a devastating mudflow which raced through the town in a wall 30 meters high. This spread out over the lowlying ground as a series of hot waves, carrying blocks as large as 10 meters high with it. At its peak, an estimated 47,000 cubic meters per second of debris sped downhill: roughly one fifth the discharge rate of the mighty Amazon River.

KEYWORDS

BENIOFF ZONE
CALDERA
EARTHQUAKE
FAULT
K/T BOUNDARY EVENT
LANDSLIDE
SHOCKED QUARTZ
TSUNAMI
VOLCANO

▷ Volcanoes are one of the most spectacular natural catastrophes. Thick clouds of ash tower as high as 20 kilometers into the air. The ash may be deposited on the ground to a depth of 15 cm (6 inches). Within a few kilometers of the blast site, everything is destroyed; the ash may carry for several hundred kilometers and disrupt weather systems all over the globe. Millions of trees may be felled. The eruption of Mount Pinatubo in the Philippines in 1991, shown here, caused no loss of life because the area was evacuated in time.

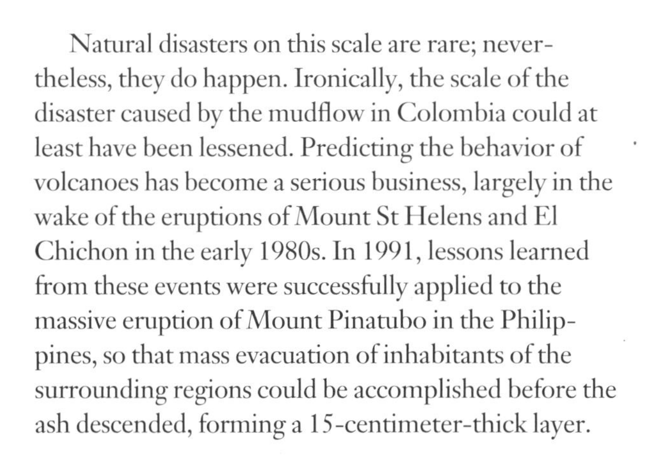

Natural disasters on this scale are rare; nevertheless, they do happen. Ironically, the scale of the disaster caused by the mudflow in Colombia could at least have been lessened. Predicting the behavior of volcanoes has become a serious business, largely in the wake of the eruptions of Mount St Helens and El Chichon in the early 1980s. In 1991, lessons learned from these events were successfully applied to the massive eruption of Mount Pinatubo in the Philippines, so that mass evacuation of inhabitants of the surrounding regions could be accomplished before the ash descended, forming a 15-centimeter-thick layer.

▽ Floods in lowlying coastal areas such as Bangladesh BELOW are common, but none the less devastating; a flood in 1988 left 30 million people homeless. By contrast, the damage in an earthquake such as the one in Los Angeles, California in 1994 BELOW LEFT may appear more severe, but actual loss of life is usually considerably less – only hundreds.

▷ A huge caldera is all that remains of a volcano on the island of Thira (Santorini) off the coast of Greece, which exploded in about 1500 BC, wiping out all life on the island. Calderas are formed when some or all of the magma drains away from the chamber below the cone of a dormant volcano, causing the cone to collapse. The destruction on Thira was so vast, and the caldera so deep, that the bottom is now filled by the sea. The slopes of the caldera, however, are rich in mineral deposits, providing fertile soil and supporting a town.

▷ An unexplained event in Siberia in 1908 felled trees and killed animals. Called the Tunguska event, it may have been due to the explosion of a comet's nucleus in the atmosphere above Siberia. Unlike meteorite impacts, the event did not leave a crater. Nor, unlike the K/T boundary event, an impact which occurred 65 million years ago, did it cause any extinctions.

Earthquakes too have caused large-scale loss of life and widespread damage to property and communications. San Francisco and neighboring settlements which lie along the San Andreas Fault in California are particularly at risk. In 1989 this fault ruptured, reactivating older fractures that had been opened as long ago as 1906. The quake, which registered 7.1 on the Richter Scale, claimed 62 lives and destroyed nearly 1000 homes. In January 1994 a quake on an associated fault caused a similar degree of damage in Los Angeles. This region is one of the most closely monitored seismic zones on Earth. Although the local and state authorities can do nothing to avert an earthquake, their catastrophic effects can be lessened by enforcing strict regulations about building structures. The death toll from earthquakes elsewhere in the world indicates how much this matters: a quake of similar strength killed 20,000 in China in 1974.

There is evidence of one natural disaster that had possible global significance. In recent years, it has been proposed that the extinction of the dinosaurs about 65 million years ago resulted from an enormous impact – possibly from a meteorite 10 kilometers in diameter striking the Earth, raising its temperature and causing darkness for several weeks. This is known as the K/T boundary event from the evidence which comes from a layer of rock at the Cretaceous-Tertiary boundary. Within it are higher than average amounts of iridium, believed to be derived from the meteorite; widespread shocked quartz, feldspar and stishovite (formed at very high pressures); remains of soot (possibly from forest fires caused by the impact); and anomalously high ratios of strontium isotopes, the result of increased weathering rates due to nitric acid rain formed from atmospheric nitrogen during the impact.

This event (or some other natural disaster) set the stage for mammalian evolution to proceed. It eventually allowed *Homo sapiens* to proliferate without competition from the highly successful predecessors which had ruled the world for millions of years.

MAN LEAVES THE EARTH

IN 1960 Vostok 1, a five-tonne spacecraft, launched Major Yuri Gagarin of the Soviet Union into Earth orbit: he thereby became the first man in space. Less than one month later, United States Commander Alan Shepard rose 170 kilometers into the outer atmosphere before returning safely to base.

The first American attempt to reach the Moon with an unmanned spacecraft had taken place during 1958; it failed. Then, in 1959 the Soviet craft Lunik 1 flew past the Moon and sent back information about its magnetic field. Ten years later a massive United States Saturn 5 rocket launched Apollo 11 on its way toward the Moon, enabling astronauts Neil Armstrong and Buzz Aldrin to step down onto the surface of Mare Tranquillitatis. Since that time, several manned and unmanned craft have landed on the Moon and sent back an immense amount of data about its geological history. Roving vehicles allowed astronauts to collect a range of rock samples and bring them back to Earth. Instruments were left on the surface to measure moonquakes, cosmic-ray intensity and magnetic fields.

KEYWORDS

APOLLO MISSIONS
MARINER-9
MARINER-10
VIKING MISSIONS
VOYAGER MISSIONS

The Earth itself has been thoroughly studied from space. During the early 1970s a series of Landsat satellites began to image the Earth's surface from orbit, a task that has continued to the present time. Using a variety of wavelengths, Landsat and its successors have provided global data on terrestrial geology, heatflow, vegetation, land use, ocean currents, weather patterns and gravity.

Much more difficult is the task of sending spacecraft to other planets. This has been accomplished several times with great success. Spectacular images of the surface of Mercury were transmitted by Mariner-10 in 1974; in the same year radar images were obtained by Venera probes, which penetrated to the surface of cloud-covered Venus and analyzed its surface rocks. Several more probes have reached Venus, the most recent being the highly sophisticated radar mapper, Magellan, which began orbiting Venus in August 1990. The data that it returned has caused the geology of that world to be rewritten.

Several missions have reached Mars, the first during 1962. Most of what we have learned has come from data returned by Mariner-9 and by the two Viking missions during the mid-1970s. They included both orbiter and lander probes, and transmitted a huge amount of imagery, supplemented by thermal, atmospheric, geochemical and geophysical data. The launch of a pair of missions, Mars-94 and Mars-96, was planned for 1994 and 1996.

Perhaps the most successful mission of all was Voyager, whose two probes, between 1979 and1989, visited Jupiter, Saturn, Uranus and Neptune, together with many of their moons. Stunning pictures of the Jovian clouds, Saturn's ring system, volcanoes on Io,

▲ **In July 1969 the Apollo 11 astronauts made their giant leap for mankind, and set up detectors on the Moon, such as this seismometer.**

▼ **Probes to the outer planets can involve complex orbits, in which the spacecraft uses the gravitational influence of one planet to accelerate it sufficiently to reach the next. The Voyager 2 craft INSET used an unusual alignment of the planets to reach Neptune in 1989, some seven billion kilometers away, and 12 years after launch. It is hoped that Voyager 2 may eventually send information about the outer limits of the solar wind, in the early 21st century. Voyager 1 passed Jupiter (1979) and Saturn (1980) before leaving the plane of the Solar System. The probe Galileo, launched in 1989, had a flightpath that took it past Venus in 1990, then back past Earth en route to the asteroid belt in October 1991. It flew by Earth again in 1992, to pass the asteroid Ida in 1993 and to reach Jupiter in December 1995.**

Voyager 1
Pioneer 1

geysers on Triton and huge scarps on Miranda gave many new insights into those worlds. Another probe, Galileo, was sent – catapulted via Venus, Earth and Moon (1990), and the asteroid Gaspra (1991) – to Jupiter. A small probe was planned to enter Jupiter's atmosphere, sending back a stream of information about lightning, the magnetic field and charged particles. An orbiting probe was planned to continue to circle Jupiter for a further 20 months.

Despite the success of these unmanned probes, a fully manned mission to Mars to allow further exploration is still a long-term goal of many space scientists. However, securing funding for such a huge project remains an enduring difficulty.

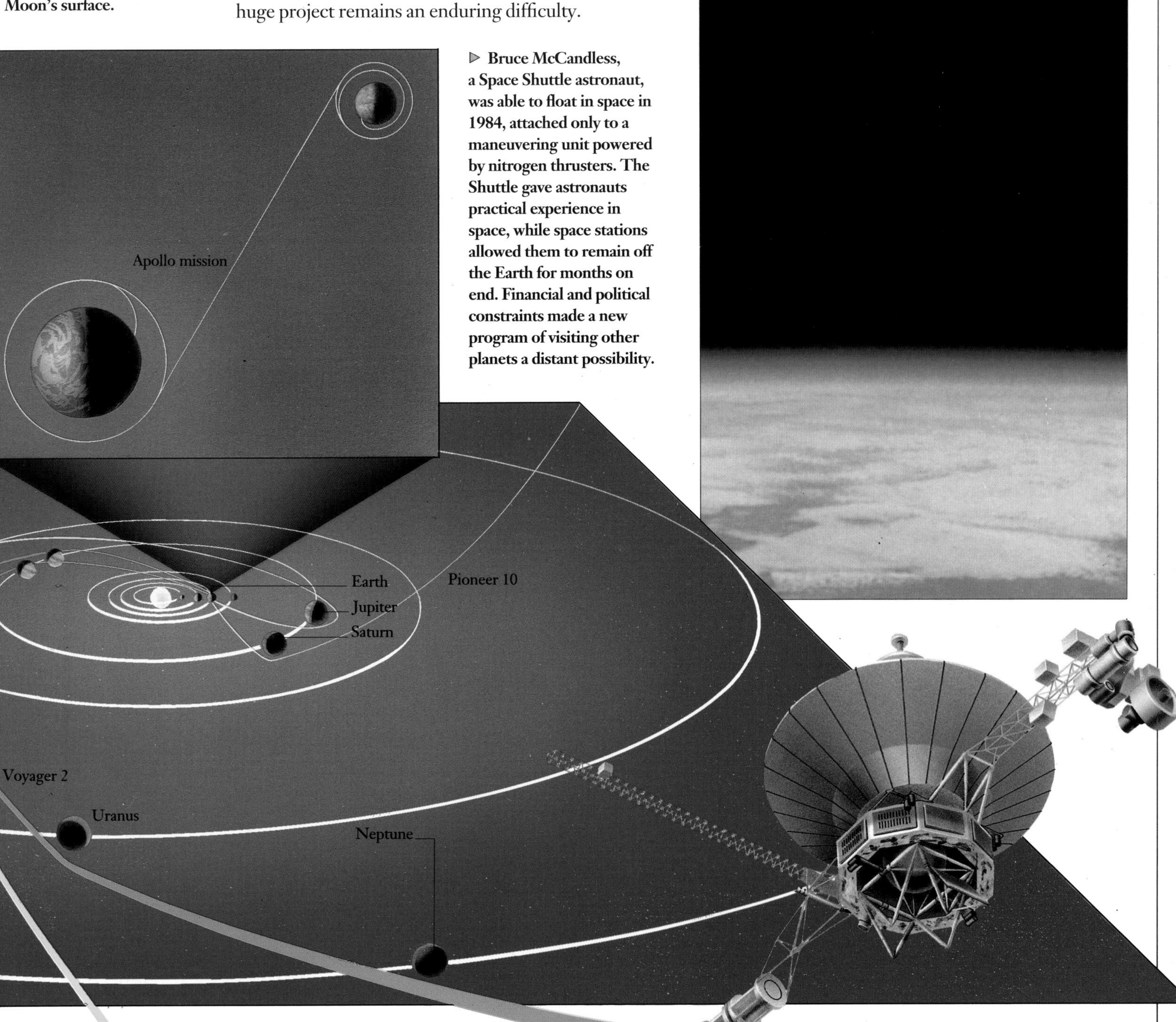

▽ The United States' Apollo missions to the Moon began with the 110m-high, three-stage Saturn rockets putting the spacecraft into orbit around the Earth. The craft was accelerated to 40,000km/h to set off on its 2.5-day, 384,000-km journey to the Moon. On its arrival, one part – the command module – remained in orbit, while a small lunar module was sent down to the Moon's surface.

▷ Bruce McCandless, a Space Shuttle astronaut, was able to float in space in 1984, attached only to a maneuvering unit powered by nitrogen thrusters. The Shuttle gave astronauts practical experience in space, while space stations allowed them to remain off the Earth for months on end. Financial and political constraints made a new program of visiting other planets a distant possibility.

Geological Stories

GEOLOGICAL time has been split into a number of major divisions: eons, eras, epochs and periods. The earliest eon is the Archean, which spans the period between the birth of the Solar System, about 4.6 billion years ago, and 2.5 billion years ago. Evidence for mountain-building in this period is to be found within Earth's rocks, as well as volcanism and deposition of marine sediments. Although the Archean was once defined as the period before life existed, simple life forms are now known to have existed at the end of this period.

KEYWORDS

ARCHEAN ERA
CAMBRIAN
CENOZOIC
CRETACEOUS
JURASSIC
MESOZOIC
PALEOZOIC
PHANEROZOIC
PRECAMBRIAN
PROTEROZOIC
QUATERNARY
TERTIARY

The Archean was followed by the Proterozoic (which means period of first life), which extended until about 590 million years ago. Again, periods of mountain-building punctuated periods of lesser activity and, during the latter part of the Proterozoic, more complex life forms developed in the primeval oceans. Together these two eons comprise the period of time known as the Precambrian. Throughout this time the Earth's continental crust was never more than 40 kilometers thick – considerably less than the modern crust, which can be up to 70 kilometers thick.

During the Archean eon the Earth's crust was repeatedly bombarded by meteoroids and asteroids. This also happened on Mercury, the Moon and Mars, and indeed, on the moons of the outer planets, which are intensely cratered. Samples of Moon rock brought back by the Apollo astronauts show that lunar activity, including volcanic filling of the Moon's maria, had virtually finished by the end of Archean times; this is probably true of Mercury too. Mars and Venus, in contrast, have remained geologically active well into more recent times.

Life on Earth began to multiply and diversify at the beginning of the Cambrian period (590-505 million years ago), the first of a series of geological periods which comprise the Phanerozoic (eon of life). By Cambrian times it appears that volcanism on Mars was well past its peak; however, resurfacing of Venus by volcanic activity seems to have continued beyond that point, and possibly continued until the present time.

Although the Universe is so large, the number of stars resembling our Sun is so high, and many scientists consider the likelihood that the conditions for life to emerge on other planets in other parts of the Universe is high, no such planets have yet been found, and no signs of life have been found on the other planets in our own Solar System.

The first mammals evolved over 200 million years ago, but their domination of the Earth may have been assisted by a collision between an asteroid and the Earth about 65 million years ago, causing the extinction of the hugely successful dinosaurs. Had this not taken place, humans might not now be in their privileged position: the earliest ancestors of the human race did not emerge, probably in East Africa, until the Pliocene epoch, no more than four or five million years ago.

■ The geological timescale is based on obvious changes in rock type or fossil groups from one period to the next. The major divisions are the eons, which are divided into eras, and further broken down into periods and, for the most

1 Shelled organisms
2 Early jawless fish
3 Tree ferns, horsetails
4 Air-breathing fish
5 Large amphibians
6 Early reptiles
7 Ginkgos, conifers
8 Early mammals
9 First birds
10 Flowering plants
11 Last dinosaurs
12 First primates
13 Early horses
14 Modern humans

recent era, covering the last 65 million years, epochs. The Precambrian era encompasses about four-fifths of the Earth's entire history. The end of the Precambrian era was marked by the appearance of the first fossils. Major life forms developed through the major periods, starting with the Cambrian; jawless fish emerged in the Ordovician period (505-438 million years ago), the first land plants were found in the Silurian (438–408) and the first amphibians during the Devonian (408–360 million years ago). The first mammals appeared in the Triassic period (248-213 million years ago). The recent Ice Age was in the Pleistocene and Holocene.

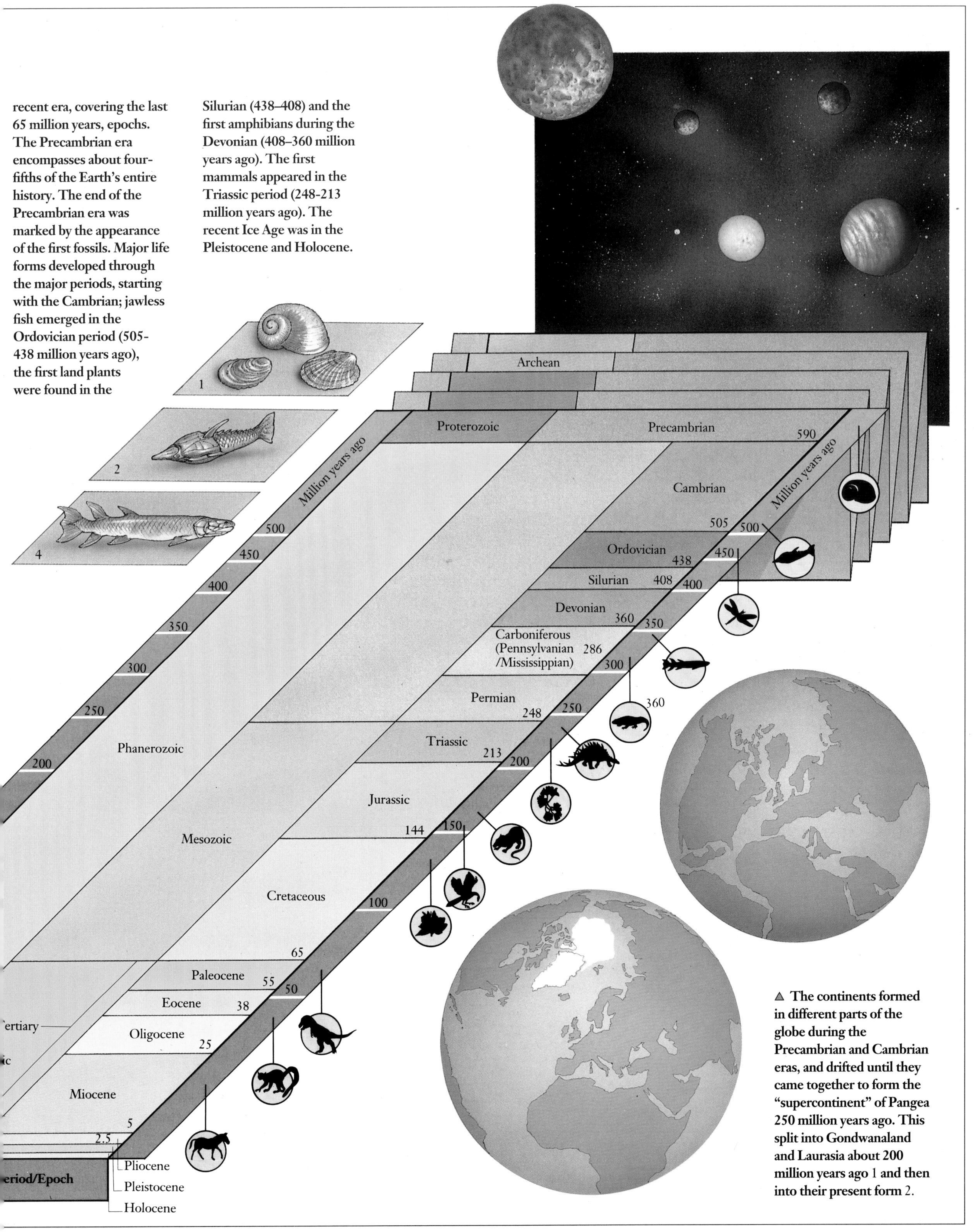

▲ The continents formed in different parts of the globe during the Precambrian and Cambrian eras, and drifted until they came together to form the "supercontinent" of Pangea 250 million years ago. This split into Gondwanaland and Laurasia about 200 million years ago 1 and then into their present form 2.

FACTFILE

PRECISE MEASUREMENT is at the heart of all science, and the several standard systems have been in use in the present century in different societies. Today, the SI system of units is universally used by scientists, but other units are used in some parts of the world. The metric system, which was developed in France in the late 18th century, is in everyday use in many countries, as well as being used by scientists; but imperial units (based on the traditional British measurement standard, also known as the foot–pound–second system), and standard units (based on commonly used American standards) are still in common use.

Whereas the basic units of length, mass and time were originally defined arbitrarily, scientists have sought to establish definitions of these which can be related to measurable physical constants; thus length is now defined in terms of the speed of light, and time in terms of the vibrations of a crystal of a particular atom. Mass, however, still eludes such definition, and is based on a piece of platinum-iridium metal kept in Sèvres, near Paris.

METRIC PREFIXES

Very large and very small units are often written using powers of ten; in addition the following prefixes are also used with SI units. Examples include: milligram (mg), meaning one thousandth of a gram, kilogram (kg), meaning one thousand grams.

Name	Number	Factor	Prefix	Symbol
trillionth	0.000000000001	10^{-12}	pico-	p
billionth	0.000000001	10^{-9}	nano-	n
millionth	0.000001	10^{-6}	micro-	µ
thousandth	0.001	10^{-3}	milli-	m
hundredth	0.01	10^{-2}	centi-	c
tenth	0.1	10^{-1}	deci-	d
one	1.0	10^{0}	–	–
ten	10	10^{1}	deca-	da
hundred	100	10^{2}	hecto-	h
thousand	1000	10^{3}	kilo-	k
million	1,000,000	10^{6}	mega-	M
billion	1,000,000,000	10^{9}	giga-	G
trillion	1,000,000,000,000	10^{12}	tera-	T
quadrillion	1,000,000,000,000,000	10^{15}	exa-	E

CONVERSION FACTORS

Conversion of METRIC units to imperial (or standard) units

To convert:	to:	multiply by:
LENGTH		
millimeters	inches	0.03937
centimeters	inches	0.3937
meters	inches	39.37
meters	feet	3.2808
meters	yards	1.0936
kilometers	miles	0.6214
AREA		
square centimeters	square inches	0.1552
square meters	square feet	10.7636
square meters	square yards	1.196
square kilometers	square miles	0.3861
square kilometers	acres	247.1
hectares	acres	2.471
VOLUME		
cubic centimeters	cubic inches	0.061
cubic meters	cubic feet	35.315
cubic meters	cubic yards	1.308
cubic kilometers	cubic miles	0.2399
CAPACITY		
milliliters	fluid ounces	0.0351
milliliters	pints	0.00176 (0.002114 for US pints)
liters	pints	1.760 (2.114 for US pints)
liters	gallons	0.2193 (0.2643 for US gallons)
WEIGHT		
grams	ounces	0.0352
grams	pounds	0.0022
kilograms	pounds	2.2046
tonnes	tons	0.9842 (1.1023 for US, or short, tons)
TEMPERATURE		
Celsius	fahrenheit	1.8, then add 32

Conversion of STANDARD (or imperial) units to metric units

To convert:	to:	multiply by:
LENGTH		
inches	millimeters	25.4
inches	centimeters	2.54
inches	meters	0.245
feet	meters	0.3048
yards	meters	0.9144
miles	kilometers	1.6094
AREA		
square inches	square centimeters	6.4516
square feet	square meters	0.0929
square yards	square meters	0.8316
square miles	square kilometers	2.5898
acres	hectares	0.4047
acres	square kilometers	0.00405
VOLUME		
cubic inches	cubic centimeters	16.3871
cubic feet	cubic meters	0.0283
cubic yards	cubic meters	0.7646
cubic miles	cubic kilometers	4.1678
CAPACITY		
fluid ounces	milliliters	28.5
pints	milliliters	568.0 (473.32 for US pints)
pints	liters	0.568 (0.4733 for US pints)
gallons	liters	4.55 (3.785 for US gallons)
WEIGHT		
ounces	grams	28.3495
pounds	grams	453.592
pounds	kilograms	0.4536
tons	tonnes	1.0161
TEMPERATURE		
fahrenheit	Celsius	subtract 32, then × 0.55556

SI UNITS

Now universally employed throughout the world of science and the legal standard in many countries, SI units (short for *Système International d'Unités*) were adopted by the General Conference on Weights and Measures in 1960. There are seven base units and two supplementary ones, which replaced those of the MKS (meter–kilogram–second) and CGS (centimeter–gram–second) systems that were used previously. There are also 18 derived units, and all SI units have an internationally agreed symbol.

None of the unit terms, even if named for a notable scientist, begins with a capital letter: thus, for example, the units of temperature and force are the kelvin and the newton (the abbreviations of some units are capitalized, however). Apart from the kilogram, which is an arbitrary standard based on a carefully preserved piece of metal, all the basic units are now defined in a manner that permits them to be measured conveniently in a laboratory.

Name	Symbol	Quantity	Standard
BASIC UNITS			
meter	m	length	The distance light travels in a vacuum in 1/299,792,458 of a second
kilogram	kg	mass	The mass of the international prototype kilogram, a cylinder of platinum-iridium alloy, kept at Sèvres, France
second	s	time	The time taken for 9,192,631,770 resonance vibrations of an atom of cesium-133
kelvin	K	temperature	1/273.16 of the thermodynamic temperature of the triple point of water
ampere	A	electric current	The current that produces a force of 2×10^{-7} newtons per meter between two parallel conductors of infinite length and negligible cross section, placed one meter apart in a vacuum
mole	mol	amount of substance	The amount of a substance that contains as many atoms, molecules, ions or subatomic particles as 12 grams of carbon-12 has atoms
candela	cd	luminous intensity	The luminous intensity of a source that emits monochromatic light of a frequency 540×10^{-12} hertz and whose radiant intensity is 1/683 watt per steradian in a given direction
SUPPLEMENTARY UNITS			
radian	rad	plane angle	The angle subtended at the center of a circle by an arc whose length is the radius of the circle
steradian	sr	solid angle	The solid angle subtended at the center of a sphere by a part of the surface whose area is equal to the square of the radius of the sphere

Name	Symbol	Quantity	Standard
DERIVED UNITS			
becquerel	Bq	radioactivity	The activity of a quantity of a radio-isotope in which 1 nucleus decays (on average) every second
coulomb	C	electric current	The quantity of electricity carried by a charge of 1 ampere flowing for 1 second
farad	F	electric capacitance	The capacitance that holds charge of 1 coulomb when it is charged by a potential difference of 1 volt
gray	Gy	absorbed dose	The dosage of ionizing radiation equal to 1 joule of energy per kilogram
henry	H	inductance	The mutual inductance in a closed circuit in which an electromotive force of 1 volt is produced by a current that varies at 1 ampere per second
hertz	Hz	frequency	The frequency of 1 cycle per second
joule	J	energy	The work done when a force of 1 newton moves its point of application 1 meter in its direction of application
lumen	lm	luminous flux	The amount of light emitted per unit solid angle by a source of 1 candela intensity
lux	lx	illuminance	The amount of light that illuminates 1 square meter with a flux of 1 lumen
newton	N	force	The force that gives a mass of 1 kilogram an acceleration of 1 meter per second per second
ohm	Ω	electric resistance	The resistance of a conductor across which a potential of 1 volt produces a current of 1 ampere
pascal	Pa	pressure	The pressure exerted when a force of 1 newton acts on an area of 1 square meter
siemens	S	electric conductance	The conductance of a material or circuit component that has a resistance of 1 ohm
sievert	Sv	dose	The radiation dosage equal to 1 joule equivalent of radiant energy per kilogram
tesla	T	magnetic flux density	The flux density (or density induction) of 1 weber of magnetic flux per square meter
volt	V	electric potential	The potential difference across a conductor in which a constant current of 1 ampere dissipates 1 watt of power
watt	W	power	The amount of power equal to a rate of energy transfer of (or rate of doing work at) 1 joule per second
weber	Wb	magnetic flux	The amount of magnetic flux that, decaying to zero in 1 second, induces an electromotive force of 1 volt in a circuit of one turn

☐ THE SOLAR SYSTEM

The Solar System comprises the planets, their satellites, numerous relatively large rocky objects within the asteroid belt, and comets, the most distant members of the Sun's family. It also includes millions of smaller particles, comprising interplanetary dust, meteoroids, tiny outer planet moons and ring particles, and small asteroids. The tables below give data relating only to the major objects. For example, the asteroid belt includes many thousand objects; there are estimated to be at least 700 asteroidal bodies with diameters greater than 1 kilometer.

The orbital period is the time taken for a planet to orbit the Sun; its rotation period is usually also the length of its day. Orbital eccentricity is the amount by which the orbit deviates from a circle; orbital inclination is the angle by which its orbit deviates from the ecliptic.

The inner rocky planets (Mercury, Venus, Earth and Mars) have shapes that depart little from spherical; however, the rapidly-rotating outer gaseous worlds are highly oblate, or squashed, the polar diameters being significantly less than the equatorial ones. The relatively long rotation periods of the inner group – measurable in days – contrast with those of the outer group, which are generally measurable in hours (Pluto being the exception). The larger planetary satellites also are spherical, but the smaller ones often have irregular shapes, like many asteroids. Comet nuclei also are believed to be irregular in shape. Because of the highly elliptical orbits of most comets, their sidereal periods are listed; this is the time it takes each object to complete one circuit around the Sun. This time is thus equivalent to the orbital periods quoted for the major planets.

THE PLANETS

	Distance from Sun (million km)		Orbital period	Rotation period	Inclination of axis	Orbital eccentricity	Orbital inclination	Equatorial diameter (km)	Escape velocity (km/sec)	Mean surface temperature (°C)	Mass (Earth =1)	Volume (Earth =1)	Density (water =1)	Surface gravity (Earth =1)
	max	min												
Mercury	69.7	45.9	87.97 days	58.65 days	0°	0.206	7°00'16"	4878	4.3	+350	0.055	0.056	5.5	0.38
Venus	109	107.4	224.7 days	243.16 days	178°	0.007	3°23'40"	12,012	10.4	+480	0.815	0.86	5.3	0.9
Earth	152	147	365.26 days	23 hr 56 min 4 sec	23.5°	0.017	0°	12,750	11.2	+22	1	1	5.5	1
Mars	249	207	686.9 days	24 hr 37 min 27 sec	23.98°	0.093	1°50'59"	6787	5.0	-23	0.107	0.15	3.9	0.38
Jupiter	816	741	11.86 years	9 hr 50 min 30 sec	3.2°	0.048	1°18'16"	143,884	60.2	-150	318	1319	1.3	2.64
Saturn	1507	1347	29.46 years	10 hr 39 min	26.7°	0.056	2°29'22"	120,536	32.3	-180	95	744	0.7	1.16
Uranus	3004	2735	84.01 years	17 hr 14 min	98°	0.047	0°46'28"	51,118	22.5	-214	15	67	1.3	1.17
Neptune	4537	4456	164.8 years	16 hr 3 min	29°	0.009	1°45'20"	50,538	23.9	-220	17	57	1.5	1.2
Pluto	7375	4425	247.7 years	6 days 9 hr	88–112°	0.248	17°12'	2445	low	-230	0.002	<0.01	2.0	low

ASTEROIDS

No.	Name	Date found	Distance from Sun (astronomical unit) min	max	Sidereal period (yr)	Diameter (km)
1	**Ceres**	1801	2.55	2.94	4.6	1003
2	**Pallas**	1802	2.11	3.42	4.60	535
3	**Juno**	1804	1.98	3.35	4.36	250
4	**Vesta**	1807	2.15	2.57	3.63	500
5	**Astrea**	1845	2.10	3.06	4.14	180
6	**Hebe**	1847	1.93	2.92	3.78	195
7	**Iris**	1847	1.84	2.94	3.69	209
8	**Flora**	1847	1.86	2.55	3.27	151
9	**Metis**	1848	2.09	2.68	3.68	151
10	**Hygeia**	1849	2.84	3.46	5.59	430
15	**Eunomia**	1851	2.15	3.14	4.30	272
16	**Psyche**	1851	2.53	3.32	5.00	250
24	**Themis**	1853	2.76	3.52	5.53	249
44	**Nysa**	1857	2.05	2.79	3.77	82
65	**Cybele**	1861	3.01	3.83	6.33	309
95	**Arethusa**	1867	2.61	3.53	5.37	230
451	**Patienta**	1899	2.82	3.30	5.34	276
611	**Patroclus**	1906	4.48	5.94	11.88	147
624	**Hector**	1907	4.99	5.25	11.59	179
944	**Hidalgo**	1920	2.02	9.68	14.04	15
1221	**Amor**	1932	1.08	2.76	2.66	5
1566	**Icarus**	1949	0.19	1.97	1.12	1
1862	**Apollo**	1932	0.65	2.29	1.78	3
2060	**Chiron**	1977	8.43	18.8	50.7	500?

COMETS

Name	Sidereal period (yr)	Inclination	Perihelion (million km)	Meteor shower
Encke	3.3	12°	50.8	Taurids
Grigg-Skjellerup	5.1	21°	14.8	
Machholtz	5.2	60°	19.5	
Wild 2	6.2	3°	223	
Kopff	6.4	5°	236	
Giacobini-Zinner	6.6	32°	154	October Draconids
Biela	6.6	13°	129	Andromedids
Whipple	8.5	10°	460	
Schwassmann-Wachmann 1	15.0	10°	815	
Hartley-IRAS	21.5	96°	191	
Halley	76	162°	88.3	Aquarids Orionids
Swift-Tuttle	120?	114°	144	Perseids
Wilk	187	26°	92.7	-
Metcalf 1919 V		46°	167	Draconids
Ikeya 1964 VIII	391	172°	123	Geminids
Ikeya-Seki 1965 VIII	880	142°	1.2	-
Bennett 1970 II	-	90°	80.7	-
Kohoutek 1973 XII	-	14°	21.0	-
Bennett 1974 XV_d	-	90°	78.8	-
Schuster 1975 II	-	112°	1025.1	-
West1976 VI	-	43°	29.2	-

THE MAJOR MOONS OF THE PLANETS

The satellite moons of the planets formed of the same stuff as the Sun and planets, and condensed out of the solar nebula at the same time. They are built from silicates (in the inner Solar System) or ices (in the outer reaches). Several, such as Titan and Ganymede, are larger than Mercury, but most are substantially smaller than even the smallest planets. Only Titan has a significant atmosphere; the rest are barren worlds with solid surfaces frequently scarred by impact craters.

In the inner Solar System, Earth's Moon is by far the largest satellite and is unique in that it is as much as quarter as large as its primary. No other moon has such a large relative size. Of the other inner planets, only Mars has moons: Phobos and Deimos. Both are irregular silicate bodies less than 30 kilometers across. They resemble asteroids in composition, size and aspect, and astronomers presume that they are in fact captured asteroids.

Triton, which is associated with Neptune, reveals the activity of geyser-like volcanoes on its surface, involving nitrogen-bearing compounds. It may have been a separate body that was captured by Neptune. Titan, the largest moon of Saturn, is also the second largest moon in the Solar System. It is unique in being surrounded by a nitrogen-rich atmosphere.

The most famous group of outer system moons are the four larger satellites of Jupiter, which were first seen by Galileo Galilei in 1610. These "Galilean" moons can be seen through small telescopes, and exhibit a complex series of eclipses, occultations and transits that have been observed for centuries. Ganymede is a little larger than Titan and is the largest moon in the Solar System, but has no atmosphere. It has a complicated geology, there being lighter and darker regions pockmarked with craters and traversed by peculiar grooves. Callisto has an icy surface and the most extensive impact record in the Solar System. Europa appears to have the smoothest surface, whereas Io, the smallest, is unique in that it has a yellow–orange surface devoid of craters. The coloration is due to sulfur compounds, and the surface features are due to active volcanism. Io, the innermost of the four moons, is the most volcanically active world in the entire Solar System and appears to be resurfacing itself every million years or so. This resurfacing explains why no impact craters are visible. All of the Galilean moons are comparable in size to Earth's Moon.

The larger moons of Saturn and Uranus have heavily cratered surfaces, some being traversed by grooves and fractures. However, the surface of Titan cannot be seen through its atmosphere. The most bizarre outer system moon is perhaps Miranda, associated with Uranus. It is broken up by immense fault scarps several kilometers high. It appears that this body may have broken apart and then reassembled.

Charon, the companion body to Pluto, is sometimes described as a moon, but the two bodies are less than 20,000 kilometers apart, and Charon is only a little less than half the diameter of Pluto, so that they are often considered as forming a binary planet system.

Name of moon	Distance from planet (km)	Orbital period (days)	Diameter (km)	Escape velocity (km/h)	Magnitude
EARTH					
Moon	384,392	27.32	3476	2.38	–12.7
MARS					
Phobos	9270	0.319	28×23×20	0.016	11.6
Deimos	23,500	1.2624	8	0.008	12.8
JUPITER					
Metis	127,960	0.295	40	0.02?	17.4
Adrastea	128,980	0.298	26×20×16	0.01?	18.9
Amalthea	181,300	0.498	262×146×134	0.16?	14.1
Thebe	221,900	0.675	110×100×90	0.1?	15.5
Io	421,600	1.769	3642	2.56	5.0
Europa	671,000	3.551	3126	2.10	5.3
Ganymede	1,070,000	7.12	5276	2.78	4.6
Callisto	1,800,000	16.689	4820	2.43	5.6
Leda	11,094,000	238.7	10	0.005?	20.2
Himalia	11,480,000	250.6	170	0.1?	14.8
Lysithea	11,720,000	259.2	24	0.01?	18.4
Elara	11,737,000	259.7	80	0.05?	16.7
Ananke	21,200,000	631	20	0.01?	18.9
Carme	22,600,000	692	30	0.02?	18.0
Pasiphaë	23,500,000	735	36	0.02?	17.7
Sinope	23,700,000	758	28	0.01?	18.3
SATURN					
Pan	133,600	0.58	20?	-	-
Atlas	137,670	0.602	37×34×27	-	18.1
Prometheus	139,350	0.613	148×100×68	-	16.5
Pandora	141,700	0.629	110×88×62	-	16.3
Epimetheus	151,420	0.694	138×110×110	-	15.5
Janus	151,470	0.695	190×194×154	-	14.5
Mimas	185,540	0.942	398	0.1	12.9
Enceladus	238,040	1.370	498	0.2	11.8
Tethys	294,700	1.888	1060	0.4	10.3
Telesto	294,670	1.888	30×26×16	-	19.0
Calypso	294,670	1.888	30×16×16	-	18.5

Name of moon	Distance from planet (km)	Orbital period (days)	Diameter (km)	Escape velocity (km/h)	Magnitude
Dione	377,000	2.737	1120	0.9	10.4
Helene	377,000	2.737	36×34×28	-	18.5
Rhea	527,040	4.518	1528	0.6	9.7
Titan	1,221,000	15.945	5150	2.47	8.4
Hyperion	1,481,100	21.277	360×280×226	0.2	14.2
Iapetus	3,561,300	79.331	1436	0.7	10(var)
Phoebe	12,954,000	550.4	220	0.1	16.5
URANUS					
Cordelia	49,471	0.330	26	very low	-
Ophelia	53,796	0.372	30	very low	-
Bianca	59,173	0.433	42	very low	-
Cressida	61,777	0.463	62	very low	-
Desdemona	62,676	0.475	54	very low	-
Juliet	64,352	0.493	84	very low	-
Portia	66,085	0.513	108	very low	-
Rosalind	69,941	0.558	54	very low	-
Belinda	75,258	0.622	66	very low	-
Puck	86,000	0.762	154	very low	-
Miranda	129,400	1.414	472	0.5?	16.5
Ariel	130,000	2,520	1160	1.2	14.4
Umbriel	266,000	4.144	1190	1.2	15.3
Titania	583,400	8.706	1580	1.6	14.0
Oberon	583,400	13.463	1526	1.5	14.2
NEPTUNE					
Naiad	48,200	0.30	50	very low	-
Thalassa	50,000	0.31	90	very low	-
Galatea	52,500	0.33	140	very low	-
Despina	62,000	0.40	160	very low	-
Larissa	73,600	0.56	200	very low	-
Proteus	117,600	1.12	420	very low	-
Triton	354,800	5.877	2705	low	13.6
Nereid	5,513,400	365.2	340	very low	18.7
PLUTO					
Charon	20,000	6.4	1200	low	19

The Earth is the planet about which we know the most. In shape it is almost spherical: unlike the outer planets, it has not been pulled out of shape much, partly because it has a rigid crust of solid silicate rock, and partly because its rotation is relatively slow (24 hours). Our planet has a layered structure, with a dense iron-nickel core enclosed in a silicate mantle that is itself surrounded by the relatively thin crust. The outer part of the core is liquid and within it is generated a self-perpetuating dynamo that produces the magnetic field. The brittle crust and the upper regions of the mantle are in constant slow motion above a region of less dense material known as the asthenosphere. The region above that layer, termed the lithosphere, has a segmented structure.

Individual segments of lithosphere are called lithospheric plates or tectonic plates. They move in response to convective motions in the mantle layer beneath; these movements are more complex than any yet found on other planets. Major geological activity is concentrated along the boundaries between the plates, and the results of this activity are manifested in many of the largest surface features such as oceans, deep-sea trenches, mountains, volcanoes and islands. The highest point on the Earth's continents is Mount Everest, on the borders of China and Nepal, which stands 8848 meters above sea level; however the greatest difference in relative relief is not to be found above the sea, but below it: Mariana Trench, in the Western Pacific, reaches 11,022 meters below sea level.

Seventy percent of the Earth's surface is today covered by water which occupies the depressed ocean basins. These are floored by thin but dense crust, which is generated at major mid-oceanic rift-ridges. These ridges are the sites of oceanic crust formation, and link up to form a global network.

The present upland regions are located on the terrestrial continents. These stand above the mean (or sea) level, are composed of material that is less dense than the ocean floors, and have grown in area through the Earth's history. Mountain ranges form as a result of collisions between tectonic plates, which deform and push up once-buried sections of predominantly

THE EARTH

Equatorial radius	6378 km
Polar radius	6357 km
Radius of sphere with Earth's volume	6371 km
Volume	1.083×10^{27} cc
Surface area	5.1×10^{18} sq cm
Average elevation of land	623 m
Average depth of oceans	3.8 km
Mass	5.976×10^{27}gm
Average density	5.517 gm/cc
Gravity at Equator	978.032 cm/sec/sec
Mass of atmosphere	5.1×10^{21} gm
Mass of ice	$25\text{–}30 \times 10^{21}$ gm
Mass of oceans	1.4×10^{24} gm
Mass of crust	2.5×10^{25}gm
Mass of mantle	4.05×10^{27} gm
Mass of core	1.90×10^{27} gm
Mean distance to Sun	1.496×10^{8} km
Rotational velocity	7.292×10^{-5} radians/sec (40,000 km/day linear velocity at Equator)
Average velocity around Sun	29.77 km/sec
Ratio: Mass of Sun: mass of Earth	3.329×10^{5}
Ratio: Mass of Earth: mass of Moon	81.303

THE WORLD AND THE CONTINENTS

	Area (sq km)	% of surface
The World	509,450,000	
Land	149,450,000	29.3
Water	360,000,000	70.7
Asia	44,500,000	29.8
Africa	30,302,000	20.3
North America	24,241,000	16.2
South America	17,793,000	11.9
Antarctica	14,100,000	9.4
Europe	9,957,000	6.7
Australasia & Oceania	8,557,000	5.7

OCEANS

	Area (sq km)	% of total ocean surface
Pacific Ocean	179,679,000	49.9
Atlantic Ocean	92,373,000	25.7
Indian Ocean	73,917,000	20.5
Arctic Ocean	14,090,000	3.9

SEAS

	Area (sq km)
Caribbean Sea	2,766,000
Mediterranean Sea	2,516,000
South China Sea	2,318,000
Bering Sea	2,268,000
Sea of Okhotsk	1,528,000
Gulf of Mexico	1,543,000
East China and Yellow Sea	1,249,000
Hudson Bay	1,232,000
Sea of Japan	1,008,000
North Sea	575,000
Black Sea	452,000
Red Sea	438,000
Baltic Sea	397,000
The Gulf	239,000
Gulf of St Lawrence	238,000
Gulf of California	162,000
Bass Strait	75,000

MOUNTAINS

Name	Location	Height (m)
Everest	China/Nepal	8848
Godwin Austen (K2)	China/Kashmir	8611
Kanchenjunga	India/Nepal	8598
Lhotse	China/Nepal	8516
Makalu	China/Nepal	8481
Cho Oyu	China/Nepal	8201
Dhaulgiri	Nepal	8172
Manaslu	Nepal	8156
Nanga Parbat	Kashmir	8126
Annapurna	Nepal	8078
Gasherbrum	China/Kashmir	8068
Broad Peak	India	8051
Gosainthan	China	8012
Disteghil Sar	Kashmir	7885
Nuptse	Nepal	7879
Masherbrum	Kashmir	7826
Nanda Devi	India	7817
Rakaposhi	Kashmir	7788
Kamet	India	7756
Namcha Barwa	China	7756
Gurla Mandhata	China	7728
Aconcagua	Argentina	6960
Illimani	Bolivia	6882
Bonete	Argentina	6872
Ojos del Salado	Argentina/Chile	6836
Tupungato	Argentina/Chile	6800
Pissis	Argentina	6779
Mercedario	Argentina/Chile	6770
Huascaran	Peru	6768
Llullaillaco	Argentina/Chile	6723
Mt McKinley	USA (Alaska)	6194
Mt Logan	Canada	6050
Kilimanjaro	Tanzania	5895
Citlaltepeti	Mexico	5700
Mt St Elias	USA/Canada	5489
Popocatepetl	Mexico	5452
Mt Foraker	USA (Alaska)	5304
Ixaccihuatl	Mexico	5286
Lucania	USA (Alaska)	5226
Mt Kenya	Kenya	5199
Ruwenzori	Uganda/Zaire	5109
Puncak Jaya	Indonesia	5029
Mt Steele	Canada	5011
Mt Bona	USA (Alaska)	5005
Mt Blackburn	USA (Alaska)	4996
Vinson Massif	Antarctica	4897
Puncak Mandala	Indonesia	4760
Puncak Trikora	Indonesia	4750
Ras Dashan	Ethiopia	4620
Meru	Tanzania	4565
Mt Kirkpatrick	Antarctica	4528
Mt Wilhelm	Papua NG	4508
Karisimbi	Rwanda/Zaire	4507
Mont Blanc	France/Italy	4807
Monte Rosa	Italy/Switzerland	4634
Dom	Switzerland	4545
Weisshorn	Switzerland	4505

sedimentary rock along narrow zones where continental and oceanic crust collide. Some ranges, such as the Himalayas, are located well away from oceanic margins. In these cases a collision between two continental plates has thrust together wedges of continental sedimentary material.

Whereas most of the Earth's highest mountains are the result of such crustal folding, some have been built by volcanic activity, which is also usually associated with plate boundaries. Thus Aconcagua, in Argentina, is a volcano set high among the Andean fold chain. Kilimanjaro, the highest mountain in Africa, is simply a huge volcano which rises up from the shoulder of the East African Rift Valley. Mount Kenya also is volcanic, as is Meru, located in northern Tanzania.

Deformation of the crust and volcanic activity also are responsible for some of the largest islands on the Earth. Papua New Guinea, Sumatra and New Zealand are all the manifestation of such forces. In effect they are mountain ranges rising out of the sea. Iceland is unusual in that it has grown at the intersection of the mid-Atlantic Ridge with a hot spot (or region where hot mantle material comes unusually close to the surface) and fracture zones.

Earth is also unique among the planets in that has widely developed oceans and river systems. Mars once had flowing water on its surface and, indeed, bodies of standing water, but the climate is now far too cold for these to exist. Martian rivers cannot have flowed for at least three billion years. Venus may once have had surface water but this has long since been boiled away. Mercury and the Moon have probably never contained water. Terrestrial river systems have flowed for billions of years and some have developed enormous catchment areas: the Amazon basin has a catchment area almost as large as Australia. Rivers are instrumental in transporting huge volumes of sedimentary material from the continents into the oceans. It has been estimated that each year 225×10^8 tonnes of detritus is carried by them. Glaciers, the wind and marine currents also move large amounts of sediment around the globe.

OCEAN DEPTHS

	Depth (m)
Mariana Trench	11,022
Tonga Trench	10,822
Japan Trench	10,554
Kuril Trench	10,542
Mindanao Trench	10,497
Kermadec Trench	10,047
Puerto Rico (Milwaukee) Deep	9200
Peru-Chile Trench	8050
Aleutian Trench	7822
Cayman Trench	7680
Java Trench	7450
Middle American Trench	6662
Molloy Deep	5608
Gulf of Mexico	5203
Mediterranean Sea	5121
Red Sea	2266
Black Sea	2211

LAND LOWS

Name	Location	Depth (m)
Dead Sea	Asia	-400
Lake Assal	Africa	-156
Death Valley	N. America	-86
Valdes Peninsula	S. America	-40
Caspian Sea	Europe	-28
Lake Eyre North	Oceania	-16

RIVERS

Name	Location of Mouth	Length (km)
Nile	Mediterranean Sea	6670
Amazon	Atlantic Ocean	6430
Yangtse	Pacific Ocean	6380
Mississippi/ Missouri	Gulf of Mexico	6020
Yenisei-Angara	Arctic Ocean	5550
Ob-Irtysh	Arctic Ocean	5410
Hwang Ho	Pacific Ocean	4840
Zaire/Congo	Atlantic Ocean	4670
Amur	Pacific Ocean	4510
Mekong	Pacific Ocean	4500
Mackenzie	Arctic Ocean	4240
Niger	Atlantic Ocean	4180
Paraná-Plate	Atlantic Ocean	4000
Murray-Darling	Indian Ocean	3720
Volga	Caspian Sea	3700
Purus	Amazon River	3350
Madeira	Amazon River	3200
Yukon	Pacific Ocean	3185
Darling	Murray River	3070
Rio Grande	Gulf of Mexico	3030
São Francisco	Atlantic Ocean	2900
Danube	Black Sea	2850
Parana	Plate River	2800
Zambezi	Indian Ocean	2740
Tocantins	Atlantic Ocean	2640
Murray	Indian Ocean	2575
Paraguay	Paraná River	2550
Ural	Caspian Sea	2535
Orinoco	Atlantic Ocean	2500

DRAINAGE BASINS

Name	Area (sq km)
Amazon	7,050,000
Congo	3,700,000
Mississippi-Missouri	3,250,000
Piraná	3,100,000
Yenisey	2,700,000
Ob'	2,430,000
Lena	2,420,000
Nile	1,900,000
Amur	1,840,000
Mackenzie	1,765,000
Ganges-Brahmaputra	1,730,000
Zambezi	1,330,000
Volga	1,380,000
Niger	1,200,000
Chang Jiang	1,175,000
Orange	1,020,000
Orinoco	945,000
Huang He	980,000
Indus	960,000
Yukon	850,000
Danube	815,000
Mekong	810,000

LAKES

Name	Location	Area (sq km)
Caspian Sea	Asia	371,000
Lake Superior	Canada/USA	82,200
Lake Victoria	East Africa	68,000
Lake Huron	Canada/USA	59,600
Lake Michigan	USA	58,000
Aral Sea	Kazakhstan	36,000
Lake Tanganyika	Central Africa	33,000
Great Bear Lake	Canada	31,500
Lake Baikal	Russia	31,500
Lake Malawi/Nyasa	East Africa	29,000
Great Slave Lake	Canada	28,700
Lake Erie	Canada/USA	25,700
Lake Chad	Central Africa	25,000
Lake Winnipeg	Canada	24,400
Tonle Sap	Cambodia	20,000
Lake Ontario	Canada/USA	19,500
Lake Balkhash	Kazakhstan	18,500
Lake Ladoga	Russia	18,400
Lake Eyre	Australia	9000
Lake Titicaca	Bolivia/Peru	8200

ISLANDS

Name	Location	Area (sq km)
Australia	Australia	7,686,849
Greenland	Greenland	2,175,600
New Guinea	Indonesia/ Papua New Guinea	780,000
Borneo	SE Asia	737,000
Madagascar	Indian Ocean	587,000
Baffin Island	Canada	508,000
Sumatra	Indonesia	425,000
Honshu	Japan	230,000
Great Britain	UK	229,880
Victoria I.	Canada	212,200
Ellesmere I.	Canada	212,000
Celebes	Indonesia	189,000
New Zealand (S)	New Zealand	150,000
Java	Indonesia	126,700
Cuba	Cuba	114,500
New Zealand (N)	New Zealand	114,400
Luzon	Philippines	104,700
Iceland	Atlantic Ocean	103,000

The rocks in the Earth's crust are mainly composed of silicate minerals, plus some carbonate rocks and certain ferruginous types. Mineral identification is achieved by applying physical and optical tests. One of the most widely used is that of hardness which is measured in relation to a set of standard minerals. Thus each mineral which sits above another on Moh's Scale can scratch any which sits below it. Color, luster and cleavage are other diagnostic features.

Rock classification is based on both mineralogy and textural characteristics. The igneous rocks are divided into coarsely-, medium- and finely-crystallized types and then further subdivided on their mineralogy. Basic rocks are typically of low silication whereas acidic rocks are of high silication. Sedimentary rocks may be clastic, chemical or organic in nature; the former include the sandstones and mudrocks, composed of rock-forming silicates in varying proportions, while the latter two groups include the limestones, dolomites, ironstones, cherts and phosphatic groups. Some limestone are precipitated inorganically from sea water whereas others are produced organically from the shells of dead sea creatures.

Metamorphic rocks are dominated by rock-forming silicates and vary according to the parent rock type from which they were produced and the degree of metamorphism to which they were subjected. In broad terms thermally metamorphosed rocks include marbles and hornfelses, while regionally metamorphosed types include slates, phyllites, schists and gneisses. The latter types tend to have their constituent minerals arranged in

MOH'S HARDNESS SCALE

Scale	Mineral	Scale	Mineral
1	Talc (softest)	6	Orthoclase
2	Gypsum	7	Quartz
3	Calcite	8	Topaz
4	Fluorite	9	Corundum
5	Apatite	10	Diamond (hardest)

SEDIMENTARY ROCKS

Type	Rocks	Composition
CHEMICAL		
Calcareous	Dolomite	Calcium and magnesium carbonates
	Oölitic limestone (part)	Calcium carbonate
	Pisolitic limestone (part)	Calcium carbonate
	Travertine	Calcium carbonate
Ferruginous	Ironstone	Various iron minerals with calcium carbonate
Phosphatic	Rock phosphate (part)	Calcium phosphate
Saline	Gypsum	Calcium sulfate
	Rock salt	Sodium chloride
Siliceous	Chert	Silicon dioxide
	Flint	Silicon dioxide
MECHANICAL		
Fine	Mudstone	Clay minerals
	Siltstone	Clay minerals with some silicon dioxide
	Shale	Clay minerals
Medium	Arkose	Silicon dioxide with alkali aluminosilicate
	Greywacke	Silicon dioxide with alkali aluminosilicate
	Grit	Silicon dioxide
	Orthoquartzite	Silicon dioxide
	Sandstone	Silicon dioxide
Coarse	Breccia	Angular rock fragments
	Conglomerate	Rounded rock fragments
	Tillite	Rock fragments in a matrix of shale or slate
ORGANIC		
Calcareous	Biochemical	Calcium carbonate limestone
	Oölitic limestone (part)	Calcium carbonate
	Pisolitic limestone (part)	Calcium carbonate
Carbonaceous	Coal	Mainly carbon
Phosphatic	Rock phosphate (part)	Calcium phosphate

IGNEOUS ROCKS

Rock	Mineral composition (from most to least SiO_2)
FINE GRAIN	
Basalt	Magnetite, olivine, plagioclase feldspar and pyroxene
Andesite	Augite, biotite mica, hornblende, plagioclase feldspar and pyroxene
Phonolite	Aegerine or amphibole, nepheline and potassium feldspar
Trachyte	Potassium feldspar with a little quartz
Rhyolite	Potassium feldspar, quartz, sometimes hornblende
MEDIUM GRAIN	
Dolerite	Olivine and/or pyroxene (dolerite) and plagioclase feldspar; sometimes also biotite, hornblende and quartz
Microsyenite	Biotite, hornblende, potassium feldspar, pyroxene, quartz
Microdiorite	Hornblende and plagioclase feldspar; sometimes also biotite, potassium feldspar, pyroxene and quartz
Microgranite	Potassium feldspar and quartz, sometimes with hornblende
COARSE GRAIN	
Dunite	Olivine
Peridotite	Olivine with some augite, biotite and hornblende
Gabbro	Plagioclase feldspar and pyroxene; sometimes also hornblende, olivine and biotite
Syenite	Potassium feldspar with some amphibole, biotite and pyroxene; there may be some quartz
Diorite	Hornblende and plagioclase feldspar; sometimes also biotite, potassium feldspar, pyroxene and quartz
Granodiorite	Potassium/sodium feldspar and quartz, sometimes with hornblende
Granite	Potassium/sodium feldspar and quartz, sometimes with hornblende, muscovite and biotite mica

METAMORPHIC ROCKS

Parent rock	Rock	Mineral composition (one or more of)
Limestone	Marble	Calcite, dolomite
Limestone (impure)	Calc-schist	Calcite, diopside, grossular (a garnet), forsterite, idocrase, tremolite, wollastonite
Basalt	Amphibolite	Plagioclase, hornblende, chlorite, garnet, biotite
	Pyroxene hornfels	Pyroxene, plagioclase, quartz, spinel
Mudstones and Shales	Hornfels	Andalusite, cordierite, biotite, sillimanite, corundum
	Slate	Mostly quartz and micas
	Phyllite	Mostly quartz and micas
	Schist	Chlorite, biotite, garnet, byanite, sillimanite, staurolite, quartz
	Gneiss	Quartz, biotite, muscovite, hornblende, plagioclase, pyroxene

coplanar layers or laminae, giving rise to the fabric termed schistosity.

Rocks themselves are made up of minerals, naturally occurring solid substances with a definite and constant chemical composition and a characteristic appearance. A few such minerals, such as anhydrite and quartz, consist of a single chemical substance (calcium sulfate and silicon dioxide in these examples). But most minerals contain a mixture of chemicals and sometimes also incorporate a definite proportion of water.

Some rocky minerals are extracted and utilized in their own right. Many are used as building stone or crushed to produce aggregate for making concrete, and clay has long been used as a raw material for making ceramics from bricks to pottery. Quartz, in the form of sand, is employed in making glass and concrete. Gypsum is made into plaster for building, and mica – an unusual semitransparent mineral that easily splits into thin plates or flakes – is a useful electrical insulator, also used for making fireproof windows for stoves and furnaces. Zeolite is the basis of some water softeners that make use of an ion-exchange process to remove the chemicals that cause hardness in natural water supplies.

Some of the minerals in the table below occur in transparent forms colored by metallic impurities; because of their beauty, hardness and rarity they are valued as gemstones. They include ruby and sapphire, which are forms of corundum, amethyst and opal, forms of quartz, and the green gemstone peridot, which is a type of the mineral olivine.

ROCK-FORMING MINERALS

Group	Variety	Composition	Color	Hardness
Amphibole	Tremolite/actinolite	Hydrated silicate of calcium, iron and magnesium	Light to dark green	5–6
	Hornblende	Complex silicate of calcium, iron, sodium, aluminum	Green/black and brown	5-6
Andalusite		Aluminum silicate	Red, brown or olive	7.5
Anhydrite		Calcium sulfate	Colorless	3–3.5
Calcite		Calcium carbonate	Colorless	3
Clay	Kaolinite	Hydrated aluminum silicate	White–dark gray	1.5–2.5
	Illite	Hydrated potassium aluminum silicate	White–dark gray	1.5–2.5
	Smectite	Complex silicate of calcium, iron, sodium and magnesium	White–dark gray	1.5–2.5
Cordierite		Magnesium and iron aluminosilicates	Dark blue or blue-gray	7
Corundum		Aluminum oxide	Brown, pink or blue; emery black	9
Dolomite		Carbonate of calcium and magnesium	Colorless	3.5-4
Epidote		Silicate of aluminum, iron	Green–gray or dark brown	6-7
Feldspar	Plagioclase feldspar	Sodium or calcium aluminum silicate	White to gray, green or yellow	6
	Alkali feldspar	Potassium aluminum silicate	White–gray pink or yellow	6
Feldspathoid	Nepheline	Silicate of aluminum, sodium and potassium	Colorless, white or gray	5.5–6
	Leucite	Potassium aluminum silicate	White–gray	5.5–6
Garnet		Silicate of calcium, iron, magnesium and aluminum	Red, green	6.5–7
Gypsum		Hydrated calcium sulfate (hydrated anhydrite)	Colorless to white	2
Halite		Sodium chloride	Colorless , pink	2.5
Idocrase (vesuvianite)		Complex aluminosilicate of calcium with magnesium	Dark green	6-7
Kyanite		Aluminum silicate	White–gray	5-7
Mica	Biotite	Complex silicate of aluminum, iron, magnesium and potassium	Black to dark brown	2.5-3
	Chlorite	Complex silicate of aluminum, iron, and magnesium	Green	2-2.5
	Muscovite	Hydrated potassium aluminum silicate	Colorless	2-2.5
Olivine		Silicate of iron and magnesium	Olive–gray-green	6.5–7
Opal-chalcedony		Silicon dioxide	Colorless to white; banded by impurities	5-6.5
Pyroxene	Aegirine	Silicate of iron and sodium	Dark green or brown	6
	Augite	Complex silicate of calcium, iron, sodium, magnesium and aluminum	Dark green to black	5–6
	Diopside	Silicate of calcium and magnesium	Light to dark green	5–6
	Enstatite-hypersthene	Silicate of iron and magnesium	Green and brown to gray	5-6
	Jadeite	Silicate of aluminum and sodium	Green or white	6
	Spodumene	Silicate of aluminum and lithium	White	6.5-7
Quartz		Silicon dioxide	Colorless to light gray, colored by impurities	7
Serpentine		Hydrated silicate of magnesium	Green, brown or gray	4-6
Sillimanite (fibrolite)		Aluminum silicate	Colorless, white or ocher	6.5–7
Staurolite		Hydrated silicate of iron and aluminum	Reddish brown or black	7
Talc		Hydrated silicate of magnesium	White to green	1
Wollastonite		Calcium silicate	White to gray	4.5-5
Zeolite		Complex hydrated silicates	Colorless/white	4-5

☐ METALLIC MINERALS

Minerals that bear metals generally are nonsilicates and are seldom present in rocks in large amounts. As a result a concentration process is required to render them economically viable, converting the disseminated metals into ore deposits. Metal-bearing solutions emanate from basic magmas that rise along fractures associated with mid-oceanic ridges, and from intermediate and acidic magmas associated with continental margins at convergent plate boundaries. Most emerge as metal chlorides but, after reaction with wall rocks through which the solutions pass, may form oxides, sulfides and carbonates. A zonal sequence occurs, with oxides at high temperature, sulfides at intermediate and carbonates at low temperatures.

The metals that are associated with granitic bodies include tungsten, molybdenum and gold; with basic magmas are associated zinc, copper and lead (although there is considerable overlap). Their transportation in hydrothermal solutions is aided by fractures and faults in the rocks through which they pass and suitable "preparation" of the rock materials into which they are eventually concentrated. Such preparation is usually achieved by brecciation (associated with faulting), or by dissolution and reprecipitation of metals within porous rocks. Because silica is readily dissolved in mobile hydrothermal fluids, quartz is often found with metal-bearing minerals in such deposits.

Name	Composition	Color	Hardness
Anatase	Titanium dioxide	Yellow, brown	5.5–6
Anglestite	Lead sulfate	Colorless to white	2.5–3
Argentite	Silver sulfide	Black	2–2.5
Arsenopyrite	Sulfide of iron and arsenic	Gray-white	5.5–6
Barite	Barium sulfate	Colorless or white	2.5–3.5
Bauxite	Aluminum hydroxide	Ochre, brown, gray	1.5–2.5
Bismuthite	Bismuth sulfide	Light gray	2
Bornite	Sulfide of iron and copper	Red-brown	3
Cassiterite	Tin oxide	Red-brown to black	6–7
Celestite	Strontium sulfate	Colorless to white	3–3.5
Cerussite	Lead carbonate	White or gray	3.5–4
Chalcocite	Copper sulfide	Dark gray to black	2.5–3
Chalcopyrite	Sulfide of copper and iron	Brassy yellow	3.5–4
Chlorargyrite	Silver chloride	Colorless to gray	1.5–2.5
Chromite	Mixed oxide of iron and chromium	Brown to black	5.5
Cinnabar	Mercury sulfide	Red to red-brown	2–2.5
Cobaltite	Sulfide of cobalt and arsenic	Gray-white	5.5
Covellite	Copper sulfide	Blue to purple	1.5–2
Crocoite	Lead chromate	Orange-red to brown	2.5–3
Cuprite	Copper oxide	Red to dark brown	3.5–4
Galena	Lead sulfide	Silver- gray	2.5
Hematite	Iron oxide	Red-brown to black	6
Ilmenite	Oxide of iron and titanium	Metallic black	5–6
Limonite	Oxide and hydroxide of iron	Yellow-brown to black	5–5.5
Magnetite	Mixed iron oxides	Metallic black	6
Malachite	Hydroxide and carbonate of copper	Bright green	3.5–4
Molybdenite	Molybdenum sulfide	Light blue-gray	1–1.5
Monazite	Phosphates of lanthanum, cerium and thorium	Brown	5–5.5
Orpiment	Arsenic sulfide	Yellow to brown	1.5–2
Pyrargyrite	Sulfide of silver and antimony	Black	2.5
Pyrite	Iron sulfide	Brassy yellow	6–6.5
Pyrrhotine	Iron sulfide	Bronze-yellow	3.5–4.5
Realgar	Arsenic sulfide	Red to orange	1.5–2
Rutile	Titanium dioxide	Red-brown or black	6–6.5
Scheelite	Calcium tungstate	White	4.5–5
Siderite	Iron carbonate	Gray to brown	3.4–4.5
Smithsonite	Zinc carbonate	Gray or brown	4–4.5
Sphalerite	Zinc sulfide	White to green, brown and black	3.5–4
Stibnite	Antimony sulfide	Gray	2
Strontianite	Strontium carbonate	White, gray or green	3.5–4
Tungstite	Hydrated tungsten oxide	Yellow to greenish	4–4.5
Uraninite	Uranium oxide	Dark brown to black	5–6
Vanadinite	Chloride and vanadate of lead	Orange, red or brown	3
Witherite	Barium carbonate	White or gray	3–3.5
Wolframite	Iron and manganese tungstates	Dark gray to black	5–5.5
Wurtzite	Zinc sulfide	Brown or black	3.5–4
Zincite	Zinc oxide	Red to orange	4–4.5
Zircon	Zirconium silicate	Light to red-brown	7.5

☐ ROCK DATING

Radiometric dating provided geologists with the first reliable method of dating rocks. Physicists have estimated the half-life of radioactive isotopes (the time it takes for half the atoms in a sample to decay into their daughter products, by emission of alpha or beta particles). By observing the isotopes locked up in them, and the proportion that have undergone radioactive decay, the date at which the rocks were formed can be calculated. Different isotopes, with different half-lifes, are used for different age ranges of rock. Some rocks can be dated by several methods, allowing "cross-checks."

Name	Parent isotope	Daughter isotope	Half-life (years)	Age range (years)	Comments
Uranium–lead	U-238	Pb-206	4.468×10^9	10 million	The most precise method for dating igneous rock
Uranium–lead	U-235	Pb-207	7.04×10^8	10 million	Used for dating igneous rock
Rubidium–strontium	Rb-87	Sr-87	4.88×10^{10}	10 million	Used to date time of strontium emplacement
Potassium–argon	K-40	Ar-40	1.28×10^{10}	50,000 and greater; generally up to 200 million	The only method used for dating young basic extrusive and minor intrusive rocks
Samarium–neodymium	Sm-147	Nd-143	1.06×10^{11}	100 million and greater	Used for dating Archean rocks
Carbon–nitrogen	C-14	N-14	5570	Recent	Used for the most recent 30,000 years only

☐ GEOLOGICAL ERAS

Geological timescales can be either relative or absolute. Until radiometric dating allowed absolute ages to be measured for rocks bearing radioactive isotopes, geologists assigned relative ages to strata according to the order in which they occur and the fossil content.

The broadest subdivisions of geological time are the four eras: Precambrian, Paleozoic, Mesozoic and Cenozoic. The first spans a huge period of time, from the Earth's beginning to 590 million years ago. This is six times as long as the succeeding Phanerozoic, a term which which includes the Paleozoic, Mesozoic and Cenozoic eras and applies to all those rocks that contain relatively abundant remains of living organisms.

Each era is subdivided into shorter epochs which may be identified with certain broad biological developments, such as the rise of fishes or spread of land mammals. The two divisions of Cenozoic time – Tertiary and Quaternary – are further subdivided into periods.

Epoch	Period	Date (years ago)	Record of life
CENOZOIC ERA			
Quaternary	Holocene	10,000 – present	Modern mankind
	Pleistocene	2. 5 million – 10,000	Early mankind
Tertiary	Pliocene	5–2.5 million	Large carnivores
	Miocene	25–5 million	Whales, apes, grazing mammals
	Oligocene	38–25 million	Large browsing mammals
	Eocene	55–38 million	Rise of modern plants
	Paleocene	65–55 million	First placental mammals
MESOZOIC ERA			
Cretaceous		144–65 million	Early flowering plants, last of dinosaurs
Jurassic		213–144 million	Flying reptiles, first mammals
Triassic		248–213 million	Rise of dinosaurs, ammonites and cycads
PALEOZOIC ERA			
Permian		286–248 million	Primitive reptiles, para-mammals, last trilobites
Carboniferous (Pennsylvanian)		320–286 million	Spread of amphibians, great coal forests
Carboniferous (Mississippian)		360–320 million	Climax of crinoids and blastoids
Devonian		408–360 million	Rise of ferns, earliest amphibians
Silurian		438–408 million	Expansion of brachiopods and corals; early plants
Ordovician		505–438 million	Appearance of primitive fishes, climax of trilobites
Cambrian		590–505 million	Many shelled invertebrates
PRECAMBRIAN ERA			
		1.2 billion–590 million	Jellyfish, worm burrows, flatworms
Proterozoic		2.1–1.2 billion	Marine algae
		2.98–2.1 billion	Abundant carbon of organic origin
		3.2–2.98 billion	Oldest stromatolites
		3.49–3.2 billion	Earliest known record of life
Archean		4.5–3.49 billion	Oldest rock

☐ EARTHQUAKE SCALES

Earthquake magnitude is usually measured according to the Richter scale, which is based on the amplitude of seismic waves recorded at seismograph stations. It was developed by the American geophysicist Charles Richter in the mid-20th century. It is a logarithmic scale, so that an increase in magnitude of one unit implies a tenfold increase in strength. The most powerful terrestrial quakes reach 8.5 on this scale.

The Richter scale is not a measurement of the intensity (surface effect) of the quake, which may depend, for example, on the depth of the epicenter. The San Fernando earthquake of 1971 was of magnitude 6.6 (relatively modest) but the damage bill was very high indeed. The Mercalli scale is also used and provides a readily identifiable subjective measure of intensity. Earthquakes are also sometimes measured in terms of their seismic moment, which is the product of the average amount of slip, the size of the rupture and the strength of the rocks affected.

THE MERCALLI SCALE

Intensity	Effects
I	Not felt at all
II	Felt only by a few people
III	Felt indoors by many, but not necessarily identified as an earthquake
IV	Felt indoors and outdoors and comparable to vibrations produced by a passing train
V	Strong enough to wake a sleeping person; small objects fall off shelves
VI	Perceptible to everyone; plaster may fall from ceilings
VII	Difficult to stand upright; waves may occur in ponds
VIII	Chimneys and smokestacks may fall; unsecured frame houses slide off their foundations
IX	Heavy damage to masonry structures and underground pipes; large cracks open in ground; panic is general
X	Numerous buildings collapse; water splashes over riverbanks
XI-XII	Virtually complete destruction

THE RICHTER SCALE

Appropriate magnitude	Characteristic effects of shallow shocks in populated areas	Earthquakes per year	Energy (ergs)
2.0–3.4	Not felt but recorded	800,000	4×10^{10}–9×10^{13}
3.5–4.2	Felt by some	30,000	$1.6–76 \times 10^{15}$
4.3–4.8	Felt by many	4800	$1.3–27 \times 10^{16}$
4.9–5.4	Felt by all	1400	$3.6–57 \times 10^{17}$
5.5–6.1	Slight damage to buildings	500	$1–27 \times 10^{19}$
6.2–6.9	Considerable damage to buildings	100	$0.5–23 \times 10^{21}$
7.0–7.3	Serious damage, railroads bent	15	$0.04–0.2 \times 10^{24}$
≥7.4	Great damage	4	$>0.4 \times 10^{24}$
≥8.0	Damage nearly total	0.1–0.2	$>10^{25}$

☐ SPACE PROBES

Space probes, manned and unmanned, have revolutionized the study of the Moon and planets and other bodies in the Solar System, which until the 1960s could be studied only through terrestrial telescopes. Only the Apollo missions to the Moon have returned and brought rock samples back to Earth, but soft landings have also been achieved and surface measurements taken on Mars and Venus. Photography at visible and invisible wavelengths, imaging radar, magnetic field instruments, spectroscopy and other instrumentation have all yielded important data from orbiting space probes and fly-bys.

In addition to the information gleaned from space probes about the planets themselves, the Pioneer series in particular has provided a great deal of information about interplanetary space, and other probes (such as Ulysses, which flew over the South Pole of the Sun in 1994) have provided important data about the Sun. Since 1991 the Hubble Space Telescope has provided important data on the planets, in addition to its imaging of objects in deeper space.

Several of the unmanned probes have left the plane of the Solar System, and it is hoped that data will be received, at least, fromVoyager 2 concerning the point, well beyond the orbit of Pluto, known as the heliopause, at which cosmic rays overcome the solar wind.

Year	Nation	Craft	Result
MOON			
1959	USSR	Luna 1	Bypasses Moon
		Luna 2	Impacts on Moon
		Luna 3	Bypasses and photographs rear
1962	USA	Ranger 4	Hard landing, cameras fail
1964	USA	Ranger 7	Bypasses, takes photographs
1965	USSR	Zond 3	Bypasses, photographs rear
1966	USA		First soft landing
	USSR	Luna 9	Soft landing
	USSR	Luna 10	First lunar orbiter
	USA	Surveyor 1	Soft landing
	USSR	Luna 11	Lunar orbiter
	USA	Lunar Orbiter 1	Detailed photographs of surface
	USA	Lunar Orbiter 2	Photographs surface
	USSR	Luna 13	Soft landing, studies rocks
1967	USA	Lunar Orbiter 3	Photographs surface
	USA	Surveyor 3	Lander, photographs surface
	USA	Lunar Orbiter 4	Photographs surface
	USA	Surveyor 5	Soft landing, studies soil
	USA	Surveyor 6	Soft landing
1968	USA	Surveyor 7	Soft landing
	USSR	Zond 5	Flies around Moon and back
	USSR	Zond 6	Flies around Moon and back
	USA	Apollo 8	Manned orbits of Moon and back
1969	USA	Apollo 9	Lander tested in Earth orbit
	USSR	Luna 15	Soft landing
	USA	Apollo 10	Lander tested in Moon orbit
	USA	Apollo 11	First manned landing
	USA	Apollo 12	Second manned landing
1970	USA	Apollo 13	Recalled from Moon orbit
	USSR	Luna 16	Returns with rock samples
	USSR	Lunokhod 1	Moon rover
1971	USA	Apollo 14	Third manned landing
	USA	Apollo 15	Fourth landing, Lunar Rover
1972	USA	Apollo 16	Fifth manned landing
	USA	Apollo 17	Sixth (last) manned landing
1994	USA	Clementine	Moon mapping from orbit
MERCURY			
1974	USA	Mariner 10	Two bypasses, takes photographs
1975	USA	Mariner 10	Makes third bypass
VENUS			
1961	USSR	Venera 1	Contact lost, bypasses planet
1962	USA	Mariner 2	Bypasses planet
1966	USSR	Venera 2	
	USSR	Venera 3	Lander
1967	USSR	Venera 4	Drops instruments in atmosphere
	USA	Mariner 5	Close bypass

☐ FURTHER READING

Banyard, P. *Natural Wonders of the World* (Orbis, London 1981)

Bridges, E.M. *World Geomorphlogy* (Cambridge University Press, Cambridge 1990)

Briggs, G. and Taylor, F. *The Cambridge Photographic Atlas of the Planets*, (Cambridge University Press, Cambridge 1982)

Brown, G. and Mussett, A. *The Inaccessible Earth* (Allen and Unwin, London 1983)

Brown, G.C., Hawksworth, C.J. and Wilson, R.C.L. *Understanding the Earth* (Cambridge University Press, Cambridge 1992)

Cadogan, P. *The Moon – Our Sister Planet* (Cambridge University Press, Cambridge 1981)

Cattermole, P. *Mars – Story of the Red Planet* (Chapman and Hall, London 1992)

Cattermole, P. *Venus – the Geological Story* (UCL Press, London 1994)

Cattermole, P. and Moore, Patrick *The Story of the Earth* (Cambridge University Press, Cambridge 1985)

Decker and Decker. *Mountains of Fire* (Cambridge University Press, Cambridge 1991)

Deer, W, Howie, R. and Zussman, J. *An Introduction to the Rock-Forming Minerals* (Longman. London 1992)

Emiliani, C. *Planet Earth* (Cambridge University Press, Cambridge 1992)

Francis, P. *Volcanoes in the Solar System.*

Gehrels, T. *Asteroids* (University of Arizona Press, Tucson, 1979)

Glass, B. *Introduction to Planetary Geology* (Cambridge University Press, Cambridge 1982)

Greeley, R. *Planetary Landscapes* (Chapman and Hall, London, 1994)

Greensmith, J. *Petrology of the Sedimentary Rocks* (Allen and Unwin, 1978)

Gregory, K. (ed) *Earth's Natural Forces* (Oxford University Press, New York, 1990)

Gregory, K. (ed). *Guinness Guide to the Restless Earth* (Guinness , London 1991)

Hall, A. *Igneous Petrology* (Longman, London 1987)

Hamblin, K. and Christiansen, E. *Exploring the Planets* (MacMillan, New York 1990)

Hunt, G. and Moore, Patrick *Atlas of Neptune* (Cambridge University Press, Cambridge, 1994)

Year	Nation	Craft	Result
1969	USSR	Venera 5	
	USSR	Venera 6	
1970	USSR	Venera 7	Soft landing
1971	USSR	Venera 8	Soft landing
1974	USA	Mariner 10	Bypasses planet
1975	USSR	Venera 9	Soft landing, photographs
	USSR	Venera 10	Soft landing, photographs
1978	USA	Pioneer Venus 1	Makes radar maps of planet
	USA	Pioneer Venus 2	Launches atmospheric probes
1980	USSR	Venera 11	Lander, analyzes atmosphere
	USSR	Venera 12	Lander, analyzes atmosphere
1982	USSR	Venera 13	Soft landing, examines rocks
	USSR	Venera 14	Soft landing
1983	USSR	Venera 15	Orbits, radar maps of surface
1984	USSR	Venera 16	Orbits, radar maps of surface
1985	USSR	Vega 1	Bypasses en route to Halley's Comet
	USSR	Vega 2	Bypasses en route to Halley's Comet
1990	USA	Magellan	Radar maps surface
MARS			
1962	USSR	Mars 1	Contact lost March 1963
1965	USA	Mariner 4	Bypasses planet, photographs
1969	USA	Mariner 6	
	USA	Mariner 7	
1971	USA	Mariner 8	Failed
	USA	Mariner 9	Orbits planet
	USSR	Mars 2	Orbits planet
	USSR	Mars 3	Lands, contact soon lost
	USA	Mariner 9	Orbits planet

Year	Nation	Craft	Result
1974	USSR	Mars 4	Lander
	USSR	Mars 5	
	USSR	Mars 6	
	USSR	Mars 7	
1976	USA	Viking 1	Soft landing, photographs
	USA	Viking 2	Soft landing, photographs
1993	USA	Orbiter	Loses contact when entering orbit
ASTEROID BELT			
1991	USA	Galileo	Photographs Gaspra
1993	USA	Galileo	Photographs Ida
JUPITER			
1973	USA	Pioneer 10	Bypasses planet
1974	USA	Pioneer 11	Bypasses planet
1979	USA	Voyager 1,2	Bypass planet, discover ring
1995	USA	Galileo	Tours Jupiter's moons
SATURN			
1979	USA	Pioneer 11	Bypasses planet
1980	USA	Voyager 1	Bypasses planet, finds moons
1981	USA	Voyager 2	Bypasses planet
URANUS			
1986	USA	Voyager 2	Bypasses planet, finds moons
NEPTUNE			
1989	USA	Voyager 2	Bypasses planet
HALLEY'S COMET			
1986	USA	ICE	Bypasses comet; no useful data
	USSR	Vega 1	Bypasses comet; no useful data
	USSR	Vega 2	Bypasses comet; no useful data
	Japan	Sakigake	Bypasses comet; no useful data
	Europe	Giotto	Passes within 600km of nucleus
	Japan	Suisei	Bypasses comet

Imbrie, J. and Imbrie, K. *Ice Ages: Solving the Mystery* (Enslow Press, New York 1979)

Kerridge, J. and Matthews, Mildred Shipley *Meteorites and the Early Solar System* (University of Arizona Press, Tucson, 1988)

Lambert, D. *The Cambridge Guide to the Earth* (Cambridge University Press, Cambridge 1988)

MacKenzie, W. and Adams, A. *A Colour Atlas of Rocks and Minerals in Thin Section* (Manson, London 1993)

Mason, R. *Petrology of the Metamorphic Rocks* (Allen and Unwin 1978)

Moore, P. *Missions to the Planets* (Cassell, London 1990)

Moore, P., and Hunt, G. *Atlas of the Solar System* (Mitchell Beazley, London 1985)

Murray, B., Malin, M. and Greeley, R. *Earthlike Planets* (Freeman, San Francisco 1981)

Phillips, K.J.H., *Guide to the Sun* (Cambridge University Press, Cambridge, 1992)

Press, F. and Siever, R. *Earth*. (Freeman, New York 1986; 2nd edition 1994)

Ringwood, A.E. *Origin of the Earth and Moon* (Springer, New York, 1979)

Rothery, D. *Satellite of the Outer Planets*. (Oxford University Press, Oxford, 1992)

Scientific American. *The Dynamic Earth*. September 1983.

Smith, P. (ed) *The Earth* (Macmillan, New York, 1986)

Stanley, S. *Earth and Life Through Time*. (Freeman, New York 1989)

Tarling, D. and Tarling, M. *Continental Drift*. (Doubleday, New York 1971).

Taylor, S.Ross, *Solar System Evolution*. (Cambridge University Press, Cambridge 1992)

Taylor, S. Ross *Lunar Science– A Post-Apollo View* (Pergamon, Oxford, 1975)

Tucker, M. *Sedimentary Petrology*. (Blackwell,Oxford 1991)

Uchupi, E. and Emery, K. *Morphology of the Rocky Members of the Solar System*. (Springer-Verlag, Berlin 1993)

Whittow, J. *The Penguin Dictionary of Physical Geography*. Penguin Books, Harmondsworth 1984)

Wood, J. *The Solar System*. (Prentice-Hall, New Jersey, 1984)

Wylie, P. *The Way the Earth Works*. (Wiley, New York 1976)

Yardley, B. *An Introduction to Metamorphic Petrology* (Longman, London 1989)

Index

Figures in *italic type* refer to the captions to the illustrations, and to Timechart entries; figures in **bold type** refer to the Keywords and Factfile articles.

A

- aa **16**
- ablation **16**
- absolute age **16**, 86, *108*
- abyssal plain **16**, 100, 129
- abyssal sediment **16**
- accretion
  - continental **16**, 116–17
  - inner (terrestrial) planets 66, 78
  - island arcs 116–17
  - planetesimal **16**, 56–57, *56*, 58–59, 78
- accretional energy **16**, 74, 76, 77, 88
- accretion disk *52*
- Aconcagua 149
- adiabatic heating **16**, 76, 77
- aeolian process 22
- aerolite **16**, 20
- aerosol spray 36
- Agassiz, Louis *13*
- air-fall deposit **16**
- air pressure *12*
- albedo **16**, 65, 70
- Aldrin, Buzz 140
- Aleutian islands 116
- algae *137*
- alluvium **16**
- Alps, European 25, *124*, *133*
- aluminum 31, 32, 40, 50, 58, 65, 81, *81*, 90
- "Alvin" research vessel 100, *115*
- Amazon 126, *127*, 149
- amino acid **16**
- ammonia **16**, 58, 85, 98
- ammonite **17**, *17*
- Amor asteroid 17–18, 70, *70*
- amphibian **17**, 137
- amphibole **17**, 41, 84, 93
- andalusite 39
- Andes 102, *103*, 118, *119*, 149
- andesite **17**
- angular momentum **17**
- anhydrite 151
- anion **17**, 124
- anorthosite **17**, 65, 90
- Antares *135*
- anticline **17**, *89*
- aphelion **17**, 62
- Apollo asteroid 17, 70, *70*
- Apollo missions **17**, 43, *58*, 97, 140, *140*, *141*
- Appalachian Mountains 25
- aquifer **17**
- aragonite 39
- Archean **17**, 28, 80, 102–03, 142
- Archimedes *12*
- argon **17**, 29, 58
- Ariel (moon of Uranus) **17**
- Ariel satellite **17**
- Aristarchus of Samos 12, *12*
- Aristotle 13
- arkose **17**
- Armero disaster 138
- Armstrong, Neil *15*, 140
- arthropod *137*
- Ascension Island 114
- ash flow deposit **17**
- asteroid **17–18**, *17*, 57, 58, 60, 70, *71*, 142, **146**
  - accretional energy **16**
  - albedo (reflectivity) 70
  - Amor 17–18, 70, *70*
  - Apollo 17, 70, *70*
  - Aten object 18, 70
  - classes 70
  - composition 70
  - orbits 70, *70*, *71*
  - Pluto and Charon 69
  - size and shape 70
  - Trojan group *70*
- asteroid belt **18**, 56, 70, *71*, 146
- asthenosphere **18**, 31, 92, *92*, 93, *93*, 96, 113, 148
- astronomical unit (AU) **18**, 62
- Aten object 18, 70
- Atlantic Ocean mapped *13*
- atmophile elements **18**
- atmosphere **18**, *18*, 23, *75*, 84–85, *84*, *85*, 98–99, *98*, *99*, 125
  - air pressure *12*
  - cloud belt **20**
  - exosphere **24**
- atoll **18**, 128
- atom **18**, 50, 86
  - ionization **29**
  - nucleus **34**
- atomic lattice 76, 124
- atomic mass 50
- atomic number 86
- augite 18, 25
- aurora **18**, *83*
- axial rift 23

B

- Bangladesh *138*
- barchan **18**, 130, *130*
- barium *81*
- Barton, Otis *14*
- basalt **18**, 31, 65, 90, *93*, 94, *95*, 97, 100, *100*, 114, 120, 124
- basin 60, 64
- batholith **18**, *94*, 118
- bathyscape *14*
- bathysphere *14*
- beach 128
- bedding plane **18**, 124
- bedform 130
- Beebe, William *14*
- Benioff, H. 18
- Benioff zone **18**, 22, 117, *119*
- benthic **18**
- Bessel, Friedrich Wilhelm *13*
- Beta Regio **18**, 27, 47, 120
- Big Bang **18**, 48, 50
- binary star **18**
- Biot, Jean *13*
- biotite 29, *125*
- black hole *50*
- black smoker **19**, 42, *115*
- blue-green algae **19**, 44, *84*, 136, *137*
- Bode, J.E. 45
- Bode's law *see* Titius-Bode rule
- Bouguer correction 27
- boulder clay **19**, 132
- bow shock **19**, 82–83
- brachiopod **19**
- Brahe, Tycho *12*, 13
- braided river channel *127*
- breccia 65
  - impact **28**
  - polymict 29
- Bromo, Mount *117*
- bryozoan 31
- Buffon, Comte George de *13*

C

- calcite 39
- calcium 31, 32, 40, 50, 58, 65, *81*, 90
- calcium carbonate 31, 39, 125
- caldera **19**, *94*, *139*
- Callisto **19**, 147
- Cambrian **19**, 137, 142, *143*
- captured rotation **19**
- carbide 136
- carbon 47, 50
  - organic molecule **35–36**
- carbonaceous chondrite **19**
- carbonate 98
- carbonatite **19**
- carbon dioxide 27, 47, 50, 84, *84*, 98, 100–101, 136
- carbon fixing 98, 125
- Carboniferous **19**, *110*, *137*
- carbon monoxide 84
- Carpentier, Johan von *13*
- Cascades 118
- Cassini, Giovani *12*
- cation 41, 124
- Cavendish, Henry *13*
- Celcius, Anders *13*
- celestial latitude 30–31
- celestial longitude 31
- celestial sphere **19**, *19*, 31
  - ecliptic **24**, 35, *62*
- cement **20**, 125
- Cenozoic **20**, *110*, **153**
- centigrade scale *13*
- cephalopod 17
- Ceres 18, **20**, 45, 70
- chalcophile element **20**, 50, 81, *81*
- chalk **20**
- *Challenger* expedition *14*
- Charon *15*, **20**, *59*, 68, *68*, 69, 147
- Chassigny Meteorite 43
- chemical fractionation 80
- chert **20**, 150
- Chicxulub Crater 30
- chlorite 29
- chlorofluorocarbon (CFC) 27, 36
- chocolate brown clay 16
- chondrite **20**, 59, 70
- chondrule **20**, 33, 70
- chromium *81*
- chromosphere *54*, 55, *55*
- cirque **20**, *132*
- clastic rock **20**
- clay **20**, 124–25, *125*
- cleavage plane **20**, 124
- cliff 128, *128*, *129*
- climate 110
  - Milankovitch theory **33**
  - seasons 62
- climatology 11
- cloud belt **20**
- coastline 128, *128*, *129*
- coastline of emergence **20**, 128
- coastline of submergence **20**, 128
- coesite **20**, 29, 91
- collapse depression 132
- collision zone **20**
- Columbia River 47, *86*
- coma **20**, 72, *72*, *73*
- comet *20*, **21**, 57, 59, 60, *139*, **146**
  - coma **20**, 72, *72*, *73*
  - composition 72, *73*
  - dust tail **23**, 72, *72*, *73*
  - ion (plasma) tail **29**, 72, *72*, *73*
  - mass 72
  - orbit *62*
  - origin 72
  - returns *62*, 72
- Comet West 72
- complex ridged terrain (CRT) *see* tessera
- composite volcano *see* stratovolcano
- compound 50
- condensation **21**, 58, 72
- conduction **21**
- confining pressure *see* hydrostatic pressure
- constellations 62
- contact metamorphism 33
- continent **21**, 102–03, *102*, *103*, *143*
  - accretion **16**
  - *see also* Gondwanaland; Laurasia; Pangea
- continental drift 14, *14*, *15*, **21**, 36, 106, *107*, 110, *110*, *111*, 112–13, 120, *121*
- continental shelf **21**, *21*, 128
- continental shield 21, *111*
- continental slope **21**, 128, 129
- convection **21**, 80–81, *81*
  - subsolidus **44**, 80–81
- convection cell *see* convection
- convergence zone *see* collision zone; convergent margin
- convergent margin **21**, 118
- Copernicus, Nicolas 12–13, *12*
- copper 152
- coral 31, 128, *137*
- core **21**, 74, 76, 77, 78–79, *78*, *79*, 80, 81, *81*, 92
- corona (of the Sun) **21–22**, *54*, 55, *55*
- corona (on Venus) **21–22**, 66, *112*
- corrie *see* cirque
- cosmic radiation 82
- cosmic rays **22**
- crater **22**
  - impact 22, 28
  - maar 38, 47
  - transient cavity **45**
- crater-age dating 86, *86*
- craton 17, **22**, 42, 102, *102*, 110
- Cretaceous **22**
- critical point **22**
- cross-bedding **22**, 23, 130
- crust **22**, 66, 80, 90–91, *90*, *91*, 97, 102–03, *103*, 116
  - isostacy **29**
  - oceanic *see* oceanic crust
  - primordial 103
  - sialic **42**
  - simatic **42**
- crustacean *137*
- cryovolcanism **22**, 69
- crystallization **22**, 27, 76, 124, 125
  - cleavage plane **20**, 124
  - fractional 23, **25**, 31, 37, 93
- cumulate rock 31
- Curie point **22**, 36, 45
- current bedding *see* cross-bedding
- cwm *see* cirque
- cyanobacteria *see* blue-green algae
- Cygnus Loop *see* Vela supernova remnant

D

- dacite 27
- dating
  - crater-age 86, *86*
  - geochronology **26**, 86
  - radiometric *14*, **16**, 17, 40, 86, *108*, 152
  - rock dating **152**
- daughter isotope **22**, 76, 77, 86
- Davis, William Morris *14*
- decay sequence **22**, 59, 86
- Deccan plateau 47
- deep-focus earthquake **22**
- Deep Sea Drilling Project *15*, 115
- deformation 117
- Deimos *14*, **22**, 147
- delta **22**, *22*, 126, *127*, 128
  - front 22
- deoxyribonucleic acid (DNA) *136*
- desert **22**, 130–31, *130*, *131*
  - dune *see* dune
  - millet-seed grain **33**
  - playa **38**
- detrital remanent magnetization (DRM) 36
- deuterium **22**, 35, 50, *53*
- Devonian **22**, 137, *143*
- diagenesis 20, **22**, 41
- diapir **22**, 93, *93*
- diatom 36, 129
- differentiation **23**, 78
- dinosaur 142, *142*
- Dione **23**
- dipole **23**, 82
  - magnetic **31**
- divergence zone *see* divergent plate margin
- divergent plate margin **23**, 32, 100, 108, *114*
- dolomite (rock) 150
- Dolomites *124*
- dreikanter **23**, 130
- drumlin **23**, *105*, 132, *133*
- ductility **23**
- dune **23**, *23*, 130–31, *130*, *131*
  - barchan **18**, 130, *130*
  - seif 130, *130*
- dune-bedding 22, **23**, *130*
- dust (extraterrestrial) **23**
- dust tail **23**, 72, *72*, *73*
- Dutton, Clarence *14*
- dyke **23**, *95*, 108

E

- Earth **23**, *58*, **146**
  - asthenosphere **18**, 31, 92, *92*, 93, *93*, 96, 113, 148
  - atmosphere 18, 23, 44, 45, 46, *75*, *81*, 84, *84*, 98–99, *98*, 125, 136
  - collision zone **20**
  - composition 60
  - convection 80
  - core 21, 74, 76, 78–79, *78*, 80, *81*, 82, 92
  - crust 39, 43, 66, 80, 81, 84, 90–91, *90*, *91*, 97, 102, 116
  - data 63
  - data listed **148–49**
  - density 64, 66
  - energy storage 74, 88
  - equator **24**
  - formation 78
  - gravitational pull 70
  - gravity anomaly **27**
  - hemispheres 24
  - hydrosphere *see* oceans *below*
  - ionosphere **29**, 55
  - life on 136–37, *136*, *137*, 142, *142*, *143*
  - lithosphere 23, 26, **31**, 38, 76, 91, 106, 112–13, 120, 148
  - lithospheric plate 31, 115, 118, 148
  - magnetic field *12*, *13*, **31**, 32, 42, 55, 82, 83, *83*
  - magnetosphere **32**, 43, 82, *83*
  - mantle *14*, **32**, 37, 66, 78–79, 80, 81, *81*, 84, 90, 92, 93, *93*, 112–13, 120, 148
  - mass determined *13*
  - mesosphere 92
  - mobile zone **34**, 108, *109*
  - Moon *see* Moon, the
  - oceans 23, **35**, *75*, 84, 99, 100–101, *100*, *101*, 148
  - orbit 62
  - origin 66
  - ozone layer *see* ozone layer
  - periods of history proposed *12*
  - polarity reversal **38**, *114*, 115
  - principal stress **39**
  - rotation and rotational axis 62, 82
  - shape 26
  - shields **42**
  - sialic crust **42**
  - simatic crust **42**
  - soil 90
  - stable zone **43**, 108
  - stratigraphic succession **43**
  - stratosphere **43**, 98–99
  - surface *76*
  - temperature 78–79, 80, 92–93, *92*
  - thermal energy 74, 76, 77, 78
  - thermosphere **44–45**, 99
  - tropopause **46**, 98, 99
  - troposphere 18, **46**, 98, 99
  - volcanism *75*, 84, 92–93, *92*, *93*, 94, 149
  - weather systems 98, 99
- earthquake *13*, 15, **23**, 96–97, 117, 118–19, *119*, *138*, 139
  - active regions *109*
  - deep-focus **22**
  - epicenter **24**
  - Mercalli scale **153**
  - mobile zones 108, *109*

  Richter scale **40**, 96, *97*, **153**
  seismograph stations *14*, 96
  shadow zone **41**
Earth Resources Technology Satellite (ERTS) *see* Landsat satellites
East African Rift Valley 19, 41, *107*, 120, *121*, 149
East Pacific Rise 19, *100*
eccentricity (orbital) **23**, 62, *62*, 146
echinoderm **23**
echo-sonar *14*
eclipse **23**
  lunar 23–24, *64*
  solar 23–24, *54*, *61*
  solar eclipse photographed *14*
  solar eclipse predicted *12*
ecliptic **24**, 35, 62
Einstein, Albert 24
Einstein's equation **24**
ejecta **24**, 28
ejecta blanket 22, **24**
El Chichon *138*
electricity, dipole **23**
electromagnetic force **24**
electron **24**, 50, 82, *83*, 86
element **24**, 50, 86
  nuclear transformation **34**, 50
Emaliani, Cesare *15*
Enceladus **24**
energy
  accretional **16**, 88
  conduction **21**
  gravitational **27**, 78
  kinetic **30**, 77
  potential **38**, 41, *50*
  rotational **40**
  solar 53, 60
  thermal 74–85
Eocene **24**
ephemeris **24**
epicenter **24**
equator **24**
equinox **24**, 62
eras, geological **153**
Eratosthenes 12, *12*
erg 130
Eros **24**
erosion **24**, 122, *123*, 124, *124*, 126, 130
  coastal 128, *129*
  glacial 132
eruption plume **24**
esker **24**, *105*, 132
estuary **24**
eukaryote **24**, 136
Europa **24**
Evening Star *12*
Everest, Mount 148
Ewing, Maurice *14*
exfoliation **24**
exosphere **24**
Explorer I satellite *14*, 82
extinction **24**

F

fault **24**, *25*, 40, 120
  escarpment 118
  thrust **45**, 119
  transcurrent **45**
  transform (strike-slip) 25, 35, 41, **45**, 114, *114*, 119
feldspar 17, 18, **25**, 27, 31, 41, 124–25, *125*, 139
  alkali *25*
  plagioclase 25, 90
  shocked quartz **42**
feldspathoid **25**, 38, 41
felsic **25**
ferrimagnetic 36
ferromagnetic 36
field reversal **25**, 36, 110
firn *132*
fissure eruption *95*
flint 20
flood deposit 126
flooding *138*
flood plain *127*
fluvial valley 130–31
fluviatile **25**
fold **25**, *25*
fold belt **25**, 36
fold mountains **25**, 102, 118–19, *118*
foraminifera 36
fossil *12*, *13*, *15*
  Glossopteris flora **26**, *111*
  index 43, *87*
  trace **45**, *45*
  see also paleontology
fractional crystallization 23, **25**, 37, 93
  layered intrusion **31**
Franklin, Benjamin *13*
free-air correction *27*
fusion **25**

G

gabbro **25**, 100
Gagarin, Yuri *15*, 140
Galapagos Islands 114
galaxy **25**, 48
  Milky Way *13*, 26, 48, *135*
  nucleus 35
Galileo Galilei *12*, 13, 19, 26, 147
Galileo mission **25**, 70, *71*, *140*, 141
Galle, Johann *13*
gamma radiation *53*
Ganymede **26**, 147
garnet 29
gas **26**
  inert (noble; rare) **29**, 47
gas-and-ice giant 57, 60, 68–69, *68*, *85*
gas giant 57, 60, 68–69, *68*, 85
Gaspra 26, 70, *71*, 141
Gay-Lussac, Joseph *13*
gemstones 151
general theory of relativity 24
geochemistry 10
geochronology **26**, 86
geoid **26**
geology, history 12–15
geomagnetics 11
geomorphology 10
geotectonic imagery 114
geothermal gradient **26**, *92*
geothermal vent *137*
Gesner, Conrad 14
geyser **26**, *94*
Gilbert, William *12*, 82
Giotto mission 15, *15*, **26**, 28, 72
glacial **26**
glaciation **26**, 34, 104–05, *104*, *105*, 110
glacier **26**, 125, 130, 132, *132*, *133*, 149
  ablation **16**
  load **31**
"Glomar Challenger" 115
Glossopteris flora **26**, *111*
gneiss **26**, 150
gold 50, 152
Gondwanaland *14*, **26**, *26*, 104, 110, *110*, *111*, 120, *143*
Gosse Bluff *73*
graben 24, **26**, 28, 41, 120
Grand Canyon *86–87*, 122
granite **26**, 27, 97, 102, 124, *125*
granodiorite **27**, 116
granulation 54
graptolite **27**
gravitational energy **27**, 78
gravity
  accretion 56
  gas giant *57*
  Moon's pull on Earth 64
  proto-planets 78
  tides 64
gravity anomaly **27**
gravity mapping 11
gravity segregation **27**
Great Bombardment **27**, 28
Great Dark Spot 68, *98*
Great Red Spot **27**, 68, *99*
Great Unconformity *86*
greenhouse effect 15, **27**, *27*, 66, 98, 99, 125
greenstone belt **27**, *102*
Greenwich meridian 31
Grigg-Skjellerup comet 26
ground ice **27**
ground moraine 34, 132, *132*
Guilin, limestone karst mountains *123*
Gulf of Aqaba *107*
Gutenberg, Beno *14*, 81, 112
gypsum 151

H

Hadley, George *13*
half-life 22, **27**, 59
Hall, Asaph *14*
Halley, Edmund *12*, 28
Halley's comet *12*, 21, **27**, *62*, 72, *72*
  Giotto mission *15*, **26**, 28, 72
harzburgite **28**
Hawaii 94, 113, *113*
headland 128
heat
  convection **21**
  see also temperature
heat flow **28**
heavy elements **28**, 50, *50*
Heezen, Bruce *14*
helium 29, 35, 50, 58, 85, 134
helium-3 *53*
helium-4 *53*
Hellenic Arc 29
Herschel, John 34, 105
Herschel, William *13*
Hess, Harry *14*
Hidalgo asteroid *70*
Himalayas *109*, *113*, 119, 149
Hipparchos *12*
Holmes, Arthur *14*
Holocene **28**, *143*
hominid **28**, 137
hornblende 17
hornfelse 150
horst **28**, 120
hot spot **28**, 94, *109*, *112*, 113, 116, 149
Hubble Space Telescope *15*
human 137, 142, *142*
Humboldt, Alexander von *13*
Hutton, James *13*, 14, 46
Huygens, Christan *12*
hydraulic action **28**
hydrocarbon **28**, 136
hydrogen 47, 58, *81*, 84, *84*, 85, *85*, 98, 136
  heavy *see* deuterium
  isotopes 50
hydrogen chloride 84
hydrogen sulfide 98
hydrolysis **28**
hydrosphere 23. *See also* ocean
hydrostatic pressure 16, **28**
hydrothermal **28**

I

ice 132, *132*, *133*
ice age *13*, **28**, 29, 34, 104–05, *104*, *105*, 132, *143*
ice cap 104, *104*, *105*, 126, 132
Iceland 114, 115, *115*, 149
Ida 26, *140*
igneous activity 108
igneous intrusion **28**
igneous rock **28**, 90, 92, *102*, **150**
imbricate structure **28**
impact basin **28**, 56, 65, 90
impact breccia **28**
impact crater 22, 28, 60, 64, 66, *66*, 67, *73*, *76*, 86, *86*, 90–91, *90*, *91*
impact event, ejecta **24**
index fossil 43, *87*
index mineral **28**
Indonesia 29, 116–17
inert gas **29**, 47
Inland Sea, Japan *129*
insolation **29**
intercrater plain **29**, 91
interglacial 26, **29**, 104
interstellar dust *see* dust
invertebrate **29**, 137, *137*
Io **29**, 69, *95*, 140, 147
ion *83*
ionization **29**, 72, *73*, 82, 99
ionosphere **29**, *29*, 55
ion tail **29**
iridium anomaly *see* K/T boundary event
iron 32, 36, *50*, 58, 78, *79*, 81, *81*, 90
iron meteorite **29**
iron oxide *125*
ironstone 150
Isidore of Seville 14
island, volcanic 94, *112*, *115*, 116
island arc 21, **29**, 33, 46, 103, 116–17, *117*, 118
isochron **29**
isostasy *14*, **29**
isotope **29**, 35, 50, 86
  daughter **22**, 76, 77, 86
  decay sequence **22**, 59, 86
  half-life 22, **28**, 59, 86
  long-lived radionuclide **31**
  parent **36**, 86
  radioactive decay 16, 59, 74, 80, 86, 88
  radiogenic **39–40**, 86
  stable **43**
isotopic ratio **29**

J

jet stream **29**
joint (within rock) **29**
Jovian planets 68–69, *68*, *69*
Jupiter 13, **29**, *58*, 59, 68, *68*
  atmosphere 68, *98*, 99, *99*
  axial inclination 62, *62*
  Callisto **19**, 147
  cloud belt **20**, 68
  composition 60, 68, *68*
  core 21, 57, 68
  data 63, **146**
  density 68
  Europa **24**
  Galileo mission **25**, 141
  Ganymede **26**, 147
  gravitational influence 70
  Great Red Spot **27**, 68, *99*
  Io **29**, 69, *95*, 140, 147
  magnetic field 83, *83*
  mass 57, 68, *68*, 69
  moons 69, **147**
  ring system 41, *58*, 69
  rotation 68
  Voyager missions **47**, 140, *140*
Jurassic **29–30**, *128*, *130*

K

Kenya, Mount 149
Kenya Dome 27
Kepler, Johannes *12*, 13, 30, 62
Kepler's laws of planetary motion *12*, 13, **30**, 62
Kilauea *113*
Kilimanjaro 120, 149
kinetic energy **30**, 77
komatiite 28, **30**
Krakatoa 46
krypton 29
K/T boundary event **30**, 139
kyanite 29, 38, 39
Kyushu *129*

L

lacolith *94*
lahar **30**, 94
lamellae 42
Landsat satellites **30**, *30*, *127*, 140
landslide **30**, 138
Langevin, Paul *14*
Large Magellanic Cloud *53*
lateral moraine **30**, 34, 132, *132*
latitude **30**, *30*
  celestial 30
Laurasia 104, 110, *110*, *143*
lava **30**, 90, 92, *93*, 94, *94*
  aa **16**
  komatiite 28, **30**
  pillow **38**
layered intrusion **31**
lead 50, 152
Leonardo da Vinci *12*
Leonid meteor shower *70*
leucite 25
light
  albedo **16**
  sunlight 60
lightning *13*
light year **31**
limestone 20, **31**, *84*, *85*, *123*, *124*, 125, 150
limnography 11
lithium *81*
lithophile **31**, 50, 81, *81*, 90
lithosphere 23, **31**, 76, 81, 91, 106, 112–13, 120, 148
  convergent margin **21**
  crust **22**
  divergent plate margin **23**, *114*, 115
  geothermal gradient **26**
  phase change 38
  tectonics (tectonism) *see* plate tectonics; tectonics
  volcanism 38
lithospheric plate **31**, 112–13
  collision zone **20–21**
  subduction zone **44**, 113
  triple junction **46**, 119
load **31**
  suspended **44**, 126, *126*
  traction 126, *126*
longitude **31**, *31*
  celestial 31
long-lived radionuclide **31**
longshore drift **31**, 128
Los Angeles *138*, 139
low-velocity zone **31**
Luna probes *14*, *15*
L waves **31**
Lyell, Charles *13*, 46

M

maar crater 38, 47
McCandless, Bruce *141*
Madagascar 120
Magellan missions *15*, **31**, *67*, 108, 140
magma **31**, 83, 84, 92–93, *92*, *93*, 94, *102*, 108, 113, 124
  crystallization *27*
  fractional crystallization **25**
  lava **30**
  radiometric dating **40**
  volcanism *see* volcanism; volcano
  see also lava
magma ocean **31**, 90
magmatism 117
magnesium 32, 58, 81, *81*, 90
magnetic dipole **31**, 82
magnetic field **31**, 82–83, *82*, *83*
  bow shock **19**, 82–83

  field reversal **25**, 36, 110
magnetic inclination **31**
magnetic striping **31**, 115
magnetism 36
  detrital remanent magnetization (DRM) 36
  dipole **23**, **31**, 82
  geomagnetics 11
  paleomagnetism **36**, 83, *108*, 110, 114–15
  polarity reversal **38**, 83, *114*, 115
  polar wandering **38**, 110
  self-exciting dynamo **41**
  thermoremanent magnetization (TRM) 35, 36, **44**
magnetite 36, *125*
magnetometer 36
magnetopause **32**, 83
magnetosheath 19, **32**, 83
magnetosphere **32**, *32*, 43, 82, *83*
main sequence **32**, *50*, *52*
major element **32**
mammal 142
manganese 32
manganese nodule **32**
Manson Crater 30
mantle *14*, **32**, 66, 78–79, *79*, 80, 81, *81*, 90, 92–93, *93*, 120
  plate tectonics 112–13, 116
  subsolidus **44**, 80
mantle nodule **32**
mapping 11
  first geological map *13*
  oceans *14*, *15*
margin (of tectonic plate) **32**, 100, 108, *109*, 114, *114*
  convergent **21**, 118
maria **32**, 65, *90*, 142
Marianas Trench *14*, 35, 116, *117*, 148
Mariner missions 15, *15*, **32**, 32, *58*, 66, *66*, 120, 131, 140
Mars-94 and Mars-96 missions 140
Mars 13, **32**, *66*, 80, 101, 125, 142, 149
  atmosphere 67, 85, 98, *98*, 99
  calderas 19
  Candor Chasma *67*
  climatic changes 39
  composition 60
  core 79
  crust 91
  data 63, **146**
  Deimos *14*, **22**, 147
  density 66, 67
  dunes 131, *131*
  ejecta 24
  energy storage 88
  faulting 120
  glacial activity 26
  ground ice 28
  Hellas basin *76*
  ice caps *105*, 132
  impact basins **28**
  intercrater plain 29, 91
  magnetic field 83
  Mariner missions *15*, **32**, 32, 120, 131, 140
  moons **147**
  orbit 67
  origin 66
  outflow channels **36**, 126
  permafrost 37
  Phobos *14*, **37**, 147
  plains 47
  precession 105
  rotational axis 62
  Schiaparelli crater *76*
  seasons 62
  SNC meteorites 42, 70
  surface 67, *67*, *76*
  temperature 98
  Tharsis Bulge 67, 120, *120*
  thermal structure 76
  Valles Marineris 67, *67*, 120, *120*
  Viking missions 15, *15*, 32, **47**, 140
  volcanism 38, 47, 94, 113, 142
marsh *129*
Mars Observer **32**
Martinque **34**
mass **32**
  atomic 50
massif **32**
mass number 86
mass wasting **32**
matrix **33**
Mauna Kea *113*
Mauna Loa *113*
Maury, Matthew Fontaine *13*
meander *127*
mélange **33**
melting point **33**
meltwater *132*, *133*, 138
Mercalli scale **153**
Mercury 13, **33**, *58*, *66*, 85, 142
  Caloris Basin *58*, *66*
  composition 60
  core *78*, 79, 80
  crust 91
  data 63, **146**
  density 79
  impact basins 28
  magnetic field 83
  Mariner missions *15*, **32**, *58*, 66, *66*, 140
  orbit 62, *62*, 66
  origin 66
  surface 66, *66*
  thermal structure 76
  volcanism *66*
Meru, Mount 149
mesosphere 92
Mesozoic **33**, 101, 118, **153**
metalic minerals, data **152**
metal phase **33**
metamorphic rock **33**, 41, 102, **150**
  cleavage plane **20**
metamorphism 28, **33**, 36, 41, 86, 117, 118
  contact (thermal) 33
  regional 33
metasomatism **33**
meteor **33**, 70, *70*, *71*
meteorite 18, **33**, *33*, 50, 56, 58, 70, 74, 80, *102*, 139
  accretional energy **16**
  aerolite **16**, 20
  chondrite **20**, 59, 70
  classes 70
  crater-age dating 86, *86*
  iron **29**, 70
  orbits 70
  shatter cone **42**
  SNC **42**, 70
meteoroid **33**, 57, 70, 142
meteorology 11
methane 27, **33**, 58, 60, 68, 84, 85, 98, 136
mica 27, **33**, 41, 84, 93, *125*
Michell, John *13*
Mid-Atlantic Ridge *14*, *100*, 114–15, *115*, 149
mid-oceanic ridge *see* oceanic ridge
Milankovitch, Milutin 33, 105
Milankovitch theory **33**
Milky Way *13*, 25, 48, *135*
millet-seed grain **33**
Milne, John *14*
Mimas **33**
mineral **34**
  data **150–52**
  index **28**
  metallic **152**
  rock-forming **151**
mineralogy 10
Miocene **34**, 120
Miranda *15*, *120*, 141, 147
mobile belt (mobile zone) **34**, 108, *109*
Mohorovicic, Andrija *14*
Mohorovicic discontinuity **34**, 35, 41, 97
Moh's Scale **150**
Moine Thrust 45
molecular cloud **34**, 52, *52*
molecule **34**, 50
  organic **35**
  photodissociation **37**
  volatile 47
mollusk 17
molybdenum 152
moon **34**, *58*, 60, 69
  major moons listed **147**
  shepherd *see* shepherd moon
Moon, the **34**, *58*, 60, 64–65, *64*, *65*, 80, 142, 147
  albedo **16**, 65
  Apollo missions **17**, 140, *140*, *141*
  captured rotation **19**
  composition 65
  core 79, 80, 97
  crater-age dating 86, *86*
  crust 80, 90, *90*, *91*, 97
  density 64
  eclipse **23**, *64*
  ejecta 24
  far side photographed *14*
  gravitational pull 64
  highlands 31, 65, 80, 90
  intercrater plain 29
  landings *15*, 140, *140*, *141*
  Luna probes *14*, *15*
  magma ocean **31**
  magnetic field 83
  mantle 97
  Mare Imbrium 65
  Mare Orientale 28
  maria **32**, 65, *90*, 142
  Moonquakes 97
  orbit 64, *64*
  origin 65
  phases 64, *65*
  rock samples 17, *86*
  rotation 64
  solar eclipse *54*, *61*
  surface 64–65, 90
moraine **34**, 104, *104*, 132, *133*
  ground 34, 132, *132*
  lateral **30**, 34, 132, *132*
  terminal 34, **44**, 132, *132*
Morning Star *12*
mountain 148–49
  building 102, 108, 113, *113*, 118–19, *118*, *119*, 142, 149
  fold **25**, 102, 118–19, *118*
mudrock 150

N

Nakhla Meteorite 42
nappes 45
Nebecula Major *50*
nebula **34**
  angular momentum **17**
  protostellar *50*
  solar *see* solar nebula
Neogene 118
neon 29
Nepheline 25
Neptune **34**, 59, 68, 69, *69*
  atmosphere *98*
  axial inclination *62*
  cloud belt **20**
  composition 60, 68, *69*
  data 63, **146**
  discovery *13*
  Great Dark Spot 68, *98*
  magnetic field 83
  moons 69, **147**
  Nereid **34**
  orbit *68*
  ring system 41, *58*
  Triton *15*, 22, **46**, 69, 147
  Voyager missions 140, *140*
Nereid **34**
neutrino *50*
neutron **34**, 35, 50, 86
neutron star *50*
Newton, Isaac *12*, 13
New Zealand 149
nickel 50, 78, 81, *81*
nickel-iron alloys **34**, 78–79, *78*
Nicol, William *13*
nitrogen 47, *81*, 84, *84*, 98
nitrous oxide 27
noble gas *see* inert gas
non-sequence **34**
Nordenskiöld 23
nuclear fusion 25, 50
nuclear transformation **34**, 50
nucleus **34**, *86*
nuée ardente **34**, 40, 94

O

obduction **35**, *35*
Oberon **35**
obliquity **35**, 62
ocean **35**, *75*, 84, 100–101, *100*, *101*, 114–15, 128–29, *128*, *129*, 148
  abyssal plain **16**, 100, 129
  abyssal sediment **16**
  Benioff zone **18**, *22*, *119*
  black smoker **19**, 42, *115*
  carbon dioxide content 101
  circulation 101
  Deep Sea Drilling Project *15*
  hydrothermal springs 100–101
  mapping *14*, *15*
  salinity 100–101, *100*
  sea-floor spreading *see* sea-floor spreading
  seamount **41**, 114
  subduction trench 35, 114
  thermocline 101
  tide 64
  trench *14*, **45**, 100, *100*, 114, 116, 118, *119*
  turbidity current 21
  see also hydrosphere
oceanic crust 18, **35**, 35, *35*, 97, 100, 103, *103*, *112*, 114–15, *114*, 116, 148
  obduction **35**
oceanic ridge *14*, 23, 31, **35**, 41, 94, 100, *100*, *108*, *109*, 112, 113, 114–15, *114*, *115*, 120, 148
oceanography 10, 12–15
Olgas *111*
Oligocene **35**
olivine 20, 25, 28, **35**, 37, 41, 151
onion-skin weathering *see* exfoliation
Oort, James 35
Oort cloud **35**, 72
ooze **35**, 129
  calcareous 16
  siliceous 16
ophiolite **35**
orbit **35**
  asteroids 70, *71*
  comets *62*
  eccentricity **23**, 62, *62*, 146
  meteorites 70
  planets 57, 60, 62, *62*
Ordovician **35**, 104, *137*, *143*
organic molecule **35**
orogenesis 25, **36**, 39, 118, 122
orogenic belt 18, **36**, 102
osmium 40
outflow channel **36**
outgassing **36**
oxbow lake *127*
oxide *56*, 58
oxygen 32, 47, 50, *81*, 84, *84*, 98, 124, 136, 137
ozone 27, 98
ozone layer **36**, 98–99, 136
ozonosphere *see* ozone layer

P

Packe, Christopher *13*
pahoehoe **36**
Paleocene **36**
paleolatitude 110
paleomagnetism **36**, 83, *108*, 110, 114–15
paleontology 11, **36**, 110, 122
paleopole **36**, 110
Paleozoic **36**, 110, *110*, **153**
Pangea 26, **36**, 101, 104, *104*, 110, *110*, *143*
Panthalassa *110*
Papua New Guinea 149
paramagnetic 36
parent isotope **36**, 86
parsec *36*, **37**
partial melting 18, 23, **37**, 93
particle-particle collision **37**, 56–57
pediment (geological) **37**
pelagic **37**
peridotite 35, **37**, 97
perihelion **37**, 62
permafrost **37**, 132
  stone polygon **43**
permeable rock **37**
Permian **37**, *110*, 137
Permo-Carboniferous 104, *104*
petrology 10
Phanerozoic **37**, 102, 142
phase change **37**
Phobos *14*, **37**, 147
phonolite **37**, 120
phosphatic rock 150
phosphorus 32
photodissociation **37**
photography
  far side of Moon photographed *14*
  solar eclipse photographed *14*
photoionization 99
photosphere **37**, *37*, 44, 54, *54*
  sunspot *see* sunspot
photosynthesis **38**, 84, *84*, 98, 136–37, *137*
phreatomagmatic eruption **38**
phyllite 150
physical geography 10, 12–15
Piazzi, Giuseppe 70
Piccard, Auguste *14*
Pilbara *102*
pillow lava **38**
Pinatubo, Mount 94, 138, *138*
pingo 132
planet **38**
  accretion **16**, 56–57, *56*, *57*
  core 74, 76, 77, 78–79, *78*, *79*, 80, 81, *81*
  formation 48, 56–59, 78–79, *78*, *79*
  gas-and-ice giant *see* gas-and-ice giant
  gas giant *see* gas giant
  inner (terrestrial) planets 58, 60, 66–67, *66*, *67*, 78–81, *78*, 84–85, *85*, 90–91, *90*, *91*
  Jovian 68–69, *68*, *69*
  mass 74
  minor *see* asteroid
  moon *see* moon
  orbits 57, 60, 62, *62*
  planetary data 63, **146**
  rotation and obliquity **35**, 62
  thermal energy 74, 76, 77
planetesimal 18, **38**, 76
  accretion **16**, 56–57, *56*, 58–59, 66, 74
planetology 10
plasma **38**, *73*, 82, *83*
  bow shock **19**, 82–83
plate tectonics *14*, 15, *15*, **38**, 83, 91, 94, 96, 100, 102–03, 106, 112–13, *112*, *113*, *124*, 148–49
  Benioff zone **18**, 22, 117, *119*
  convergent margin **21**, 118

island arcs *see* island arc
  margin **32**, 100, 108, *109*, 114, *114*
  measurement *15*
  mobile belt (mobile zone) **34**, 108, *109*
  orogenesis *see* orogenesis
  trench **45**
playa **38**, 131
Pleistocene **38**, 104, 126, 132, *143*
Pliny 14
Pliocene 142
plume **38**, *93*, 94, *112*, 113, *114*, *121*
  eruption **24**
  hot spot **28**, 94, *109*, *112*, 113
Pluto **38**, *59*, 68, *68*, 69
  Charon *15*, **20**, *59*, 62, 68, *68*, 69, 147
  composition 60
  data 63, **146**
  discovery *14*
  orbit 62, *62*
point bar **38**, 126
polarity reversal **38**, 83, *114*, 115
polarizing prism *13*
polar wandering **38**, 110
polymict breccia 29
polymorph **38**
potassium 32, 76, *81*
potential energy **38**, 41, *50*
Pre de Bar *133*
Precambrian **38**, 80, *111*, 136, *137*, 142, *143*, **153**
precession *12*, **39**, 62, 105
pressure
  air pressure in atmospheres *12*
  critical point 22
  hydrostatic (confining) **28**
pressure wave *see* P waves
primary wave *see* P waves
prime meridian 31
principal stress **39**
projection **39**, *39*
Proterozoic **39**, 103, 142
proton 35, **39**, 50, 82, *83*, 86
proto-planet 76, 78
protoplanet *see* planetesimal
protostar **39**, 43, 48, 52, *52*
protostellar nebula *50*
proto-Sun 66
Ptolemy *12*
pumice **39**, 138
push-pull waves *see* P waves
P waves **39**, 96–97, *96*, *97*
pyroclast **39**, 94, *94*
pyroxene 18, 20, **39**, 41
Pythagoras 12, *12*

Q

Qattara basin 130
quartz 27, **39**, 41, 102, 124, 125, *125*, *131*, 151, 152
  coesite **20**
  shocked **42**
Quaternary **39**

R

radar *15*, **39**
radiation, Van Allen belts *14*, **46–7**, *47*, 82
radiative layer *54*
radioactive decay 74, 76, 80, 86
  decay sequence **22**, 59, 86
  half-life 22, **27**, 59, 86
  isochron **29**
  long-lived radionuclide **31**
radioactive nuclide **31**, 76, 77
radiogenic isotope **40**, 86
radiolaria 36
radiometric dating *14*, 17, **39–40**, 86, 152
  absolute age **16**, *108*
raised beach **40**
Raleigh Taylor instability *93*
rarefaction wave **40**
rare gas *see* inert gas
red giant **40**, *50*
Red Sea *107*
reef 128
refractory element **40**, 58
refractory inclusion **40**
regional metamorphism 33
regolith 28, 37, **40**, 57, 90
rejuvenation **40**
relativity, general theory 24
remote sensing 11, **40**
retrograde rotation **40**
reversing layer 55
Reykjanes ridge *115*
Rhea **40**
rhenium 40
rhyolite 27
Richter scale **40**, 96, *97*, **153**
ridge belt **40**
rift valley **40**, *67*, 120–21, *120*, *121*
Ring of Fire 116
ring system **40**, 68, *69*
  shepherd moon **42**, 68
Rio Negro *127*
river 126–27, *126*, *127*, 149
  delta **22**, 22, 126, *127*, 128
  estuary **24**
  life cycle proposed *14*
  load **31**, **44**, **45**, 125, 126, 128
roche moutonnée **40**, 132, *133*
rock **40**, 122, 124–25, *124*, *125*
  data **150–52**
  dating **152**
  rock-forming minerals **151**
rock-forming silicates **40**
rotational energy **40**
rubidium *81*

S

Sahara 130, *130*
St Helens, Mount 94, *138*
salinity **401**
salt dome 23
San Andreas Fault 45, 119, *138*, 139
sand sea 130
sandstone **40–41**, *124*, *128*, *130*, 151
San Francisco 139
saprolite **41**
satellite, artificial *14*
satellite, natural 60
Saturn 13, **41**, *58*, 59, 68
  atmosphere 68, *98*
  axial inclination *62*
  Cassini divison *12*
  cloud belt **20**
  composition 60, 68, *68*
  data 63, **146**
  Dione **23**
  Enceladus **24**
  magnetic field 83
  Mimas **33**
  moons 69, **147**
  Rhea **40**
  ring system *12*, 34, 41, 42, *58*, 68, *69*
  shepherd moons (guardian satellites) 34, 42, 68, *69*
  Tethys **44**
  Titan 34, **45**, 69, 147
  Voyager missions **47**, 68, 140, *140*
scablands **41**, 126
schist **41**, 150
sea-floor spreading *14*, *15*, 21, 32, **41**, 103, 114–15
seamount **41**, 114
Seasat satellite *15*
sea stack *129*
sediment **41**, 125, 126, 128, *129*, 149
  abyssal **16**
  air-fall deposit **16**
  beach 128
  bedding plane **18**
  delta **22**
  diagenesis **23**
  load **31**, **44**, **45**, 125, 126
  longshore drift 128
  sorting **43**
  turbidity current 21, **46**
sedimentary rock **41**, 124, **150**
sedimentology 10
seif dune 130, *130*
seismic discontinuity **41**, *96*, 97, *103*
  Mohorovicic **34**, 35, 41, 97
seismic reflection profiling **41**
seismic wave 96–97, *96*, *97*, 108, *109*, 112, 128
  L wave **31**
  low-velocity zone **31**
  P wave **39**, 96–97, *96*, *97*
  S wave **44**, 96–97, *96*, *97*
Seismograph *12*, 96, 97
seismograph stations *14*
seismology 10
seismometer **41**, *96*
self-exciting dynamo **41**
serac *133*
shadow zone **41**, *96*, 97, *97*
shatter cone 29, **42**
Shepard, Alan 140
shepherd moon 34, **42**, *42*, 68, *69*
Shergotty Meteorite 43
shield 21, **42**, 47, 67, *67*, 94, 102, *111*, *112*, 113, *113*
shocked quartz **42**, 139
Shoemaker-Levy 9 comet *99*
shooting star *see* meteor
sialic crust **42**
Siberia, Tunguska event *139*
sidereal period **42**
siderophile **42**, 50, 81, *81*
Sierra Nevada 118
silica 90, 152
silicate **42**, 50, 56, *56*, 58, 66, 78, 80, 81, 84, 92, 93, 94, 102, 124, 125
  rock-forming **40**
siliciclastic rock 41
silicon 32, *81*, 124
sillimanite 29, 39
Silurian **42**, 110, *143*
silver 50
simatic crust **42**
6344P-L asteroid 70
Slapin, Loch *128*
Smith, William *13*
smoker **42**
  black **19**, 42, *115*
SNC meteorite **42**, 70
snow line 69
sodium 31, 32, *81*
soil 40, **42**, 90
solar nebula 34, **42**, 52–53, *52*, 56, *56*, 57, 58, 59, 72, *85*
  particle-particle collision **37**
solar prominence **42**, *42*, 55, *55*
solar radiation 98, 136
Solar System *12*, 13, 60–73, 134, *135*, **146**
  origins 52–53, *52*, *53*
  Sun *see* Sun
solar wind **42–43**, 55, *57*, 72, *72*, *73*, 82–83, *83*, *140*
  magnetopause **32**
solstice **43**, 62
sonar *14*
sorting (of sediment) **43**
space probes, data listed **154–55**
space science, history 12–15
Space Shuttle *15*
spectrum, solar 54–55
spit **43**
spring line **43**, 126
Sputnik I satellite *14*
stable isotope **43**
stable zone **43**, 108
stalactite **43**
stalagmite **43**
star **43**, 134
  binary *see* binary star
  contracting phase *50*
  expansion phase *50*
  luminous phase *50*
  main sequence **32**, *50*, *52*
  neutron *50*
  protostar *see* protostar
  red giant *see* red giant
  shooting *see* meteor
  Sun *see* Sun
  supergiant *see* supergiant
  supernova *see* supernova
  white dwarf *50*
stellar wind 53
Steno, Nicolaus *12*
stishovite 29, 139
stone polygon **43**, 130, 132
stratification *90*, *91*, 124
stratigraphic succession **43**
  unconformity **46**, *86*, 122
stratigraphy 11, **43**, 122
stratosphere **43**, 98–99
stratovolcano **43**, 47, 120
stratum **43**
strike-slip fault *see* transform fault
stromatolite **43**, 136, *137*
subduction **45**, 118, 119, *119*
subduction trench 35, 114
subduction zone 29, **43**, 103, 113, 114
sublimation 16
subsolidus **44**, 80–81
Suess, Eduard *14*
sulfide 50
sulfur *15*, 81, *94*
sulfur dioxide 50
Sumatra 149
Sun **44**, 54–55, *54*, *55*, 134
  chromosphere *54*, 55, *55*
  core 54, *54*
  corona **21–22**, *54*, 55, *55*
  eclipse **23**, *54*, *61*
  eclipse photographed *14*
  eclipse predicted *12*
  energy generation 53, 60
  granulation 54
  gravitational pull 62, 70
  insolation **29**
  magnetic field *55*
  mass 53
  nuclear fusion 53, *53*
  origins 52–53, *52*, *53*
  photosphere *37*, **37**, 44, 54, *54*
  proto-Sun 66
  radiative layer *54*
  reversing layer 55
  solar nebula *see* solar nebula
  solar prominence *see* solar prominence
  solar spectrum 54–55
  solar wind *see* solar wind
sunlight 60
sunspot 38, **44**, *55*
supergiant *50*, *51*, *135*
Supernova 1987a *50*, *51*
supernova **44**, *44*, *49*, 50, *50*, *51*
  first recorded explosion *12*
supernova remnant *49*
Surtsey *115*
suspended load **44**, 126, *126*
S waves **44**, 96–97, *96*, *97*
syncline **44**

T

talus **44**, 131
Tapeats Sandstones *86*
Tarantula Nebula *50*
tectonics (tectonism) 11, **44**, 93, 94
  orogenesis **36**, 39
  plate *see* plate tectonics
tektite **44**
telescope 13, 19
temperature
  accretional energy 76, 77
  adiabatic heating **16**, 76, 77
  centigrade scale *13*
  critical point *22*
  Curie point **22**, *22*, 36, 45
  geothermal gradient **26**
  heat convection **21**
  liquidus 44
  melting point **33**
  solidus 44
terminal moraine 34, **44**, 132
terra australis incognita 12
terrestrial planets *see* planet
Tertiary **44**, 119, 120
tessera **44**, 66
Tethys **44**
Tethys Sea 26, 110, *110*, *124*
Thales *12*
thermal energy 74–85
thermal metamorphism 33
thermal radiation 98
thermocline 101
thermometer *12*
thermoremanent magnetization (TRM) **35**, 36, **44**
thermosphere **44–45**, 99
Thira *139*
thorium 76
309 Doradus nebula *53*
threshold wind speed **45**
thrust 119
thrust fault **45**, 119
Tibetan Plateau *113*
tide **45**, 64, 128
till 132
tillite **45**, 104, *104*, 132
Tiros I satellite *14*
Titan 34, **45**, 69, 147
Titania **45**
titanium 32, 40
Titus, J.D. 45
Titus-Bode rule **45**
Tombaugh, Clyde *14*
Torricelli, Evangelista *12*
trace element **45**
trace fossil **45**, *45*
traction load **45**, 126, *126*
transcurrent fault **45**
transform fault 25, 35, 41, **45**, 114, *114*, 119
transient cavity **45**
trench *14*, **45**, *100*, 114, 116, 117, *117*, 118, *119*
Triassic **45**, 110, *143*
tributary 126
trilobite **45**, *87*, *137*
triple junction **45**, 119
tritium *15*, 35
Triton 22, **46**, 69, 141, 147
Trojan asteroids *70*
tropopause **46**, 98, 99
troposphere 18, **46**, 98, 99
tsunami **46**
T Tauri activity **46**, 53
tufa **46**
tundra **46**
tungsten 58, 152
Tunguska event *139*
turbidite **46**, 129
turbidity current 21, **46**, 46, 128–29, *129*

U

ultraviolet radiation (UV) **46**, 55, 99, 136
Umbriel **46**
unconformity **46**, *86*, 122
Uniformitarianism **46**
uranium 76
Uranus **46**, 59, 68, *69*
  Ariel **17**
  atmosphere 68, 99
  axial inclination 62, *62*
  cloud belt **20**

composition 60, 68
data 63, **146**
discovery *13*, 68
magnetic field 83, *83*
Miranda *15*, *120*, 141, 147
moons 69, **147**
Oberon **35**
ring system 41, *58*, 69, *69*
Titania **45**
Umbriel **46**
Voyager missions 47, 140
Urey, Harold *15*
Ussher, Bishop 14

V

Vacquier, Victor 14
Valles Marineris **46**
Van Allen belts *14*, **46–47**, *47*, 82
Vela supernova remnant *49*
Venera missions 15, *15*, 66, *91*, 140
Venus *12*, 13, **47**, *66*, 80, 101, 108, 125, 142, 149
Aine Corona *112*
atmosphere 66, 85, *85*, 98–99, *98*
Beta Regio **18**, 27, 47, 120
calderas 19
composition 60
convection 80
core *78*, 79, 80
corona **21–22**, 66, *112*
crust 91, *91*, 103
data 63, **146**
dunes 131
ejecta 24
energy storage 88
faulting 120
fold belts 25
Galileo mission *140*, 141
greenhouse effect 15, 27, 66, 98, 99, 125
Ishtar Terra 66
Magellan missions *15*, **31**, *67*, 140
magnetic field 83
Mariner-10 **32**
Maxwell Montes 66
orbit *62*
origin 66
plumes 38, *112*, 113, 120
radioactive heat 74
ridge belt **40**
rotation 62, 66
Sapas Mons *67*
surface 66, *67*
temperature 80, 98
tesserae (complex ridged terrain CRT) **44**, 66
thermal structure 76
Venera missions 15, *15*, 66, *91*, 140
volcanism **47**, 66, *67*, 94, *112*, 113, 120, 142
vesiculation 39, **47**
Vesta 18
Victoria Falls *126*
Viking missions 15, *15*, 33, **47**, 140
volatile 50, 56, 65, 67, *81*, 84, *85*, 92
volatile element **47**, 58, 81, 93
outgassing **36**
volatile molecule 47
volcanic dome 47
volcanic rise **47**
volcanism 15, **47**, *75*, 92–93, *92*, *93*, 94–95, *94*, *95*, 96, 103, 108, 118–19, *119*, 138, *138*, *139*, 142, 149
active submarine 23
air-fall deposit **16**, 129
ash flow deposit **17**
caldera **19**, *94*
creation of atmospheres and oceans 84, *85*
cryovolcanism **22**, 69
eruption plume **24**
faults 120
geyser **26**, *94*
intercrater plain **29**, 91
island arcs *see* island, volcanic; island arc
lahar **30**, 94
lava *see* lava
lithosphere 38
Mars 38, 47, 94, 113, 142
Mercury *66*
mobile zones 108, *109*
moons 69
nue ardente **34**, 40, 94
oceanic ridges 114–15
rift valleys 120–21, *121*
shields 19, 42, 47, 67, *67*, 94, *112*, 113, *113*
and temperature changes 105
tsunami **46**
Venus **47**, 66, *67*, 94, *112*, 113, 120, 142
volcano **47**, 94–95, *94*, *95*
stratovolcano (composite volcano) *see* stratovolcano
Vostok missions *15*, 140
Voyager missions 15, *15*, **47**, 68, 69, *69*, *120*, 140, *140*
vulcanism *see* volcanism

W

wadi **47**, 131
water 50, 84, 92, 98
water table **47**
wave 128, *129*
weather **47**
weathering **47**, *123*, 124, *124*, *125*
weather satellite *14*
Wegener, Alfred 14, *14*, 21, 106
white dwarf *50*
Widmanstatten figures 34
Wilson, J. Tuzo *15*
wind 130–31, *130*, *131*
jet stream **29**
solar *see* solar wind
stellar 53
threshold wind speed **45**
Wyoming *89*

X

xenolith **47**
xenon 29
X-ray fluorescence **47**
X-ray radiation 55

Z

Zambezi *126*
zeolite **47**, 151
Zhang Heng *12*
zinc 50, 152
zirconium 40
zodiac 62

ACKNOWLEDGMENTS

Picture credits

1 SPL/NASA **2-3t** TRIP/NASA **2-3b** GSF **5** SPL/John Reader **6** TCL/Masterfile **7** SPL/David Parker **48-9** SPL/Royal Observatory, Edinburgh **50-1** SPL/Royal Observatory, Edinburgh **51** SPL/Space Telescope Science Institute/NASA **52-3** NASA **54** SPL/Roger Ressmeyer, Starlight **55t** SPL/NASA **55b** SPL/Dr Carey Fuller **58** SPL/NASA **58-9** TRIP/NASA **59** SPL/NASA **60-1** SPL/George Post **64-5** SPL/NASA **66** SPL/NASA **66-7** SPL/NASA **67** SPL/US Geological Survey **68-9** SPL/NASA **69l** SPL/NASA **69r** TRIP **70-1** SPL/NOAO **71** SPL/NASA **72** SPL/ ESA **72-3** SPL/NOAO **74-5** Images Colour Library **76-7** SPL/US Geological Survey **79** IP/Alain le Garsmeur **80-1** RHPL **82-3** TCL/Masterfile **85** SPL/ NASA **86** TRIP/NASA **86-7** RHPL/Tony Gervis **88-9** GSF **90** RHPL **91** NASA **93** RHPL **95** SPL/NASA **98** SPL/NASA **98-9** SPL/NASA **99** SPL/MSSSO, ANU **100-1** TSW **101** SPL/David Vaughan **102** Open University/Steve Drury **104** IP/Cedri/Xavier Desmier **105** SPL/US Geological Survey **106-7** SPL/NASA **109** SPL/NASA **111** Images Colour Library, **111 inset** SPL/Martin Land **112** SPL/NASA **112-3** SPL/NASA **115b** SPL/Scripps/Peter Ryan **115t** Images Colour Library/Landscape Only **115r** NHPA/Brian Hawkes **116-7** RHPL/Sassoon **118-9** Victor Engelbert **120c** SPL/NASA **120b** SPL/US Geological Survey **120-1** Natural Science Photos/A Sutcliffe **122-3** RHPL/Nigel Gomm **124** SPL/David Parker **124-5** GSF **126** Images Colour Library **127** SPL/Earth Satellite Corporation **128** AOL/Peter Cattermole, Bodge Witterer Enterprises **128-9** NHPA/Orion Press **129** Geophotos/Tony Waltham **130-1** SPL/Earth Satellite Corporation **131b** Geophotos/Tony Waltham **131r** SPL/NASA **132** Z **134-5** SPL/Reverend Ronald Royer **137l** SPL/John Reader **137c** SPL/Scripps/Peter Ryan **137r** SPL/Dr Morley Read **138t** Katz Pictures/A Garcia **138b** TCL/Colorific/Black Star/J Eyerman **138-9** HL/Trevor Page **139t** RHPL **139c** Sovfoto **140** TRIP/NASA **141** SPL/NASA

Abbreviations

b = bottom, **t** = top
l = left, **c** = centre, **r** = right

AOL	Andromeda Oxford Limited, Abingdon, UK
GSF	Geoscience Features, Kent, UK
HL	Hutchison Library, London, UK
IP	Impact Photos, London, UK
RHPL	Robert Harding Picture Library, London, UK
SPL	Science Photo Library, London, UK
TCL	Telegraph Colour Library, London, UK
TRIP	TRIP, Surrey, UK
TSW	Tony Stone Worldwide, London, UK
Z	Zefa Picture Library, London, UK

Artists

Rob and Rhoda Burns, Mick Gillah, Ron Hayward, Trevor Hill/Vennor Art, Joshua Associates, Pavel Kostell, Colin Rose, Leslie D. Smith, Ed Stewart

Editorial assistance

Peter Lafferty, Ray Loughlin, Dr Keith Moseley, Lin Thomas, Claire Turner

Index

Ann Barrett

Origination by

HBM Print Ltd, Singapore;
ASA Litho, UK